ANNALS OF DISCRETE MATHEMATICS

annals of discrete mathematics

Managing Editor
Peter L. HAMMER, University of Waterloo, Ont., Canada

Advisory Editors
C. BERGE, Université de Paris, France
M.A. HARRISON, University of California, Berkeley, CA, U.S.A.
V. KLEE, University of Washington, Seattle, WA, U.S.A.
J.H. VAN LINT, California Institute of Technology, Pasadena, CA, U.S.A.
G.-C. ROTA, Massachusetts Institute of Technology, Cambridge, MA, U.S.A.

NORTH-HOLLAND PUBLISHING COMPANY – AMSTERDAM • NEW YORK • OXFORD

ANNALS OF DISCRETE MATHEMATICS 4

DISCRETE OPTIMIZATION I

Proceedings of the ADVANCED RESEARCH INSTITUTE ON
DISCRETE OPTIMIZATION AND SYSTEMS APPLICATIONS
of the
Systems Science Panel of NATO
and of the
DISCRETE OPTIMIZATION SYMPOSIUM
co-sponsored by
IBM Canada and **SIAM**
*Banff, Alta, and Vancouver, B.C.
Canada, August 1977*

Edited by

P.L. HAMMER, University of Waterloo, Ont., Canada
E.L. JOHNSON, IBM Research, Yorktown Heights, NY, U.S.A.
B.H. KORTE, University of Bonn, Federal Republic of Germany

1979

NORTH-HOLLAND PUBLISHING COMPANY – AMSTERDAM • NEW YORK • OXFORD

© NORTH-HOLLAND PUBLISHING COMPANY – 1979

All rights reserved. No part of this publication may be reproduced, stored in a retrieval system, or transmitted, in any form or by any means, electronic, mechanical, photocopying, recording or otherwise, without the prior permission of the copyright owner.

Submission to this journal of a paper entails the author's irrevocable and exclusive authorization of the publisher to collect any sums or considerations for copying or reproduction payable by third parties (as mentioned in article 17 paragraph 2 of the Dutch Copyright Act of 1912 and in the Royal Decree of June 20, 1974 (S. 351) pursuant to article 16 b of the Dutch Copyright Act of 1912) and/or to act in or out of Court in connection therewith.

PRINTED IN THE NETHERLANDS

PREFACE

Discrete optimization has experienced a tremendous growth during the last several decades. This period has seen an explosion in computer technology, the development of mathematical programming from its origins to its present state, and attempts to deal with increasingly complex management and decision problems using quantitative methods. A common theme running through many of these applications is the central role played by discrete aspects of the problems, severely limiting the applicability of those mathematical tools developed during the last couple of centuries to successfully handle physical science and engineering problems.

The area of optimization experienced a revolution with the development of the simplex method for solving linear programming problems. With the availability of this new technique, decision makers learned to formulate linear programming models in a surprising variety of contexts. The areas of applicability seemed enormous. However, the originator of the method, George Dantzig, recognized early that dealing with the frequent presence of discreteness in practical problems required finding ways to solve linear programs when some or all variables are required to take on integer values.

In the late 1950's, Gomory developed his method for solving such problems based on an adaptation of the linear programming simplex procedure. From these early linear programming origins, discrete optimization rapidly grew into an area attracting practitioners and theoreticians from a large variety of fields. As a result, the theoretical, methodological, and practical work being done encompasses a wide diversity of approaches. The theory draws upon mathematics as diverse as linear programming, logic, group theory, and combinatorics. The methodology has grown from its cutting plane and linear programming-branch and bound origins to include a vast array of techniques for special problems, heuristics for general and special problems, and frameworks for handling the general integer programming problem. A great variety of problems arising in industry, government, and the military have been formulated, and the size of those problems which can be solved efficiently is constantly increasing, due not only to computer developments but to more appropriate formulations and to better techniques.

The Systems Science Panel of NATO decided that it was an appropriate time to attempt to assess the current state of this diverse subject and to examine its main trends of development. Accordingly, a NATO Advanced Research Institute on Discrete Optimization and Systems Applications (ARIDOSA) took place in August 1977 at Banff, Alberta. This event provided an ideal opportunity to bring together a group of leading contributors representing different areas within the broad heading of discrete optimization.

These two volumes of proceedings of the NATO ARIDOSA are organized by subject areas and contain two different kinds of papers: surveys on certain sub-fields of discrete optimization, and reports from working sessions reflecting the positions taken and the views expressed at the meeting regarding the major trends perceived and the developments needed in discrete optimization and its applications.

Leading experts had been asked one year in advance to prepare survey papers on pre-assigned sub-fields within their main area of expertise. These surveys had been distributed to all participants beforehand.

During the week preceding the NATO ARIDOSA a symposium on discrete optimization (DO) had taken place in Vancouver, offering an opportunity to present these surveys in the form of lectures to a broad audience consisting of 230 participants from 19 countries and representing industries, governments, and universities. Most of the participants of the Research Institute had the chance of attending these lectures, which laid the groundwork for a meaningful interchange of views and ideas in the following week at Banff. The final versions of the survey papers as they have been included here were written after the Research Institute had taken place, and they have been influenced to a large extent by the evaluation of the major trends of discrete optimization as perceived in the working sessions.

Preparation of these reports from ARIDOSA was only made possible by the unselfish work of the associate editors who listened to tape recordings of the sessions and spent much time and effort in preparing preliminary reports. The final texts, as they appear here, have been prepared by the session chairmen from these preliminary reports.

It is our pleasant duty to extend grateful thanks to all those who have helped make possible these two meetings and these subsequent proceedings. The credit for organizing ARIDOSA goes to the Systems Science Panel of NATO and its director, Dr. A. Bayraktar.

During the stages of planning ARIDOSA, the Systems Science Panel of NATO has included several of our distinguished colleagues. We are particularly thankful to Professor Donald Clough for his valuable encouragement and continued support and to Professors Guy deGhellinck and Jakob Krarup for their active interest and participation in the two meetings.

Without the understanding support of IBM Canada, DO could not have taken place. In addition, IBM Research helped in various ways to sponsor DO. IBM Germany and IBM/E/ME/A provided travel support for some participants. Further travel support was obtained from IBM Belgium, IBM Denmark, and IBM Italy.

SIAM's sponsorship of DO is gratefully acknowledged.

Support received from the National Science Foundation (USA) and the National Research Council (Canada) was instrumental in the organization of DO.

The support provided by the University of British Columbia, Simon Fraser University, the University of Bonn, and the University of Waterloo has helped in many important ways in making possible the meetings. We are grateful to the Faculty of Commerce of the University of British Columbia for hosting DO and to the Institute of Applied Mathematics and Statistics for helping in its organization. In particular, we thank Brian Alspach, Daniel and Frieda Granot, Jack Mitten, Fred Wan, and the graduate students of these universities for their active roles in assisting in the organization of DO

The pleasant working environment of the Banff Centre coupled with the efficient and unselfish assistance offered by Mr. Imo von Neudegg and Mrs. Katherine Hardy have made substantial contribution to the success of ARIDOSA.

We appreciate the continued interest of North-Holland Publishing Company in the two meetings and their sustained enthusiasm for the publication of these proceedings.

We thank the American Mathematical Society for valuable publicity, Canadian Pacific Airlines for arranging special group flights, the Royal Bank of Canada for providing briefcases for the participants of the two meetings, and the Provincial Government of British Columbia for hosting the luncheon banquet of DO.

We express our thanks to Mrs. Inge Moerth for her efficiency and enthusiasm in carrying out in Vancouver the organizational side of DO.

Last but far from least, our most special thanks go to Mrs. Maureen Searle without whose work, beginning long before the meetings, extending over lunch breaks, evenings, and weekends, and carrying through long after the meetings, there would not have been the two meetings or these two volumes.

The Editors

CONTENTS

Contents of Volume I

Preface	v
Contents	ix

1. Combinatorial and polyhedral aspects of discrete optimization

Surveys

C. BERGE, Packing problems and hypergraph theory: a survey	3
J. EDMONDS, Matroid intersection	39
P.L. HAMMER, Boolean elements in combinatorial optimization	51
A.J. HOFFMAN, The role of unimodularity in applying linear inequalities to combinatorial theorems	73
B. KORTE, Approximative algorithms for discrete optimization problems	85
J.K. LENSTRA and A.H.G. RINNOOY KAN, Computational complexity of discrete optimization problems	121
L. LOVÁSZ, Graph theory and integer programming	141
J. TIND, Blocking and antiblocking polyhedra	159

Reports

Complexity of combinatorial problems (R.L. GRAHAM)	175
Structural aspects of discrete problems (P. HANSEN)	177
Polyhedral aspects of discrete optimization (A.J. HOFFMAN)	183

2. Some fundamental classes of problems

Surveys

R.E. BURKARD, Travelling salesman and assignment problems: a survey	193
P.C. GILMORE, Cutting stock, linear programming, knapsacking, dynamic programming and integer programming, some interconnections	217
V. KLEE and D. LARMAN, Use of Floyd's algorithm to find shortest restricted paths	237
E.L. LAWLER, Shortest path and network flow algorithms	251
M.W. PADBERG, Covering, Packing and Knapsack problems	265

Reports

Network flow, assignment and travelling salesman problems (D. DE GHEL-
LINCK) 289
Algorithms for special classes of combinatorial optimization problems (J.K.
LENSTRA) 295

Contents of Volume II

3. Methodology

Surveys

E. BALAS, Disjunctive programming	3
P. HANSEN, Methods of nonlinear 0–1 programming	53
R.G. JEROSLOW, An introduction to the theory of cutting planes	71
E.L. JOHNSON, On the group problem and a subadditive approach to integer programming	97
J.F. SHAPIRO, A survey of lagrangian techniques for discrete optimization	113
K. SPIELBERG, Enumerative methods in integer programming	139

Reports

Branch and bound/implicit enumeration (E. BALAS)	185
Cutting planes (M. GONDRAN)	193
Group theoretic and lagrangean methods (M.W. PADBERG)	195

4. Computer codes

Surveys

E.M.L. BEALE, Branch and bound methods for mathematical programming systems	201
A. LAND and S. POWELL, Computer codes for problems of integer programming	221

Reports

Current state of computer codes for discrete optimization (J.J.H. FORREST)	271
Codes for special problems (F. GIANNESSI)	275
Current computer codes (S. POWELL)	279

5. Applications

Surveys

R.L. GRAHAM, E.L. LAWLER, J.K. LENSTRA and A.H.G. RINNOOY KAN, Optimization and approximation in deterministic sequencing and scheduling: a survey	287
J. KRARUP and P. PRUZAN, Selected families of location problems	327
S. ZIONTS, A survey of multiple criteria integer programming methods	389

Reports

Industrial applications (E.M.L. BEALE)	399
Modelling (D. KLINGMAN)	405
Location and distribution problems (J. KRARUP)	411
Communication and electrical networks (M. SEGAL)	417
Scheduling (A.H.G. RINNOOY KAN)	423

Conclusive remarks 427

PART 1

**COMBINATORIAL AND POLYHEDRAL ASPECTS
OF DISCRETE OPTIMIZATION**

CONTENTS OF PART 1

Surveys

C. BERGE, Packing problems and hypergraph theory: a survey	3
J. EDMONDS, Matroid intersection	39
P.L. HAMMER, Boolean elements in combinatorial optimization	51
A.J. HOFFMAN, The role of unimodularity in applying linear inequalities to combinatorial theorems	73
B. KORTE, Approximative algorithms for discrete optimization problems	85
J.K. LENSTRA and A.H.G. RINNOOY KAN, Computational complexity of discrete optimization problems	121
L. LOVÁSZ, Graph theory and integer programming	141
J. TIND, Blocking and antiblocking polyhedra	159

Reports

Complexity of combinatorial problems (R.L. GRAHAM)	175
Structural aspects of discrete problems (P. HANSEN)	177
Polyhedral aspects of discrete optimization (A.J. HOFFMAN)	183

PACKING PROBLEMS AND HYPERGRAPH THEORY: A SURVEY

Claude BERGE
University of Paris VI, Paris, France

Introduction

By 1955, Graph Theory has appeared as a tool to solve a large class of combinatorial problems, and his range has become wider. By 1970, Hypergraph Theory started to generalize as well as to simplify the classical theorems about graphs.

This paper deals with a particular type of problems that arise in Graph Theory and in Integer Programming; though some of the results can be proved by purely algebraic methods and polytope properties, (Edmonds, Fulkerson, etc.), the "hypergraph" context permits an elegant presentation which does not require a particular algebraic knowledge and may look more appealing to the reader familiar with Graph Theory.

Most of the classical concepts about graphs are in fact (0, 1) solutions of a linear program; if instead we consider the solutions with fractional coordinates, we get simpler results. This is called the "fractional" graph theory as suggested by Nash–Williams; may be, the first paper with this point of view is due to Rosenfeld [32] (the "fractional stability number" of a graph is sometimes called the "Rosenfeld number").

This survey is divided into two parts: the first one is a brief summary of the "classical hypergraph theory", and the second one is devoted to "fractional hypergraph theory". No proof will be given in Part 1, since the reader can find them in [3]. List of the basic definitions and notations which will be used throughout:

hypergraph – $H = (E_i/i \in M)$ collection of non-empty subsets.
vertex set of $H - X = \bigcup_{i \in M} E_i$
order of $H - n(H) = |X|$
rank of $H - r(H) = \max_{i \in M} |E_i|$
number of edges – $m(H) = |M|$
dual hypergraph – H^* (dual in the sense of the incidence structures)
subhypergraph induced by $A \subseteq X - H_A = (E_i \cap A/i \in M, E_i \cap A \neq \emptyset)$
partial hypergraph generated by $J \subseteq M - H' = (E_j/j \in J)$
section hypergraph by $A \subseteq X - H \times A = (E_i/i \in M, E_i \subseteq A)$

multiplication of H by $k \geq 1 - kH$: each edge of H is repeated k times
matching – partial hypergraph of H whose edges are pairwise disjoint
k-matching – partial hypergraph H' of kH with maximum degree $\Delta(H') \leq k$
bundle – partial hypergraph of H whose edges are pairwise intersecting
transversal – set $T \subseteq X$ which meets all the edges
k-transversal – weight $p(x) \in \{0, 1, \ldots, k\}$ such that the total weight of each edge is $\geq k$
stable set – set $S \subseteq X$ which contains no edge of H.
weak k-coloring – partition of X into k classes so that no edge (with cardinality >1) is contained in only one class
strong k-coloring – partition of X into k classes so that each edge have all its vertices in different classes
$\nu(H)$ – *matching number* – maximum number of edges in a matching
$\tau(H)$ – *transversal number* – minimum cardinality of a transversal
$\gamma(H)$ – *strong chromatic number* – minimum number of classes for a strong k-coloring
$\chi(H)$ – *(weak) chromatic number* – minimum number of classes for a weak k-coloring
$q(H)$ – *chromatic index* – strong chromatic number of the dual H^*
$\alpha(H)$ – *stability number* – maximum cardinality of a stable set.

1. Transversals and matchings

1.1. *The Helly property*

Let $H = (E_1, E_2, \ldots, E_m)$ be a hypergraph (family of non-empty sets); the elements of $X = \cup E_i$ will be called the *vertices*, each E_i is called an *edge*. Put $\{1, 2, \ldots, m\} = M$. The *rank* of H is

$$\max_{i \in M} |E_i| = r(H).$$

The *anti-rank* of H is

$$\min_{i \in M} |E_i| = s(H).$$

An edge with cardinality one is also called a *loop*; if all the edges have cardinality 2, H is called a *graph*. If all the edges have the same cardinality, H is *uniform*. The *degree* of a vertex $x \in X$ is the number of edges in H containing x, i.e.

$$d_H(x) = |\{i / i \in M, x \in E_i\}|.$$

The *maximum degree* is denoted

$$\Delta(H) = \max_{x \in X} d_H(x).$$

Now, let $J \subseteq M$; the family $H' = (E_j / j \in J)$ is called a *bundle* of H (or: *intersecting family*) if any two of its edges meet. H is said to have the **Helly Property** if every bundle of H has a non-empty intersection.

For instance, if H is a simple graph, a bundle can be either a "star" (i.e. $K_{1,p}$), or a triangle (i.e. K_3). So, *a graph has the Helly Property if and only if it has no triangles*.

Example 1. The intervals.

An "interval hypergraph" is a hypergraph whose vertices $x_1, x_2, \ldots, x_n$ are points on the line, each edge E_i being defined by those points lying in an interval of the line. Such a hypergraph has the Helly Property, by a famous theorem of Helly.

Example 2. The subtrees of a tree.

Consider a tree G (connected graph with no cycles) on a vertex set

$$X = \{x_1, x_2, \ldots, x_n\}.$$

A "tree-hypergraph" of G is a hypergraph whose vertices are $x_1, x_2, \ldots, x_n$, each edge E_i being a set such that the subgraph of G induced by E_i is also a tree.

(To show that such a hypergraph has the Helly Property, the reader may use Theorem 1.1' below.)

This example generalizes the previous one, because an interval hypergraph is also the tree-hypergraph of an elementary chain G.

Example 3. The rectangular polyominoes.

Consider a $n \times n$ chessboard. A $(a \times b)$-*polyomino* is a rectangular shape of size a by b which can cover exactly $a \times b$ squares of the chessboard.

Suppose we have an infinite number of rectangular polyominoes with sizes $a_1 \times b_1$, $a_2 \times b_2$, etc.... What is the maximum number of these polyominoes we can pack into the chessboard? Consider a hypergraph H whose vertices are the squares of the board, and whose edges are all the rectangular arrays of vertices with sizes $a_1 \times b_1$, $a_2 \times b_2$, etc.; the problem is to find a "maximum matching" of H. Clearly, H has the Helly Property, because if $H' = (E_1, E_2, \ldots, E_n)$ is a bundle, and if A_i and B_i denote the projections of the edge E_i on the two sides of the chessboard, then $(A_1, A_2, \ldots, A_m)$ has the Helly Property (by Example 1), so there exists a point $a \in \cap A_i$; also, there exists a point $b \in \cap B_i$, and all the edges of H' contain the vertex (a, b).

Example 4. Let $(X, \leq)$ be a modular lattice, and consider a hypergraph H with vertex-set X, each edge being an "interval":

$$[a, b] = \{x / x \in X, a \leq x \leq b\}.$$

Then one can show that H has the Helly Property. This generalizes the Examples 1 and 3; and also the "Chinese Theorem" (when X is the set of positive integers, and $a \leq b$ means: a is a divisor of b"). See: Ore, Am. Math. Monthly 59 (1952), 365–370.

We now generalize the above structures. Let p be a positive integer and $H = (E_1, E_2, \ldots, E_m)$ be a hypergraph. A *p-bundle* of H is a family $H' = (E_j/j \in J)$ with at least $p+1$ edges, such that any $p+1$ of them have a non-empty intersection. H is said to have the *Helly Property for dimension p* if every p-bundle of H has a non-empty intersection. Clearly, the "Helly Property" defined earlier is the "Helly Property for dimension 1".

To see when the Helly Property holds, we may use the following result:

Theorem 1. *The hypergraph $H = (E_1, E_2, \ldots, E_m)$ has the Helly Property for dimension p if and only if, for every set $A \subseteq X$ with cardinality $p+2$, the family*

$$(E_j/j \in M, |E_j \cap A| \geq p+1)$$

has a non-empty intersection. (Berge and Duchet, [5]).

In particular:

Theorem 1.1'. *The hypergraph H has the Helly Property (for dimension 1) if and only if, for any three vertices a_1, a_2, a_3 of H, the family of the edges containing at least two of them has a non-empty intersection.*

(The reader may use this result to show the Helly Property in the Examples 1 and 2.)

Let H be a hypergraph with vertex set X. A set $S \subseteq X$ is said to be *stable* if it contains no edge of H; a set $T \subseteq X$ is said to be a *transversal* (or: *cover*) if it meets all the edges of H.

Clearly, the complement of a transversal set is a stable set, and vice versa. The problem of the maximum stable set of a graph is very well known, and has many practical applications (see [2]).

P. Gilmore gave an axiomatic characterization for the family of the maximal stable sets is a graph. The Helly Property will permit a similar characterization for hypergraphs.

The *dual H^** of the hypergraph $H = (E_1, E_2, \ldots, E_m)$ on $X = \{x_1, x_2, \ldots, x_m\}$ is a hypergraph with vertex set $\{e_1, e_2, \ldots, e_m\}$, and edges

$$X_i = \{e_j/j \in M, x_i \in E_j\}.$$

Theorem 1.2. *Let H be a hypergraph such that no edge contains another. Then H is the family of the maximal stable sets in some hypergraph having all edges of size $\leq p+1$ if and only if the dual H^* has the Helly Property for dimension p.* (Berge and Duchet, [5]).

(Combining Theorems 1.1 and 1.2, we get a generalization of the Gilmore Theorem, see [3].)

1.2. The König property

Let $H = (E_1, E_2, \ldots, E_m)$ be a hypergraph. A partial hypergraph $H' = (E_i / i \in I)$ is called a *matching* if its edges are pairwise disjoint. The matching number $\nu(H)$ is the maximum number of edges in a matching. The transversal number is the minimum size of a transversal set. We have always $\nu(H) \leq \tau(H)$, because for every matching H' and for every transversal set T, we have $m(H') \leq |T|$, so $\max_{H'} m(H') \leq \min_T |T|$. If a hypergraph H satisfies $\nu(H) = \tau(H)$, we say that H has the *König Property*.

By the König theorem, every bipartite graph has the König Property. For the dual H^* of H, the König Property has an interesting meaning. $\nu(H^*)$ represent the maximum number of vertices of H such that no two of them lie in a same edge. $\tau(H^*)$ is the minimum number of edges of H needed to cover the vertex set X. The dual H^* has the König Property when these two numbers are equal. The problem of finding a maximum matching is also sometimes called a *packing problem*; when H has the König Property, it is easier to check if a matching H' is maximum.

Example 1. Some types of planes need a crew of three pilots. If we draw a graph G whose vertices represent the pilots who are available, and if we joint two vertices if the two pilots can cooperate, the maximum number of planes that can be operated simultaneously is the matching number of the hypergraph defined by the triangles of G.

Example 2. Given a chessboard of size $p \times q$ and a supply of bricks of size $a \times b$, what is the largest number of bricks that can be placed on the chessboard, each one covering exactly $a \, b$ squares?

This problem is solved by the matching number of a hypergraph H whose vertices are the pq unit squares, whose edges are the sets of ab squares that can be covered by only one brick.

1. A variation of this packing problem is: given the truncated chessboard of 5×5 unit squares, with the central square removed, show that it is impossible to cover the 24 remaining squares with 6 bricks of size 1×4? Here we have to show that the 4-uniform hypergraph H whose vertices are the 24 unit squares satisfies $\nu(H) < 6$. Since that exists a transversal set T with $|T| = 5$ (represented by the crosses in Fig. 1), we have:

$$5 \leq \nu(H) \leq \tau(H) \leq |T| = 5.$$

Therefore the tiling with 6 is impossible and $\nu(H) = \tau(H) = 5$. Thus, H satisfies the König Property.

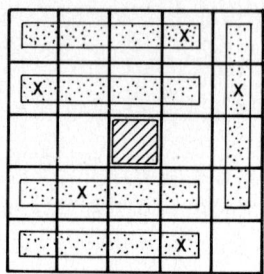

Fig. 1.

2. A companion problem is: what is the least number of tiles with size 1×4 needed to cover the 24 squares of the truncated chessboard? The tiles may overlap each other, but may not cover the removed squares. Here, we have to find $\tau(H^*)$, that is the least number of edges of H needed to cover X.

There exists a set S with $|S \cap E_i| \leq 1$ for all i ("strongly stable set") such that $|S| = 8$ (pictured by crosses in Fig. 2). So

$$\tau(H^*) \geq \nu(H^*) \geq |S| = 8$$

Since there exists a covering with 8 tiles (See Fig. 2), the least number of tiles needed for a covering of the 24 squares is exactly 8.

Thus, H^* satisfies the Helly Property.

3. If we consider the hypergraph H arising in the general packing problem of $p \times q$ chessboard with bricks of size $a \times b$, Brualdi and Foregger [8] have shown:

If a is a divisor of b, then H has the König Property, and

$$\nu(H) = \left[\frac{p}{b}\right]\left[\frac{q}{b}\right]\frac{b}{a} + \left[\frac{q}{b}\right]\left[\frac{r}{a}\right] + \left[\frac{p}{b}\right]\left[\frac{s}{a}\right] + \left[\left[\frac{r}{a}\right] + \left[\frac{s}{a}\right] - \frac{b}{a}\right]*$$

where

$$r = p - [p/b], \quad s = q - [q/b]b.$$

If a is not a divisor of b, there exists a (p, q) such that H has not the König Property.

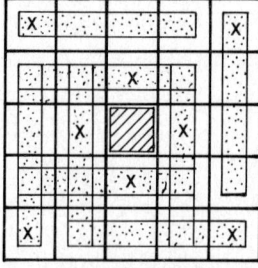

Fig. 2.

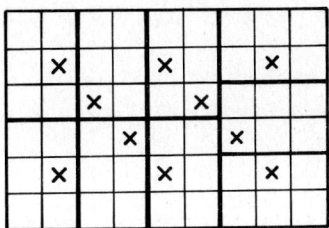

Fig. 3.

In Fig. 3, we consider the case $p = 6$, $q = 9$, $a = 2$, $b = 3$. Then, $m = 67$, $\Delta = 12$, and the crosses represent an optimal transversal with cardinality 10. Hence $\tau(H) = 10$, $\nu(H) = 9$, and H has not the König Property.

Example 3. The *complete r-uniform hypergraph* K_n^r is a hypergraph on a set X of n vertices, whose edges are all the r-tuples of X. Clearly,

$$\nu(K_n^r) = \left[\frac{n}{r}\right]$$

$$\tau(K_n^r) = n - r + 1$$

Hence the complete hypergraph has not the König Property. The hypergraph of the minimal transversals of K_n^r is K_n^{n-r+1}.

Example 4. Consider r disjoint sets $X_1, X_2, \ldots, X_r$ with $|X_i| = n_i$, $n_1 \leq n_2 \leq \cdots \leq n_r$.

The *complete r-partite hypergraph* $K_{n_1, n_2, \ldots, n_r}^r$ is a hypergraph with vertex-set $X_1 \cup X_2 \cup \cdots \cup X_r$ whose edges are all the r-tuples $\{x_1, x_2, \ldots, x_r\}$ with $x_1 \in X_1$, $x_2 \in X_2, \ldots, x_r \in X_r$.

Clearly, there is a matching with n_1 edges, and the set X_1 is a transversal with cardinality n_1, so $n_1 \leq \nu \leq \tau \leq n_1$.

Thus, the complete r-partite hypergraph has the König Property.

Now let us show that the hypergraph of the minimal transversals is $\{X_1, X_2, \ldots, X_r\}$. Clearly, $X_1, X_2, \ldots, X_r$ are minimal transversals. Let T be a transversal which does not contain any of them and let $a_i \in X_i - T$; then $\{a_1, a_2, \ldots, a_r\}$ is an edge of $K_{n_1, \ldots, n_r}^r$ which does not meet T; the contradiction follows.

The following are some important classes of hypergraphs having the König Property.

Class 1. Hypergraphs with no odd cycles

A cycle of H is a sequence $(x_1, E_1, x_2, E_2, \ldots, E_k, x_k)$, with all the E_i's distincts, all the x_i's distinct, (excepting $x_1 = x_k$) and $E_i \supseteq \{x_{i-1}, x_i\}$. The *length* of this cycle is k.

Theorem 1.3. *A hypergraph with no cycle of odd length has the König Property.*

Class 2. Unimodular hypergraphs

A hypergraph H is *unimodular* if every set A has a bipartition (S_1, S_2) such that, for $i \leq n$, and $k = 1, 2$,

$$\left[\frac{|E_i \cap A|}{2}\right] \leq |S_k \cap E_i| \leq \left[\frac{|E_i \cap A|}{2}\right]^*$$

(S_1, S_2) is sometimes called an *equitable bicoloring* of H_A.

For instance, an interval hypergraph (example 1) is unimodular, because we can color its vertices with red and blue alternatively from left to right on the line (clearly, this is an equitable bicoloring). Also, the hypergraph whose vertices are the vertices of a tree T, and whose edges are defined by chains of T, is also unimodular (by the same argument). Recently, Las Vergnas has proved: *Let G be a connected graph, and T be a spanning tree of G. Let $e(T)$ be the fundamental cycle basis associated with T. Then the hypergraph whose vertices are the edges of G, whose edges are the cycles of $e(T)$, is unimodular if and only if there is an orientation of the edges of T such that the cycles in $e(T)$ are directed circuits.* This result can be extended to orientable matroids.

Theorem 1.4. *A unimodular hypergraph has the König Property.*

This is a special case of the Hoffman Kruskal Theorem.

Theorem 1.5. *A hypergraph with no odd cycles is unimodular.*

The converse is not true; for instance an interval hypergraph is unimodular, but may have odd cycles.

Class 3. Balanced hypergraphs

A hypergraph is *balanced* if every odd cycle has at least one edge which contains three vertices of the cycle.

Theorem 1.6. *A unimodular hypergraph is balanced.*

The converse is not true; for instance, the hypergraph with vertices a, b, c, d, and edges $ab, ac, ad, abcd$ is balanced; since there exists no equitable bicoloring, it is not unimodular. Frank [16] has considered a class of balanced hypergraphs which are not unimodular (the "bi-path-hypergraph" associated with a directed graph).[1]

---

[1] Recently, A.E. Brouwer and A. Schrijver have proved: Let G be a simple graph on X; the hypergraph on X whose edges are the stars of G is balanced iff: every subgraph of G induced by a cycle of length ≥ 4 has a vertex of degree ≥ 3, and every subgraph induced by a cycle of length $4k+2$ has at most $2k$ vertices of degree 2.

Theorem 1.7. *A hypergraph H is balanced if and only if for every partial hypergraph $H' \subseteq H$, and for every set $A \subseteq X$, the subhypergraph H'_A has the König Property.* (Berge and Las Vergnas [7].)

A hypergraph H is said to have the *edge-coloring property* if the chromatic index (i.e. the least number of colors needed to color the edge-set so that not two intersecting edges have the same color) is equal to the maximum degree.

Theorem 1.8. *A hypergraph H is balanced if and only if for every partial $H' \subseteq H$ and for every set $A \subseteq X$, the subhypergraph H'_A has the edge coloring property.* (Berge [2]).

Class 4. Normal hypergraphs

A hypergraph H is *normal* if every partial hypergraph of H has the edge-coloring property.

It follows from Theorem 1.8. that a balanced hypergraph is normal. The converse is not true. For example, the hypergraph defined by four vertices a, b, c, d and the edges abd, acd, bcd is normal but not balanced. The normal hypergraphs, introduced by Lovász to prove the weak form of the perfect graph conjecture, have the following properties.

Theorem 1.9. *A hypergraph is normal if and only if every partial hypergraph has the König Property.* (Lovász, [22].)

Theorem 1.10. *Every normal hypergraph has the Helly-Property.*

In fact, H has the Helly Property iff
$$H' \subseteq H, \nu(H') = 1 \Rightarrow \tau(H') = 1.$$
The result follows.

Note that the four classes of hypergraphs above generalize the bipartite graphs. There are two other classes of hypergraphs generalizing the bipartite graphs.

Class 5. Quasi-normal hypergraphs

A hypergraph is *quasi-normal* if every odd cycle has three edges with a non-empty intersection. It is easy to show that *every normal hypergraph is quasi-normal*. The converse is not true. For instance, a hypergraph defined by the five vertices a, b, c, d, e, and the edges $\{a, b, c\}, \{a, b, d\}, \{a, c, d, e\}, \{b, c, d, e\}$ is quasi-normal, but not normal.

The quasi-normal hypergraphs have been introduced by Fournier and Las Vergnas to generalize the normal hypergraphs; they have not necessarily the König Property.

Theorem 1.11. A *hypergraph is quasi-normal if and only if it has a bicoloring, and this remains if we remove edges and contract boolean atoms.* (Fournier and Las Vergnas [15].)

Class 6. Semi-normal hypergraphs

A hypergraph is *semi-normal* if every section-hypergraph has the König Property. Clearly, a normal hypergraph is also semi-normal, but the converse is not true; Several semi-normal hypergraphs arise with a directed graph (see [43, 28, 25, 13]).

In fact, this class of hypergraphs was first considered by Jack Edmonds (who called them "exquisite hypergraphs" during the Hypergraph Seminar, Colombus 1972).

2. Fractional graph theory

2.1. *Fractional transversals*

A *fractional transversal* of a hypergraph $H = (E_1, E_2, \ldots, E_m)$ on X is a weight $p(x)$ associated with each vertex x of H such that $p(x) \geq 0$ for every vertex x, and such that $\sum_{x \in E} p(x) \geq 1$ for every edge E. Clearly, the characteristic vector $\Phi_T(x)$ of a transversal set is also a fractional transversal. The *value* of a fractional transversal $p(x)$ is

$$\sum_{x \in X} p(x) = p(X)$$

The minimum of $p(X)$ when p ranges over all the fractional transversals is called the *fractional transversal number*, and is denoted by:

$$\tau^*(H) = \min_p p(X)$$

A *k-transversal* is a function $p(x)$ on X with values in $\{0, 1, 2, \ldots, k\}$, such that $p(E) \geq k$ for every edge E. Clearly, the characteristic vector of a transversal is an 1-transversal. The *k-transversal number* denoted by $\tau_k(H)$, is the minimum of $p(X)$ when p range over all the possible k-transversals. Thus $\tau_1(H) = \tau(H)$. Let k be a positive integer, and let denote by kH a hypergraph whose vertices are the same as in H, whose $km(H)$ edges are obtained by repeating k times each edge of H. A *k-matching* H' is a partial hypergraph of kH with maximum degree $\Delta(H') \leq k$. The maximum number of edges for a k-matching is denoted by $\nu_k(H)$.

For instance, consider the graph C_5 (see Fig. 4) and the "polyomino-hypergraph" (see Fig. 5), whose vertices are the white squares, and whose edges are any three consecutive white squares of the diagram. In both cases, we see that $\tau = 3$, $\nu = 2$. But $\tau_2 = 5$, $\nu_2 = 5$, and a fractional transversal with value 5/2 is indicated by the numbers in Fig. 4 and Fig. 5. (The reader will check later that this fractional transversal is optimal, that is $\tau^* = 5/2$.)

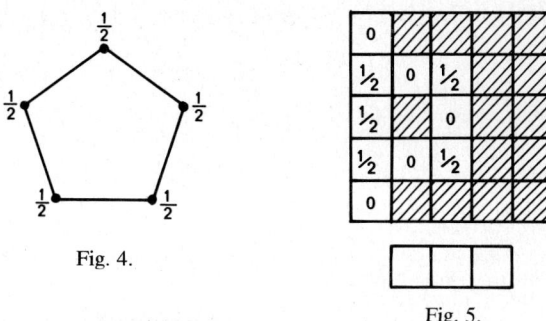

Fig. 4.

Fig. 5.

Now, let us associate with every hypergraph $H = (E_1, E_2, \ldots, E_m)$ of order n, a $(0, 1)$ matrix A as follows: the columns of the matrix A will correspond to the m edges of H, and the rows of the matrix will correspond to n vertices of H. At the intersection of the ith column and the jth row, the entry a_i^j will be 0 if $x_j \notin E_i$, and 1 if $x_j \in E_i$. Such a matrix is called the *incidence matrix* of H. Consider the two polytopes:

$$P = \{y : y \in R^n, y \geq 0, yA \geq 1\}$$
$$Q = \{z : z \in R^m, z \geq 0, Az \leq 1\}.$$

Clearly every transversal T of H determines a $(0, 1)$-vector of P and every $(0, 1)$-vector of P defines a transversal of H. Every matching of H defines a $(0, 1)$-vector of Q and conversely every $(0, 1)$-vector of Q defines a matching of H. Clearly:

$$\tau(H) = \min_{y \in P \cap \{0,1\}^n} \sum_{i=1}^{n} y_i = \min_{y \in P \cap \mathbf{Z}^n} \sum_{i=1}^{n} y_i,$$

$$\nu(H) = \max_{z \in Q \cap \{0,1\}^m} \sum_{i=1}^{m} z_i = \max_{z \in Q \cap \mathbf{Z}^m} \sum_{i=1}^{m} z_i$$

(where $\mathbf{Z}$ is the set of all integers).
For any integer $k \geq 1$ we write:

$$P_k = \{y : y \in R^n, y \geq 0, yA \geq k\},$$
$$Q_k = \{z : z \in R^m, z \geq 0, Az \leq k\}.$$

A k-*transversal* of H is defined by an integer vector $(y_1, y_2, \ldots, y_n)$ of P_k and

$$\tau_k(H) = \min_{y \in P_k \cap \{0,1,\ldots,k\}^n} \sum_{y=i}^{n} y_i = \min_{y \in P_k \cap \mathbf{Z}^n} \sum_{i=1}^{n} y_1.$$

A k-matching of H is defined by an integer vector $(z_1, z_2, \ldots, z_m)$ of Q_k and:

$$\nu_k(H) = \max_{z \in Q_k \cap \{0,1,\ldots,k\}^m} \sum_{i=1}^m z_i = \max_{z \in Q_k \cap Z^m} \sum_{i=1}^m z_i.$$

A *fractional transversal* and a *fractional matching* are the vectors of P and Q respectively. By the duality principle of linear programming, we have:

$$\tau^*(H) = \min_{y \in P} \sum_{i=1}^n y_i = \max_{z \in Q} \sum_{i=1}^m z_i.$$

Theorem 2.1. [6, 22]. *We have*

$$\nu(H) = \min_{k \geq 1} \frac{\nu_k(H)}{k} \leq \max_{H' \subseteq H} \frac{m(H')}{\Delta(H')} \leq \max_{k \geq 1} \frac{\nu_k(H)}{k} = \tau^*(H)$$

$$= \min_{k \geq 1} \frac{\tau_k(H)}{k} \leq \min_{A \subseteq H} \frac{|A|}{s(H_A)} \leq \max_{k \geq 1} \frac{\tau_k(H)}{k} = \tau(H).$$

(These inequalities are called the fundamental inequalities.)

Proof. 1. Let H' be a maximum matching of H. For every integer $k \geq 1$ we see that kH' is a k-matching of H. Hence

$$k\nu(H) = km(H') = m(kH') \leq \nu_k(H)$$

Thus for any integer $k \geq 1$, $(\nu_k(H)/k) \geq \nu(H)$, and

$$\nu(H) \leq \inf_k \frac{\nu_k(H)}{k} \leq \nu_1(H) = \nu(H),$$

It follows that

$$\nu(H) = \min_k \frac{\nu_k(H)}{k}.$$

2. Let H'' be a maximum matching of H. Then $m(H'') = \nu(H)$ and $\Delta(H'') = 1$. Hence

$$\nu(H) = \frac{m(H'')}{\Delta(H'')} \leq \max_{H' \subseteq H} \frac{m(H')}{\Delta(H')}$$

3. Let H'' be a partial hypergraph of H with $(m(H'')/\Delta(H'')) = \max_{H' \subset X} (m(H')/\Delta(H'))$ and put $k = \Delta(H'')$. The set H'' gives rise to an integral-vector $z \in Q_k$ such that $\sum_{i=1}^m z_i = m(H'')$. Since H'' is a k-matching,

$$\max_{H' \subseteq H} \frac{m(H')}{\Delta(H')} = \frac{m(H'')}{\Delta(H'')} \leq \frac{\nu_k(H)}{k} \leq \sup_k \frac{\nu_k(H)}{k}.$$

4. For an integer $k \geq 1$, let z^* be a k-matching of H with $\sum_{i=1}^m z_i^* = \nu_k(H)$. Then z^*/k is a fractional matching of H, and

$$\frac{\nu_k(H)}{k} = \sum_{i=1}^m \frac{z_i^*}{k} \leq \max_{z \in Q} \sum z_i = \tau^*(H).$$

Thus,

$$\sup_k \frac{\nu_k(H)}{k} \leq \nu^*(H).$$

Now let $\bar{z}$ be an optimal fractional matching of H such that

$$\sum_{i=1}^m \bar{z}_i = \tau^*(H).$$

Then since $\bar{z}$ is an extreme point of the polytope Q and the coordinates of an extreme point of Q are rationals there exist integers $\bar{y}_i$ ($i = 1, 2, \ldots, m$) and q such that

$$\bar{z}_i = \frac{\bar{y}_i}{q}.$$

Clearly $\bar{y}$ is a q-matching of H, and

$$\tau^*(H) = \sum_{i=1}^m \frac{\bar{y}_i}{q} \leq \frac{\nu_q(H)}{q}.$$

Thus

$$\sup_k \frac{\nu_k(H)}{k} \leq \nu^*(H) \leq \frac{\nu_q(H)}{q} \leq \sup_k \frac{\nu_k(H)}{k}.$$

So,

$$\nu^*(H) = \sup_k \frac{\nu_k(H)}{k} = \max_k \frac{\nu_k(H)}{k}.$$

5. Let $\bar{y}$ be an optimal fractional transversal of H. Then $\sum_{i=1}^n \bar{y}_i = \tau^*(H)$. Since $\bar{y}$ is an extreme point of the polytope P and the coordinates of an extreme point of P are rationals, we can write: $\bar{y}_i = (\bar{z}_i/p)$ for $i = 1, 2, \ldots, n$, where $\bar{y}_i$ and p are integers. Clearly $\bar{z}$ is a p-transversal of H and hence

$$\tau^*(H) = \frac{1}{p} \sum_{i=1}^n \bar{z}_i \geq \frac{\tau_p(H)}{p}.$$

For any integer $s \geq 1$ let y^* be a s-transversal of H such that $\sum_{i=1}^n y_i^* = \tau_s(H)$. Clearly y^*/s is a fractional transversal, and

$$\sum_{i=1}^n \frac{y_i^*}{s} = \frac{\tau_s(H)}{s} \geq \tau^*(H).$$

Thus,
$$\inf_k \frac{\tau_k(H)}{k} \geq \tau^*(H) \geq \frac{\tau_p(H)}{p} \geq \inf_k \frac{\tau_k(H)}{k}.$$

So
$$\tau^*(H) = \inf_k \frac{\tau_k(H)}{k} = \min_k \frac{\tau_k(H)}{k}.$$

6. Let us show that
$$\min_k \frac{\tau_k(H)}{k} \leq \min \frac{|A|}{s(H_A)}.$$

Given a set A, we have by definition
$$s(H_A) = \min_i |E_i \cap A| = s.$$

The characteristic vector of A is a s-transversal of H. Thus $\tau_s(H) \leq |A|$, and
$$\min_k \frac{\tau_k(H)}{k} \leq \frac{\tau_s(H)}{s} \leq \frac{|A|}{s(H_A)}.$$

The inequality follows.

7. Let us show that
$$\min_A \frac{|A|}{s(H_A)} \leq \max_k \frac{\tau_k(H)}{k}.$$

Let T be a minimum transversal. We have
$$\frac{|T|}{s(H_T)} \leq |T| \leq \tau(H) = \frac{\tau_1(H)}{1} \leq \max_k \frac{\tau_k(H)}{k}.$$

The inequality follows.

8. Let $t = (t_1, t_2, \ldots, t_n)$ be a minimum transversal of H (given by its characteristic vector). Then $\sum_{i=1}^n t_i = \tau(H)$. For an integer $k \geq 1$, kt a k-transversal of H. Thus $\tau_k(H) \leq \sum_{i=1}^n k t_i = k\tau(H)$. Hence

$$\tau(H) \geq \sup_k \frac{\tau_k(H)}{k} \geq \tau_1(H) = \tau(H).$$

Thus
$$\tau(H) = \max_k \frac{\tau_k(H)}{k}.$$

This achieves the proof.

Corollary 1. *Let H be a hypergraph such that $\tau(H) = \nu(H)$ (e.g. interval hypergraph, unimodular hypergraph, balanced hypergraph, etc...).*

Then a necessary and sufficient condition for the existence of a matching with k edges is that, for every $A \subseteq X$,

$$ks(H_A) \leq |A|.$$

The existence of a matching having k edges is equivalent to: $\nu(H) \geq k$. This is equivalent to

$$\min \frac{|A|}{s(H_A)} \geq k,$$

or:

$$ks(H_A) \leq |A| \quad (A \subseteq X).$$

Corollary 2. *Let H be a hypergraph such that $\tau(H^*) = \nu(H^*)$ (e.g. balanced hypergraphs, etc....).*

Then a necessary and sufficient condition for the existence of a covering of X with k edges is that, for every $A \subseteq X$,

$$kr(H_A) \geq |A|.$$

It is possible to cover X with k edges if and only if $\tau(H^*) \leq k$, or:

$$\max_{H' \subset H^*} \frac{m(H')}{\Delta(H')} \leq k.$$

This is equivalent to

$$\max_{A \subseteq X} \frac{|A|}{r(H_A)} \leq k,$$

or:

$$|A| \leq kr(H_A) \quad (A \subseteq X).$$

Theorem 2.2. *Let $H = (E_1, E_2, \ldots, E_m)$ be a hypergraph of order n, with $s = \min |E_i|$, let H' be the partial hypergraph of H formed by the edges of H with size s. If H' is regular, then*

$$\tau^*(H) = \frac{n}{s}.$$

Proof. The vector $t = \vec{1} = (1, 1, \ldots, 1)$ is a s-transversal of H. Hence $\tau_s(H) \leq \sum_{i=1}^n t_i = n$. This together with Theorem 2.1 gives

$$\frac{m(H')}{\Delta(H')} \leq \tau^*(H) \leq \frac{\tau_s(H)}{s} \leq \frac{n}{s}. \tag{1}$$

Let G be the bipartite graph whose vertices are the vertices and edges of H'; two vertices of G are adjacent if and only if one is a vertex x of H' and the other is an

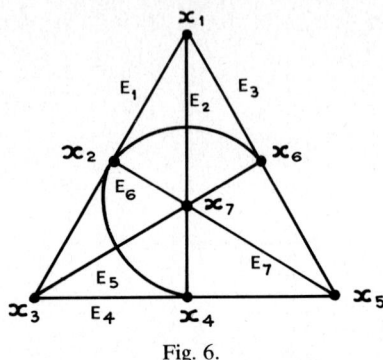

Fig. 6.

edge E of H' with $x \in E$. Counting the number of edges in G in two ways we get $(m(H')/\Delta(H')) = n/s$. Hence all the inequalities in (1) are equalities and Theorem 2.4 is proved.

Example 1. Let H be the lines of the finite projective plane with 7 points (or "Fano configuration") also noted PG(2, 2). Then H is of order 7, 3-uniform and regular. By Theorem 2.2, $\tau^*(H) = (m(H)/\Delta(H)) = 7/3$; so $\nu(H) = 1 \leq \tau^*(H) = 7/3 \leq \tau(H) = 3$.

Now consider a finite projective plane PG(2, $r-1$). It is defined by a set of $r^2 - r + 1$ 'points' and a set of $r^2 - r + 1$ 'lines' satisfying the following conditions:

(i) every point is contained in exactly r lines,
(ii) every line contains exactly r points,
(iii) any two distinct points occur together in exactly one line,
(iv) any two distinct lines have exactly one point in common.

It is evident that every line of the projective plane H is a minimal transversal of H; in PG(2, 2) there is no other minimal transversal. For the projective plane PG(2, $r-1$) with $r > 3$, Pelikan [31] has proved that edges are the only minimum transversal sets and all the other minimal transversal sets have size $\geq r + 2$. In fact the minimum size of a transversal set T which is not an edge is given in the third line of the following table:

$r = 3$	4	5	6	–	8	9
$n = 7$	13	21	31	–	57	73
$\|T\| = -$	6	7	9	–	12	?

For every projective plane H of rank r,

$$\nu(H) = 1 \leq \tau^*(H) = \frac{r^2 - r + 1}{r} \leq \tau(H) = r.$$

Example 2. Let H be a (n, k, λ)-design, i.e. a k-uniform hypergraph on n vertices such that any two vertices of H are contained in exactly λ edges. Clearly, H is regular of degree $\Delta(H) = \lambda(n-1/k-1)$ and $m(H) = \lambda(n(n-1)/k(k-1))$. Hence, we have $\tau^*(H) = (m(H)/\Delta(H)) = (n/k)$ (by Theorem 2.4). For the main known (n, k, λ)-designs we get

(n, k, λ) =	$(13, 3, 1)$	$(10, 4, 2)$	$(9, 4, 3)$	$(11, 3, 3)$	$(12, 4, 3)$
τ =	7	4	4	7	6
τ^* =	$\dfrac{13}{3}$	$\dfrac{5}{2}$	$\dfrac{9}{4}$	$\dfrac{11}{3}$	3

Example 3. The *complete r-uniform hypergraph* K_n^r is a hypergraph on a set X of n elements whose edges are all the subsets of size r. Then $\Delta(K_n^r) = \binom{n-1}{r-1}$, $n(K_n^r) = n$. Hence

$$\nu(K_n^r) = \left[\frac{n}{r}\right] \leq \tau^*(K_n^r) = \frac{n}{r} \leq \tau(K_n^r) = n - r + 1.$$

Proposition 2.3. *We have*:

$$\lim_{k \to \infty} \frac{\tau_k(H)}{k} = \tau^*(H) = \lim_{k \to \infty} \frac{\nu_k(H)}{k}.$$

Proof. Let z^* be an s-transversal and y^* be a t-transversal with $\sum_{i=1}^n z_i^* = \tau_s(H)$ and $\sum_{i=1}^n y_i^* = \tau_t(H)$ for two positive integers s and t. Then it is easy to see that $z^* + y^*$ is an $(s+t)$-transversal. Thus $\tau_{s+t}(H) \leq \tau_s(H) + \tau_t(H)$, so $\tau_k(H)$ is a sub-additive function of k. Hence by a theorem of Fekete we have

$$\lim_{k \to \infty} \frac{\tau_k(H)}{k} = \inf_k \frac{\tau_k(H)}{k} = \tau^*(H).$$

Similarly it is easy to see that $-\nu_s(H)$ is a sub-additive function of s. Hence

$$\lim_{k \to \infty} \frac{\nu_k(H)}{k} = \sup_k \frac{\nu_k(H)}{k} = \tau^*(H).$$

Proposition 2.4. *If $(\tau_k(H)/k) = \tau(H)$ for an integer $k \geq 1$ then $(\tau_p(H)/p) = \tau(H)$ for all positive integers $p \leq k$.*

Proof. By Theorem 2.1, $(\tau_p(H)/p) \leq \tau(H)$ for any integer $p \geq 1$. If the proposition were not true we could find an integer $p < k$ with $(\tau_p(H)/p) < \tau(H)$. Since $\tau_s(H)$ is a sub-additive function of s,

$$k\tau(H) = \tau_k(H) \leq \tau_{k-p}(H) + \tau_p(H) < (k-p)\tau(H) + p\tau(H) = k\tau(H).$$

The contradiction follows.

Proposition 2.5. *If $(\tau_k(H)/k) = \tau^*(H)$ for some integer $k \geq 1$, then $(\tau_{ks}(H)/ks) = \tau^*(H)$ for any integer $s \geq 1$.*

Proof. Since $\tau_p(H)$ is a sub-additive function of p, $\tau_{ks}(H) \leq s\tau_k(H)$. Hence

$$ks\tau^*(H) \leq \tau_{ks}(H) \leq s\tau_k(H) = ks\tau^*(H).$$

So

$$\frac{\tau_{ks}(H)}{ks} = \tau^*(H).$$

2.2. Application: The Erdös problem

For given integers n, k, r with $r \geq 2, k \geq 2$, Erdös [13] asked: What is the minimum number $M(n, r)$ of edges in a r-uniform hypergraph H of order $\leq n$ which has no bicoloring?

1. Erdös has proved that $M(n, r) \geq 2^{r-1}$. A short proof follows from Theorem 2.1. Let X be a set of size n. For a bipartition $\pi = (S_1, S_2)$ of the set X, let E_π denote the set of all r-tuples of X contained either in S_1 or in S_2. Then $H = (E_\pi : \pi$ is a bipartition of $X)$ is a hypergraph on the set of all r-tuples of X. The set of edges of a r-uniform hypergraph with no bicoloring is a transversal of H, and conversely. Hence $M(n, r) = \tau(H)$. Since $m(H) = 2^{n-1}$, $\Delta(H) = 2^{n-r}$, we get by Theorem 2.1,

$$M(n, r) = \tau(H) \geq \frac{2^{n-1}}{2^{n-r}} = 2^{r-1}.$$

2. Let $s(H) = \min_{E \in H} |E|$. If $\pi = (S_1, S_2)$ is a bipartition of X with $|S_1| = n_1$ and $|S_2| = n_2$, then by a classical property of the Pascal triangle,

$$|E_\pi| = \binom{n_1}{r} + \binom{n_2}{r} \geq \binom{[\frac{n}{2}]}{r} + \binom{[\frac{n}{2}]^*}{r}.$$

(In fact, the minimum of $\sum_{i=1}^k \binom{n_i}{r}$ for $n_1 + n_2 + \cdots + n_k = r$ is attained when $[n/k] \leq n_i \leq [n/k]^*$ for all i.) For the sake of simplicity let us suppose that n is even, $n = 2p$. Then $s(H) = 2\binom{p}{r}$.

Let $H' \subseteq H$ be the partial hypergraph of H whose edges are all the edges E_π of H with size $s(H)$. Then H' is a $2\binom{p}{r}$-uniform regular hypergraph. Now, by

Theorem 2.2,
$$\tau^*(H) = \frac{\binom{2p}{r}}{2\binom{p}{r}}.$$

Hence
$$M(n, r) = \tau(H) \geq \tau^*(H) = \frac{\binom{2p}{r}}{2\binom{p}{r}},$$

or:
$$M(2p, r) \geq \frac{2p-1}{p-1} \cdot \frac{2p-2}{p-2} \cdots \frac{2p-r+1}{p-r+1}.$$

For instance, consider the hypergraph H determined by the lines of the finite projective plane with 7 points (Fig. 6). It is easy to check that H has no bicoloring. Hence
$$M(7, 3) \leq 7$$

Now, by using the formula above, we get:
$$M(7, 3) \geq M(8, 3) \geq \frac{8-1}{3} \cdot \frac{8-2}{2} = 7.$$

This shows that $M(7, 3) = 7$, and that H is an optimal hypergraph.

3. Now, put $M(r) = \text{Lim}_{n=\infty} M(n, r)$.

For large values of r, the above results can be improved by looking at the particular properties of the considered hypergraph. The best result obtained so far is obtained by Jozsef Beck [2], who has proved that for large r,
$$M(r) \geq r^{\frac{1}{2}} 2r.$$

2.3. The number $\tau^0(H)$.

Theorem 2.1 is one of the possible improvements of the inequality $\nu(H) \leq \tau(H)$. We shall now give another one.

Let us denote by:
$\Delta^0(H)$: the maximum number of edges in a bundle,
$\tau^0(H)$: the minimum number of bundles needed to cover the edge-set of H,
$\tau^0_k(H)$: the minimum size of a collection of bundles such that each edge of H belongs to at least k of them.

Put:
$$\tau^{0*}(H) = \min_{k \geq 1} \frac{\tau_k^0(H)}{k}.$$

If H satisfies the Helly Property, then $\tau^0 = \tau$, $\Delta^0 = \Delta$. Chvátal [11] has conjectured that: *if H satisfies*
$$E \in H, F \subseteq E \Rightarrow F \in H,$$
then $\Delta^0(H) = \Delta(H)$.

Theorem 2.6. *We have*:
$$\nu(H) \leq \max_{H' \subseteq H} \frac{m(H')}{\Delta^0(H')} \leq \tau^{0*}(H) = \min_{k \geq 1} \frac{\tau_k^0(H)}{k}$$
$$\leq \max_{k \geq 1} \frac{\tau_k^0(H)}{k} = \tau^0(H) \leq (H).$$

Proof. Associate with the hypergraph $H = (E_i / i \in M)$ another hypergraph $\bar{H} = (\bar{E}_i / i \in M)$ whose vertices are the maximal bundles of H, the edge $\bar{E}_i$ being the set of all maximal bundles of H containing E_i.

Clearly
$$\nu(\bar{H}) = \nu(H)$$
$$\Delta(\bar{H}) = \Delta^0(H)$$
$$\tau(\bar{H}) = \tau^0(H)$$
$$\tau_k(\bar{H}) = \tau_k^0(H)$$
$$\tau^*(\bar{H}) = \tau^{0*}(H).$$

Thus, the required inequalities follow immediately from Theorem 2.1.

Example 1. The complete graph K_n with $n \geq 4$. A maximal bundle is either a K_3 (triangle) or a $K_{1,n-1}$ (star), and
$$\Delta^0(K_n) = \Delta(K_n) = n - 1.$$

Furthermore, it is always possible to cover the edges of K_n with one triangle and

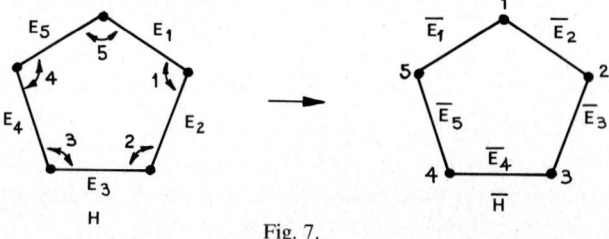

Fig. 7.

$n-3$ stars, and consequently:

$$\tau^0(K_n) \leq n-2.$$

The value of τ^0 is clearly 1 if $n=3$, or 2 if $n=4$. For $n>4$, it is impossible to cover the edges of K_n with $n-2$ triangles, since $1/3\, m(K_n) = (n(n-1)/6) > n-2$. Therefore a minimum covering must have at least one star, and:

$$\tau^0(K_n) = 1 + \tau^0(K_{n-1}) = 1 + 1 + \cdots + 1 + \tau^0(K_4) = n-2.$$

This shows that $\tau^0(K_n) = n-2$. The value of $\tau^{0*}(K_n)$ is $n/2$, since, by Theorem 2.6,

$$\frac{m(K_n)}{\Delta^0(K_n)} = \frac{n(n-1)}{2(n-1)} = \frac{n}{2} \leq \tau^{0*}(K_n) \leq \frac{\tau_2^0(K_n)}{2} \leq \frac{n}{2}$$

(because the edges of K_n are covered twice by a collection of n stars).

Example 2. The complete r-uniform hypergraph K_n^r, with $n \geq 2r$. There is a famous theorem of Erdös, Chao-Ko and Rado [14], which says that $\Delta^0(K_n^r) = \Delta(K_n^r) = \binom{n-1}{r-1}$. Hence:

$$\frac{m}{\Delta^0} = \binom{n}{r}\binom{n-1}{r-1}^{-1} = \frac{n}{r} \leq \tau^{0*} \leq \frac{\tau_r^o}{r} \leq \frac{n}{r}$$

Thus, $\tau^{0*}(K_n^r) = n/2$.

The problem of finding $\tau^0(K_n^r)$ was a famous open problem for twenty years due to Kneser [21] who conjectured that

$$\tau^0(K_n^r) = n - 2r + 2.$$

For $r=3$, this was proved by J.C. Meyer [29] (and later by Stahl [36]).

By using some combinatorial topological arguments, Lovász [24] proved it for all r in 1977. Clearly, K_n^r can be covered by one K_{2r-1}^r and $(n-2r+1)$ maximum stars (a "star" being here a set of $\binom{n-1}{r-1}$ edges containing a vertex a) so, $\tau^0(K_n^r) \leq n - 2r + 2$.

A very short proof of the above equality was given recently by Bárány [1], who improved Lovász's proof by using only Borsuk's theorem and a theorem of D. Gale:

Borsuk's theorem. *If the unit sphere $S_k = \{x/x \in R^{k+1}, \|x\|=1\}$ is the union $(k+1)$ sets which are open in S_k, then one of these sets contains two antipodal points.*

Gale's theorem [18]. *If r and k are non-negative integers, and if for $a \in S_k$, we denote by $S_k(a)$ the set $\{x/a \in S_k, \langle a, x \rangle > 0\}$, then there exists a set $X \subseteq S_k$ with $2r+k$ elements such that $|X \cap S_k(a)| \geq r$ for every $a \in S_k$.*

[Now suppose that a set X with $|X|=2r+k$ have all its r-tuples split into $k+1$ classes (bundles). This defines a $(k+1)$-coloring of S_k as follows: if $a \in S_k(a)$, then the point a gets the color of each r-tuple in $X \cap S_k(a)$. By Gale's theorem, every $a \in S_k$ has a color. Now the result follows from the Borsuk theorem.]

2.4. Upper bounds for the transversal number

Theorems 2.1 and 3 give lower bounds for $\tau(H)$. If H has the Helly Property then $\tau(H) = \tau^0(H)$, and there are many known upper bounds for $\tau^0(H)$ (because this is also the chromatic number of the complement of the line-graph $L(H)$), e.g.:

Theorem 2.7. *If H satisfies the Helly Property, then*

$$\tau(H) \leq \max_{i \in M} m[H \times (X - E_i)],$$

except if the complement of the line graph $L(H)$ possesses as a connected component either a clique with $\Delta(\overline{L(H)})+1$ vertices, or an odd cycle (and $\Delta(\overline{L(H)})=2$). In both cases,

$$\tau(H) = \max_{i \in M} m[H \times (X - E_i)] + 1$$

Another upperbound for $\tau(H)$ can be given. The "Greedy algorithm" to obtain a transversal $T = \{x_1, x_2, \ldots,\}$ with small cardinality, is an iterative procedure as follows:

First step. Choose a vertex x_1 with a maximum degree in $H_1 = H$, and remove from H the set $H(x_1)$ of all edges containing x_1.

Second step. Choose a vertex x_2 with maximum degree in $H_2 = H_1 - H_1(x_1)$, etc.

ith step. Choose a vertex x_i with maximum degree in $H_i = H_{i-1} - H_{i-1}(x_{i-1})$.

This procedure is continued until all the vertices have degree 0. Then $\{x_1, x_2, \ldots,\}$ is a transversal of H.

Let l be the set of all transversals of H obtained by a greedy algorithm. Define $\tilde{\tau}(H) = \max_{T \in l} |T|$. Then $\tilde{\tau}(H) \geq \tau(H)$. $\tilde{\tau}(H)$ is called the *greedy transversal number* of H.

Theorem 2.8. (Lovász [27], Stein [37].) *We have*:

$$\tau(H) \leq \tilde{\tau}(H) \leq \left(1 + \frac{1}{2} + \cdots + \frac{1}{\Delta(H)}\right) \max_{H' \subseteq H} \frac{m(H')}{\Delta(H')} \leq [1 + \log \Delta(H)] \tau^*(H).$$

Proof. Let $T = \{x_1, x_2, \ldots,\}$ be a transversal of $H = (E_1, E_2, \ldots, E_m)$ obtained by a greedy algorithm that satisfies $|T| = \tilde{\tau}(H)$. For the sake of simplicity, put:

$$\max_{H'} \frac{m(H')}{\Delta(H')} = \xi.$$

For an integer $\lambda \geq 1$, let t_λ be the number of steps where a vertex of degree λ is chosen. Thus

$$\tilde{\tau}(H) = |T| = t_\Delta + t_{\Delta-1} + \cdots + t_1.$$

After the step $k = t_\Delta + t_{\Delta-1} + \cdots + t_{\lambda+1}$, we choose a vertex in a hypergraph H_{k+1} with maximum degree $\leq \lambda$. Since $H_{k+1} \subseteq H$,

$$m(H_{k+1}) \leq \xi \Delta(H_{k+1}) \leq \xi \lambda.$$

Hence

$$\frac{m(H_{k+1})}{\lambda(\lambda+1)} = \frac{1}{\lambda(\lambda+1)} (t_1 + 2t_2 + \cdots + \lambda t_\lambda) \leq \frac{\xi}{\lambda+1}.$$

Thus

$$\frac{1}{1 \cdot 2} t_1 \leq \frac{\xi}{2}$$

$$\frac{1}{2 \cdot 3} (t_1 + 2t_2) \leq \frac{\xi}{3}$$

$$\cdots \cdots \cdots \cdots$$

$$\frac{1}{\lambda(\lambda+1)} (t_1 + 2t_2 + \cdots + \lambda t_\lambda) \leq \frac{\xi}{\lambda+1}$$

$$\cdots \cdots \cdots \cdots$$

$$\frac{1}{\Delta} (t_1 + 2t_2 + \cdots + \Delta t_\Delta) \leq \xi.$$

Now, we add the above inequalities; the coefficient of t_λ, at the left hand-side, is

$$\lambda \left[\frac{1}{\lambda(\lambda+1)} + \frac{1}{(\lambda+1)(\lambda+2)} + \cdots + \frac{1}{\Delta} \right]$$

$$= \lambda \left[\left(\frac{1}{\lambda} - \frac{1}{\lambda+1}\right) + \left(\frac{1}{\lambda+1} - \frac{1}{\lambda+2}\right) + \cdots + \left(\frac{1}{\Delta-1} - \frac{1}{\Delta}\right) + \frac{1}{\Delta} \right] = 1$$

So, the left-hand side reduces to:

$$t_1 + t_2 + \cdots + t_\Delta = |T| = \tilde{\tau}(H).$$

The first inequality follows. Since by Theorem 2.1, $\tau^*(H) \geq \xi$, the proof is complete.

Now, consider the maximum size of a transversal which is minimal (with respect to the inclusion). This number, denoted $\tau'(H)$ is called the *lazy transversal number*. Clearly, $\tau(H) \leq \tau'(H)$.

Theorem 2.9. Let $H = (E_1, E_2, \ldots, E_m)$ be a hypergraph of order n, let $s = \min_i |E_i|$ and $\Delta = \Delta(H)$. Then,

$$\tau(H) \leq \tau'(H) \leq \left[\frac{n\Delta}{\Delta + s - 1}\right].$$

Proof. Let T be a minimal transversal of H with $|T| = \tau'(H)$. Since T is minimal, for every $x \in T$, there is an edge E_x of H such that $E_x - \{x\} \subseteq X - T$. Let $H' = (E_x / x \in T)$. Then

$$|X - T|\Delta \geq \sum_{a \in X-T} d_{H'}(a) = \sum_{x \in T} |E_x - \{x\}| \geq (s-1)|T|.$$

Hence

$$\tau'(H) = |T| \leq \left[\frac{\Delta n}{\Delta + s - 1}\right].$$

Remark 2.10. If $s = 2$ the above Theorem 2.9 gives the best possible bound for $\tau(H)$. Write $n = p(\Delta + 1) + q$, $0 \leq q \leq \Delta$.

Let G be the union of p disjoint $(\Delta + 1)$-cliques and one q-clique. Then $\Delta(G) = \Delta$, and

$$\tau(G) = \Delta p + q - 1 = \left[\frac{\Delta n}{\Delta + 1}\right].$$

Remark 2.11. If H is a 3-uniform Δ-regular hypergraph of order n, Henderson and Dean [17] have shown that

$$\tau(H) \leq n\left(\frac{\Delta}{3} - \frac{\Delta - 1}{4} + \frac{1}{8}\sum_{t=3}^{\Delta-1}\frac{4.6 \cdots (2t-2)}{9.11 \cdots (2t+3)}\right).$$

Thus, for $\Delta = 3$, we get

$$\tau(H) \leq \left[\frac{n}{2}\right].$$

This bound is the best possible when $\Delta = 3$. In Fig. 8 we describe a 3-uniform hypergraph of order n, regular of degree 3 which attains the bound. Let us assume $n = 2k$. Let the vertices $1, 2, \ldots, 2k$ of H be arranged on the circle as shown in Fig. 8.

The edges of H are all the possible triples obtained from the triangle $\{1, k+1, 2k\}$ by clockwise rotations. Clearly, $\Delta(H) = 3$, and H has a transversal $\{1, 2, \ldots, k\}$ with cardinality $k = [n/2]$.

If there is a smaller transversal T_0, then we may assume that 1 and $k+1$ are not in T_0. Then $2k$ and k are both in T_0. Thus $|T_0| \geq k$, which is a contradiction. So, $\tau(H) = [n/2]$.

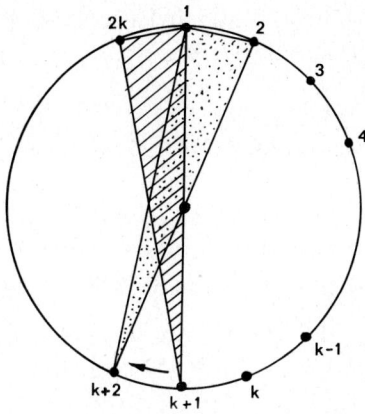

Fig. 8.

Remark 2.12. The maximum value $f(n, r)$ for the transversal number of a r-uniform hypergraph of order n with maximum degree 2 has been recently determined by Sterboul [41], who has proved:

If $\lambda = [2n/r]$, we have

For r even: $f(n, r) = [2\lambda/3]$

For r odd: $f(n, r) = [2\lambda/3]$ or $[4n/3r+1]$ or $[4n/3r+1]-1$, according to the values of λ.

(The different values of λ giving these numbers are well defined and are described in [41].)

2.5. Applications to graphs

From now, we shall give more specific results for the hypergraphs which are the most commonly used in combinatorial problems.

First consider the case where the hypergraph is a simple graph G; i.e. a 2-uniform hypergraph with no repeated edges.

Theorem 2.13. *For a graph G,*

$$\tau^*(G) = \frac{\tau_2(G)}{2} = \frac{\nu_2(G)}{2}$$

Furthermore, there exists an optimal 2-matching H of G such that each connected component of H is either a single vertex, or a pair of parallel edges ('double-edge'), or an odd cycle. For such a 2-matching, there exists an optimal 2-transversal $t(x)$ with values 0 for a singleton vertex of H; (0, 2) or (1, 1) for the vertices belonging to a double edge of H; and 1 for the vertices belonging to an odd cycle of H.

The fact that $(\nu_2(G)/2) = \tau^*(G)$ is due to Tutte: Lovász [26] has given the above equalities (with some generalizations). We shall give here a simple proof of the complete theorem.

Proof. Let $H_1 \subseteq 2G$ be an optimal 2-matching. It is easy to see that every connected component of H_1 which is an even chain or an even cycle can be replaced by a set of double-edges, so that we can get an optimal 2-matching H of the type described above.

Now, we shall label each vertex with 0, 1 or 2, by an iterative procedure described by the following rules:

Rule 1. Label with 0 each vertex which is a singleton of H.
Rule 2. Label with 2 each vertex which is adjacent in G with a vertex previously labelled with 0.
Rule 3. Label with 0 every vertex which is adjacent in H with a vertex previously labelled with 2.
Rule 4. Every vertex which cannot be labelled by the iterative procedure described by rules 1, 2 and 3 will be labelled with 1.

No odd chain consisting alternately of edges of $G-H$ and of double-edges of H connects two singletons of H because such a chain would constitute a connected component of a 2-matching H' with $m(H') = m(H) + 1$. Similarly, no odd chain of that kind connects a singleton of H to an odd cycle of H, since in this case we could replace the double-edges in the chain and the odd cycle by a set of new double-edges to get a 2-matching H' with $m(H') = m(H) + 1$. This contradicts the maximality of H.

No odd chain of the above kind can cross itself in a vertex labelled with 0 (because then we could get a better 2-matching by replacing the double-edges in the chain by an odd cycle and a set of new double-edges).

Thus only one label $t(x)$ can be given to a vertex x by the above rules, and
$t(x) = 0$ if x is a singleton of H,
$t(x) = 0$ and $t(y) = 2$ if $[x, y]$ is a double-edge of H (or vice versa) connectable to a singleton.
$t(x) = t(y) = 1$ if $[x, y]$ is a double-edge of H not connectable to a singleton,
$t(x) = 1$ if x belongs to an odd cycle of H.
The rule 2 also shows that $t(x)$ is a 2-transversal of G. Furthermore,

$$\frac{m(H)}{2} = \frac{\nu_2(G)}{2} \leq \tau^*(G) \leq \frac{\tau_2(G)}{2} \leq \frac{1}{2} \sum_{x \in X} t(x) = \frac{m(H)}{2}.$$

Thus equalities hold in the above inequalities. This completes the proof.

A hypergraph H is said to be *regularisable* if by replacing each edge E by a *non-empty* set of edges equal to E, we get a new hypergraph with all vertices having the same degree. A hypergraph H is said to be *quasi-regularisable* if by replacing each edge E by a set (possibly empty) of parallel edges we get a regular hypergraph. From Theorem 2.13, it follows that a graph G is *quasi-regularisable* if and only if the vertex set can be covered by pairwise disjoint classes consisting of either one edge, or an odd cycle.

Theorem 2.14. *For a graph G, the weight $\frac{1}{2}$ on each vertex is always a fractional transversal; it is an optimal one if and only if the vertex set can be covered by edges and odd cycles (pairwise disjoint). This optimal fractional transversal is unique if and only if G is regularisable and has no bipartite connected component.* (Berge, [4]).

Theorem 2.15. (*Lovász*, [23].) $\tau^*(G) \leq \frac{1}{2}(\nu(G) + \tau(G))$.

Proof. By Theorem 2.13, it suffices to show that

$$\tau_2(G) \leq \nu(G) + \tau(G).$$

Let S be a stable set of G with maximum size. Consider a bipartite graph H whose vertex set is the same as that of G and whose edges are the edges of G between S and $X - S$, where X is the vertex set of G. Then by König's Theorem,

$$\nu(H) = \min_{B \subseteq S} (|S - B| + |\Gamma_H(B)|).$$

Let us write $\nu(H) = k$. Then $k \leq \nu(G)$. Take a set $B \subseteq S$ with

$$\nu(H) = |S - B| + |\Gamma_H(B)|.$$

We define a 2-transversal t of G as follows:

$$t_i = \begin{cases} 0 & \text{if } x_i \in B, \\ 1 & \text{if } x_i \in S - B \text{ or } x_i \in X - S - \Gamma_H(B), \\ 2 & \text{if } x_i \in \Gamma_H(B). \end{cases}$$

Since S is a maximum stable set, $X - S$ is a minimum transversal and $|X - S| = \tau(G)$. Thus

$$\sum_{i=1}^{n} t_i = |S - B| + 2|\Gamma_H(B)| + |X - S - \Gamma_H(B)|$$
$$= |S - B| + |X - S| + |\Gamma_H(B)|$$
$$= k + \tau(G) \leq \nu(G) + \tau(G).$$

Now, since t is a 2-transversal of G,

$$\tau_2(G) \leq \sum_{i=1}^{n} t_i = \nu(G) + \tau(G). \quad \square$$

An easy corollary of this result is: $\nu(G) = \tau(G)$ if and only if $\tau^*(G) = \tau(G)$. As quoted by Lovász, this would not be true for r-uniform hypergraphs with $r > 2$.

2.6. Application to duals of graphs

Let $G = (X, E)$ be a graph of order n with no isolated vertices and with m edges. Then the *dual* G^* of G is a hypergraph whose vertices e_j represent the edges of G and whose edges are

$$X_i = \{e_j : x_i \in E_j\}.$$

Example 1. Let $G = (X, E)$ be the graph with $X = \{x_1, x_2, x_3, x_4\}$ and $E = \{e_1, e_2, e_3, e_4, e_5\}$ where $e_1 = (x_1, x_2)$, $e_2 = (x_2, x_3)$, $e_3 = (x_2, x_3)$, $e_4 = (x_4, x_2)$ $e_5 = (x_4, x_3)$. Then the dual $G^* = (X_1, X_2, X_3, X_4)$ is defined by $X_1 = \{e_1\}$, $X_2 = \{e_1, e_2, e_3\}$, $X_3 = \{e_2, e_3, e_5\}$ and $X_4 = \{e_4, e_5\}$.

For any real number λ, we write $\vec{\lambda} = (\lambda, \lambda, \ldots, \lambda)$.
For two vectors x and y we write $\langle x, y \rangle = \sum_i x_i y_i$

Lemma 2.16. $\nu_k(G^*) = nk - \tau_k(G)$.

Proof. If t is a k-transversal of G then $t' = \vec{k} - t$ is a k-matching of G^* because

$$\langle t', a_j \rangle = \langle \vec{k}, a_j \rangle - \langle t, a_j \rangle \leq 2k - k = k$$

(and conversely). Hence

$$\nu_k(G^*) = \max_{t'} \sum t'_i = kn - \min_t \sum t_i = nk - \tau_k(G).$$

Example 2. $\tau_k(G^*) = nk - \nu_k(G)$.

Proof. Let z^* be a maximum k-matching of G. Then $\nu_k(G) = \sum_{i=1}^m z_i^*$ and $\langle z^*, a^i \rangle \leq k$, for $i = 1, 2, \ldots, n$. Define $\varepsilon_i = k - \langle z^*, a^i \rangle$ for $i = 1, 2, \ldots, m$, and $S = \{x_i : \varepsilon_i > 0\}$. Then S is a stable set of G: otherwise we can get a k-matching $\bar{z}$ of G with $\sum_{i=1}^m \bar{z} > \sum_{i=1}^m z_i^*$, a contradiction.

Now, for each vertex $x_i \in S$, define a new weight $z'_j = z_j + \varepsilon_i$ for one e_j of the edges incident to x_i (for the other edges, put $z'_j = z_j$). Then z' is a k-transversal of G^*, because

$$\sum_{j : x_i \in e_j} z'_j = \sum_{j : x_i \in e_j} z_j + \varepsilon_i = \langle z, a^i \rangle + k - \langle z, a^i \rangle = k$$

Hence

$$\tau_k(G^*) \leq \sum_j z'_j = \sum_j z_j + \sum_i (k - \langle z, a^i \rangle)$$

$$= \nu_k(G) + nk - 2\nu_k(G) = nk - \nu_k(G). \quad (1)$$

Conversely, if z' is a minimum k-transversal of G^*, we have $\langle z', a^i \rangle \geq k$ for all j, and no edge e_j of G has a weight $z' > 0$ if the two end points x_p and x_q of e_j satisfy

$\langle z', a^p \rangle > k$ and $\langle z', a^q \rangle > k$. Hence, the edges of G which are incident to $S = \{x_i : \langle z', a^i \rangle > k\}$ have an excess weight of $\sum_i (\langle z', a^i \rangle - k)$, and by subtracting this total weight, we get a new vector z such that $\langle z, a^i \rangle = k$ for all $x_i \in S$. Thus, z is a k-matching of G. Hence

$$\nu_k(G) \geq \sum_j z_j = \sum_j z'_j - \sum_i (\langle z', a^i \rangle - k)$$
$$= kn - \sum z'_i = kn - \tau_k(G^*). \tag{2}$$

By comparing (1) and (2), we get $\nu_k(G) = kn - \tau_k(G^*)$.

Theorem 2.17. *We have*:
$$n - \tau^*(G) = \tau^*(G^*) = \frac{\tau_2(G^*)}{2} = \frac{\nu_2(G^*)}{2}.$$

This follows immediately from the previous lemmas and Theorem 2.13.

Theorem 2.18. $\tau^*(G^*) \geq \frac{1}{2}[\nu(G^*) + \tau(G^*)]$

This follows from Lemma 2.16 and Theorem 2.15.

2.7. Application to matroids

Let M be a matroid on a set E of elements. This is also a hypergraph with vertex-set E, and whose edges are all the maximal independent sets (bases).

Theorem 2.19 (J. Edmonds). *We have*:

$$\tau^*(M) = \min_{B \subseteq E} \frac{|B|}{s(H_B)} \quad \text{and} \quad \nu(M) = [\tau^*(M)].$$

2.8. Application to duals of matroids

Let us consider now the *dual* M^* of a matroid $M = (E, B)$ – with the sense of duality defined in Part 1, Session 1.

Theorem 2.20 (J. Edmonds). *We have*:

$$\tau^*(M^*) = \max_{A \subseteq E} \frac{|A|}{r(H_A)}$$

and

$$\tau(M^*) = [\tau^*(M^*)]^*$$

where $[z]^*$ denotes the least integer $\geq z$.

2.9. Application to edge-colorings

We shall associate with each hypergraph $H = (E_i/i \in M)$ another hypergraph $\bar{H} = (\bar{E}_i/i \in M)$ whose vertices are the maximal matchings of H, the edge $\bar{E}_i$ being the set of all maximal matchings of H containing E_i (see example on Fig. 9). Clearly, $\bar{E}_i$ and $\bar{E}_j$ intersect if and only if E_i and E_j do not intersect.

In this section, we are going to apply the Theorem 2.1 to $\bar{H}$ when H is a graph.

Let G be a multigraph with no loops, and let $k \geq 1$ be an integer. A *k-tuple coloring* of G is an assignment of k colors to each edge of G such that no two adjacent edges have a color in common. We denote by $q_k(G)$ the least number of colors needed for a k-tuple coloring of G. Let kG denote the multigraph obtained from G by replacing each edge by k parallel edges. Thus by definition, $q_k(G) = q(kG)$, the chromatic index of kG, and $q_1(G) = q(G)$, the chromatic index of G. The *fractional chromatic index* $q(G)$ is

$$q^*(G) = \min_k \frac{q(kG)}{k}.$$

Theorem 2.21. *If G is a multigraph without loops,*

$$\Delta(G) \leq \Delta_0(G) \leq \max_{G' \subseteq G} \frac{m(G')}{\nu(G')} \leq q^*(G) = \min_k \frac{q(kG)}{k} \leq \max_k \frac{q(kG)}{k} = q(G).$$

Proof. Let $\bar{G}$ be a hypergraph whose vertices are the maximal matchings of G and, for each edge E of G, we define an edge $\bar{E}$ of $\bar{G}$ by:

$$\bar{E} = \{\text{all maximal matchings of } G \text{ containing } E\}.$$

Clearly,

$$m(\bar{G}) = m(G),$$
$$\tau(\bar{G}) = q(G),$$
$$\tau_k(\bar{G}) = q(kG)$$
$$\tau^*(\bar{G}) = q^*(G),$$
$$\Delta(\bar{G}) = \nu(G),$$
$$\nu(\bar{G}) = \Delta_0(G).$$

The proposition follows from Theorem 2.1.

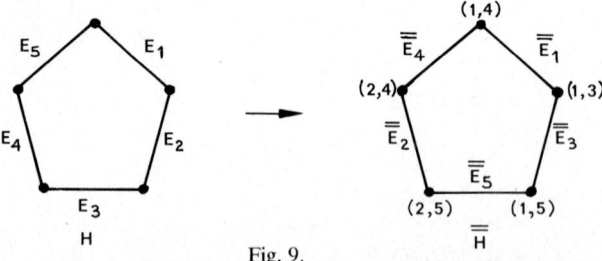

Fig. 9.

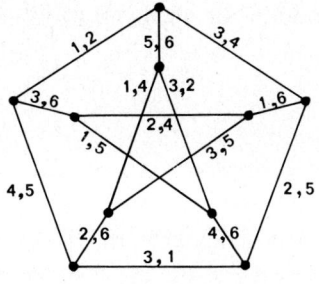

Fig. 10.

Example 1. $G = C_{2p+1}$, the odd cycle of length $2p+1$, i.e. an odd cycle with $2p+1$ edges; we get a p-tuple coloring of the edges by coloring the first edge with colors $1, 3, 5, \ldots, 2p-1$, the second edge with $2, 4, \ldots, 2p$, the third edge with $3, 5, \ldots$, etc. Thus $q(pG) \leq 2p+1$, and therefore

$$\frac{m(G)}{v(G)} = \frac{2p+1}{p} \leq q^*(G) \leq \frac{q(pG)}{p} \leq \frac{2p+1}{p}.$$

Hence

$$q^*(C_{2p+1}) = 2 + \frac{1}{p}.$$

Example 2. $G = P_{10}$, the Petersen graph with 10 vertices.
Since then exists a 2-tuple coloring with 6 colors, represented in Fig. 10, $q(2G) \leq 6$, so that

$$\Delta(G) = 3 \leq q^*(G) \leq \frac{q(2G)}{2} \leq \frac{6}{2} = 3.$$

Hence

$$q^*(P_{10}) = 3.$$

Example 3. $G = K_n$, the complete graph with n vertices. If n is even,

$$n - 1 = \Delta(G) \leq q^*(G) \leq q(G) = n - 1.$$

Thus $q^*(K_n) = n - 1$, when n is even.
If n is odd, $n \geq 3$,

$$n = \frac{\frac{n(n-1)}{2}}{\frac{n-1}{2}} = \frac{m(G)}{v(G)} \leq q^*(G) \leq q(G) = n$$

Thus $q^*(K_n) = n$, when n is odd. For all n, we can also write:

$$q^*(K_n) = 2\left[\frac{n}{2}\right]^* - 1$$

Theorem 2.22. *Let G be a Δ-regular multigraph on X without loops. Then the following are equivalent*:

  (i) *for some integer k, it is possible with Δk colors, to assign k colors with each edge such that at each vertex, each of the Δk colors occurs exactly once* ("*uniform multicoloring*"),
  (ii) $q^*(G) = \Delta(G)$,
  (iii) *for every set A of vertices with odd cardinality, the number of edges between A and $X - A$ is $\geq \Delta$.*

The only non-trivial part is (iii) $\Rightarrow$ (i), but this can be done by induction on n and Δ.

Theorem 2.23. *Let G be a multigraph without loops, then*

$$q^*(G) = \max_{G' \subseteq G} \frac{m(G')}{\nu(G')}$$

Furthermore, we can restrict the partial graphs G' to the ones of the following kinds:

  (i) *the maximum stars of G, i.e. the sets of edges incident to a vertex with maximum degree*,
  (ii) *the section-graphs $G \times A$, where A is a set such that $|A| \equiv 0$ (2), $|A| \neq 1$, $\nu(G_A) = \frac{1}{2}(|A|-1)$ and $\Delta(G_A) = \Delta(G)$.*

The fact that

$$q^*(G) = \max_{\substack{G' \subseteq G \\ |A|=1}} \left\{ \Delta(G), \max_{\substack{A \subseteq X \\ |A|=1}} \frac{2m(G_A)}{|A|-1} \right\}$$

follows also from a theorem of J. Edmonds [12] on extremal points of the matching polytop. See also Seymour [35], Stahl [36].

2.10. Application to chromatic number of r-uniform hypergraphs

In this section, we are going to apply the Theorem 2.1 to a particular hypergraph H_1 which satisfies the Helly Property for dimension r (cf. Section 1.1).

Let $H = (E_1, E_2, \ldots, E_m)$ be a r-uniform hypergraph on X. A *stable set* is a set $S \subseteq X$ which does not contain any edge of H. A *coloring of H with p colors* is a partition of X into p stable sets. The *chromatic number* $\chi(H)$ is the least p for which there exists a coloring with p colors.

A *k-tuple coloring* of H is an assignment of k colors to each vertex such that all the vertices receiving a color i constitute a stable set (no edge is "monochromatic" with i). The least number of colors needed for a k-tuple coloring is denoted by $\chi_n(H)$. Clearly, $\chi_1(H) = \chi(H)$.

Now, let H_2 be a hypergraph whose vertices are the maximal stable sets of H, each edge X_i being the set of maximal stable sets of H which contain a vertex x_i.

Clearly,

$$m(H_2) = n(H),$$
$$\Delta(H_2) = \beta(H),$$
$$\tau(H_2) = \chi(H),$$
$$\tau_k(H_2) = \chi_k(H).$$

The *fractional chromatic number* $\chi^*(H)$ is

$$\chi^*(H) = \min_k \frac{\chi_k(H)}{k} = \min_k \frac{\tau_k(H_2)}{k}.$$

Thus by Theorem 2.1,

$$\chi^*(H) = \tau^*(H_2).$$

By Theorems 2.1 and 2.8, we have immediately:

Theorem 2.24.

$$\max_{A \subseteq X} \frac{|A|}{\alpha(H_A)} \leq \chi^*(H) = \min_k \frac{\chi_k(H)}{k} \leq \max_k \frac{\chi_k(H)}{k} = \chi(H).$$

Many results [42, 34, 9, 20] are intended to improve some of these inequalities.

For instance, let G be a graph, and let $(S_1, S_2, \ldots, S_k)$ be a coloring of its vertices such that, for $j = 1, 2, \ldots, k$, the set S_j has at least one of its vertices adjacent to *all the other colors*. The worst value of k for such a coloring is called the *lazy chromatic number* $\chi'(G)$.

Also, let $(S_1, S_2, \ldots, S_k)$ be a coloring obtained by a maximum stable set S_1, and a maximum stable set S_2 of $X - S_1$, etc.... The worst value of k for such a coloring is called the *greedy chromatic number* $\tilde{\chi}(G)$. The best upper bounds for $\chi'(G)/\chi(G)$ for $\tilde{\chi}(G)/\chi(G)$ are not known, not even when G is a line-graph, or when G is an interval graph (the "Woodall conjecture").

References

[1] I. Bárány, A short proof of Kneser's conjecture (to appear).
[2] J. Beck, On a combinatorial problem of P. Erdos and L. Lovász, Discrete Math. 17 (1977) 127–132.
[3] C. Berge, Graphs and Hypergraphs, 2nd edition (North-Holland, Amsterdam, 1976).
[4] C. Berge, Regularisable graphs, Discrete Math. 23 (1978) 85–95.

[5] C. Berge and D.K. Ray-Chaudhuri, eds., Hypergraph Seminar 1972, Lecture Notes 411 (Springer-(1975) 117–123.
[6] C. Berge and M. Simonovits, The coloring numbers of the direct product of two hypergraphs, in: C. Berge and D.K. Ray-Chaudhuri, eds., Hypergraph Seminar 1972, Lecture Notes 411 (Springer-Verlag, Berlin, 1972) 21–33.
[7] C. Berge and M. Las Vergnas, Sur un théorème du type König, Annals N.Y. Acad. Sci. 175 (1970) 32–40.
[8] R. Brualdi and T. Foregger, J. Comb. Theory Ser. B, 17 (1974) 81–115.
[9] C.C. Chen and A. Hilton, A 4-color conjecture for planar graphs (to appear).
[10] V. Chvátal, On finite polarised relations, Canad. Math. Bull. 12 (1969) 321–326.
[11] V. Chvátal, Intersecting families of edges in hypergraphs having the hereditary property, in: C. Berge and D. K. Ray-Chaudhuri, eds., Hypergraph Seminar 1972, Lecture Notes 411 (Springer-Verlag, Berlin, 1972) 61–66.
[12] J. Edmonds, Maximum matchings and a polyhedron with $(0, 1)$ vertices, J. Res. Nat. Bur. Standards Ser. B 69 (1965) 125–130; 67–70.
[13] P. Erdös, On extremal problems of graphs and generalised graphs, Israel J. Math. 2(1964) 183–190.
[14] P. Erdös, Chao-Ko and R. Rado, Intersecting theorems for systems of finite sets, Quart. J. Math. 12 (1961) 313–320.
[15] J.C. Fournier and M. Las Vergnas, Une classe d'hypergraphes bichromatiques, Discrete Math. 2 (1972) 407–410.
[16] P. Frank, private communication.
[17] D.R. Fulkerson, Antiblocking pairs of polyhedra, J. Comb. Theory Ser. B 12 (1972) 50–71.
[18] D. Gale, Neighboring vertices on convex polyhedron, in: Linear Inequalities (Princeton University Press, Princeton, NJ, 1956).
[19] J.R. Henderson and R.A. Dean, The 1-width of $(0, 1)$–matrices having constant sum three, J. Comput. Theory Ser. 16 (1974) 355–370.
[20] A.J.W. Hilton, private communication.
[21] M. Kneser, Aufgabe 300, Jber. Deut. Math. Verl. 58 (1955).
[22] L. Lovász, Normal hypergraphs and the perfect graph conjecture, Discrete Math, 2 (1972) 253–267.
[23] L. Lovász, Minimax theorems for hypergraphs, in: C. Berge and D.K. Ray-Chaudhuri eds., Hypergraph Seminar 1972, Lecture Notes 411 (Springer-Verlag, Berlin, 1972) 111–126.
[24] L. Lovász, Kneser's conjecture, chromatic number and homotopy (to appear).
[25] L. Lovász, On two minimax theorems in graphs, J. Comb. Theory Ser. B 21 (1976) 96–103.
[26] L. Lovász, 2-matchings and 2-covers of hypergraphs, Acta Math. Ac.Sc. Hung. 26 (1975) 433–444.
[27] L. Lovász, On the ratio of optimal integral and fractional covers, Discrete Math. 13 (1975) 383–390.
[28] C. Lucchesi, D.H. Younger, A minimax theorem for directed graphs (to appear).
[29] J.C. Meyer, Quelques problemes concernant les cliques des hypergraphs h-complets et q-parti h-complets, in: C. Berge and D.K. Ray-Chaudhuri eds., Hypergraph Seminar 1972, Lecture Notes 411 (Springer-Verlag, Berlin, 1972) 127–139.
[30] J.C. Meyer, Problems de colorations dans les hypergraphes, Thesis, University of Paris (1976).
[31] J. Pelikan, Properties of balanced incomplete block designs, in: P. Erdös, Renyi and Sös eds., Combinatorial Theory and its Applications (North-Holland Publishing Co. Amsterdam, 1971) 869–890.
[32] M. Rosenfeld, On a problem of C.E. Shannon in graph theory, Proc. A.M.S. 18 (1967) 315–319.
[33] J. Schönheim, Hereditary systems and Chvátal's conjecture, in: Nash-Williams and Sheehan eds., Proc. Fifth British Combinatorial Conference, 1975 (Utilitas Math. Publ. Winnipeg, 1976) 537–544.
[34] S.H. Scott, Multiple node colorings of finite graphs, Doctoral dissertation, University of Reading, England (1975).
[35] P.D. Seymour, On multicolorings of cubic graphs and conjectures of Fulkerson and Tutte (to appear).
[36] S. Stahl, Multicolorings of outerplanar and line graphs (to appear).

[37] S.K. Stein, Two combinatorial covering theorems, J. Comb. Theory Ser., A.16 (1974) 391–397.
[38] F. Sterboul, On the chromatic number of the direct product of hypergraphs, in: C. Berge and D. K. Ray-Chaudhuri eds., Hypergraph Seminar 1972, Lecture Notes 411 (Springer-Verlag, Berlin, 1972) 165–174.
[39] F. Sterboul, Les paramètres des hypergraphes et les problèmes exterieurs associés, Thesis, Paris (1974).
[40] F. Sterboul, Sur une conjecture de V. Chvátal, in: C. Berge and D.K. Ray-Chaudhuri eds., Hypergraph Seminar 1972, Lectures Notes 411 (Springer-Verlag, Berlin, 1972). 152–164.
[41] F. Sterboul, Sur le nombre transversal des hypergraphes uniformes, Discrete Math. 22 (1978) 91–96.
[42] D.R. Woodale, Property B and the 4-color conjecture (to appear).
[43] D.H. Younger, Maximum families of disjoint directed cut-sets, Recent Progress in Combinatorics (Academic Press, New York, 1969) 329–333.

MATROID INTERSECTION

Jack EDMONDS
University of Waterloo, Ont., Canada

A *matroid* $M = (E, F)$ can be defined as a finite set E of elements and a non-empty family F of subsets of E, called *independent* sets, such that
  (1) every subset of an independent set is independent; and
  (2) for every $A \subseteq E$, every maximal independent subset of A, i.e. every M-*basis* of A, has the same cardinality, called the M-*rank* $r(A)$ of A. (This definition is known to be equivalent to many others).

An M-basis of E is called a *basis* of M; the M-rank of E is called the *rank* of M.
  (3) A 0, 1-valued vector $x = [x_j]$, $j \in E$, is called *the (incidence) vector* of the subset $J \subseteq E$ of elements j for which $x_j = 1$.
  (4) Let $M_a = (E, F_a)$ and $M_b = (E, F_b)$ be any two matroids on E, having rank functions $r_a(A)$ and $r_b(A)$, respectively.
  (5) For $k = a$ and b, let

$$P_k = \{x = [x_j],\ j \in E : \forall j,\ x_j \geq 0,\ \text{and}\ \forall A \subseteq E,\ \sum_{j \in A} x_j \leq r_k(A)\}.$$

  (6) **Theorem.** (i) *The vertices (extreme points) of polyhedron P_k are the vectors of the members of F_k.* (ii) *The vertices of polyhedron $P_a \cap P_b$ are the vectors of the members of $F_a \cap F_b$.*

  (7) Let $c = [c_j]$, $j \in E$, be any real-valued vector on E. For any other such vector $x = [x_j]$, $j \in E$, let $cx = \sum_j c_j x_j$, $j \in E$. For any $J \subseteq E$, let $c(J) = \sum c_j$, $j \in J$. Where x is the vector of J, $cx = c(J)$.
  (8) A *(convex) polyhedron* P in the space of real-valued vectors $x = [x_j]$, $j \in E$, can be defined as the set of solutions of some finite system L of linear inequalities and linear equations in the variables x_j.

A *vertex* of a polyhedron P can be defined as a vector $x^0 \in P$ such that, for some c, cx is maximized over $x \in P$ by x^0 and *only* x^0. Not every P has vertices. However, it is well-known that
  (8a) if P has at least one vertex then there are a finite number of them and every cx which has a maximum over $x \in P$ is maximized over $x \in P$ by at least one of them; if P is non-empty and bounded, as is P_k, then it has vertices and every cx is maximized over $x \in P$ by at least one of them;
  (8b) where L is a finite linear system whose solution-set is polyhedron P, a

vector x^0 is a vertex of P if and only if it is the *unique* solution of some linear system, L_0, obtained from L by replacing some of the "$\leq$" 's in L by "$=$" 's.

(9) It is well-known that for any finite set V of vectors x, there exists a unique bounded polyhedron P containing V and such that the vertices of P are a subset of V. P is called the *convex hull* of V. Where L is a finite linear system whose solution-set P is the convex hull of V, the problem of maximizing cx over $x \in V$ is equivalent to the linear program:
maximize cx over solutions x of L by a "basic feasible solution" of L, i.e., by a vertex of P. In particular, where V is the set of vectors of a family F of sets $J \subseteq E$, maximizing $c(J)$ over $J \subseteq F$ is equivalent to this l.p.

(10) A problem of the form, maximize cx over $x \in V$, where V is a prescribed finite set of vectors, we call a *loco problem*. "Loco" stands for "linear-objective combinatorial". The equivalence of a loco problem with a linear program can be quite useful if one can readily identify the linear program. (The linear program is not unique since the L whose P is the convex hull of V is not unique.) In particular, every loco problem has a "dual", namely the dual of a corresponding linear program.

The dual of the linear program,
(11) maximize $cx = \sum_j c_j x_j$, $j \in E$, subject to
(12) $x_j \geq 0$, $j \in E$; and $\sum_j a_{ij} x_j \leq b_i$, $j \in E$, $i \in F$; is the l.p.,
(13) minimize $by = \sum_i b_i y_i$, $i \in F$, subject to
(14) $y_i \geq 0$, $i \in F$; and $\sum_i a_{ij} y_i \geq c_j$, $i \in F$, $j \in E$.

For any x satisfying (12) and any y satisfying (14), we have

(15) $$\sum_j x_j \left(\sum_i a_{ij} y_i - c_j \right) + \sum_i y_i \left(b_i - \sum_j a_{ij} x_j \right) \geq 0.$$

(16) Cancelling we get $c \cdot x \leq b \cdot y$.

(17) Thus, $cx = by$ implies that cx is maximum subject to (12) and by is minimum subject to (14).

(18) The linear programming duality theorem says that if cx has a maximum value subject to (12) then it equals by for some y satisfying (14) — that is, it equals the minimum of by subject to (14).

(19) Theorems (6), (8a) and (18), immediately yield a useful criterion for the maximum of $c(J)$, $J \in F_k$; or the maximum of $c(J)$, $J \in F_a \cap F_b$.

(20) For a given polyhedron P, and a given set $V \subseteq P$, one way, in view of (8), to prove that the vertices of P are a subset of V is, for any linear function cx, either show that cx has no maximum over $x \in P$ or else produce some $x^0 \in V$ which maximizes cx over P.

(21) A way to show that x^0 maximizes cx over P, where say (12) is the linear system defining P, is to produce a solution y^0 of (14) such that $cx^0 = by^0$.

(22) Showing that $x^0 \in V$ maximizes cx over $x \in P$ is, at the same time, a way of showing that x^0 maximizes cx over V.

(23) A good illustration of this technique, which we also use on more difficult

loco problems, is provided by the problem, maximize $c(J) = \sum c_j$, $j \in J$, over $J \in F$, where $M = (E, F)$ is a matroid. The following algorithm, called *the greedy algorithm*, [1], solves this problem:

(24) Consider all the elements $j \in E$, such that $c_j > 0$, in order, $j(1), \ldots, j(m)$, such that $c_{j(1)} \geq c_{j(2)} \geq \cdots \geq c_{j(m)} > 0$. Let $J_0 = \emptyset$. For $i = 1, \ldots, m$, let $J_i = J_{i-1} \cup \{j(i)\}$ if this set is a member of F. Otherwise, let $J_i = J_{i-1}$.

(25) **Theorem.** J_m maximizes $c(J)$ over F.

(26) For $i = 1, \ldots, m$, let $A_i = \{j(1), \ldots, j(i)\}$. It follows immediately, by induction, from the definition of matroid that the vector $x^0 = [x_j^0]$, $j \in E$, of the set $J_m \subseteq E$ is $x_{j(1)}^0 = r(A_1)$; $x_{j(i)}^0 = r(A_i) - r(A_{i-1})$ for $i = 2, \ldots, m$; and $x_j^0 = 0$ for every other $j \in E$.

(27) It is fairly obvious that the vector of any $J \in F$ is a vertex of the set P of solutions to the system:

(28) $\quad x_j \geq 0$, $j \in E$; $\sum_j x_j \leq r(A)$, $A \subseteq E$, $j \in A$.

(29) The dual of the l.p., maximize cx subject to (28), is the l.p.:
(30) minimize $ry = \sum_A r(A) \cdot y(A)$, $A \subseteq E$, subject to:
(31) $y(A) \geq 0$, $A \subseteq E$; $\sum_A y(A) \geq c_j$, $j \in E$, $j \in A \subseteq E$.
(32) Let $y^0 = [y^0(A)]$, $A \subseteq E$, be

$$y^0(A_i) = c_{j(i)} - c_{j(i+1)} \quad \text{for } i = 1, \ldots, m-1;$$

$$y^0(A_m) = c_{j(m)}; \text{ and } y^0(A) = 0 \quad \text{for every other } A \subseteq E.$$

(33) **Theorem.** ry is minimized, subject to (31), by y^0.

(34) Theorems (33), (25), and the tricky part of (6i), i.e. that every vertex of the P of (28) is the vector of a $J \in F$, are all proved by verifying that y^0 satisfies (31) and $cx^0 = ry^0$.

Proving (25) this way is using a hammer to crack a peanut. However, a coconut is coming up — a similarly good algorithm for maximizing $c(J)$ over $J \in F_a \cap F_b$.

First we do a few preliminaries on matroids — in particular, what I believe to be a simplified approach to basic ideas of Whitney and Tutte on matroid duality and contraction.

(35) For any matroid $M = (E, F)$, with rank function $r(A)$, let $F^* = \{J \subseteq E : r(E-J) = r(E)\}$; then $M^* = (E, F^*)$ is a matroid, called the *dual* of M.

(35a) It follows immediately from (35) that $B \subseteq E$ is an M^*-basis of E if and only if $E-B$ is an M-basis of E.

Clearly, M^* satisfies (1). We show that it satisfies (2) by showing that, for any $A \subseteq E$, any $J \in F^*$, $J \subseteq A$, can be extended to a $J' \in F^*$, $J' \subseteq A$, having cardinality,

(36) $\quad r^*(A) = |A| - r(E) + r(E-A)$.

By definition of rank, there is a $J_0 \in F$, $J_0 \subseteq E - A$, $|J_0| = r(E - A)$. By property (2) for matroid M, there is a $J_1 \in F$, $J_0 \subseteq J_1 \subseteq E - J$, $J_1 - J_0 \subseteq A$, $|J_1| = r(E - J) = r(E)$. Let $J' = A - (J_1 - J_0)$. Then $J \subseteq J' \subseteq A$; $J_1 \subseteq E - J' \Rightarrow r(E - J') = r(E) \Rightarrow J' \in F^*$; and $|J'| = |A| - (|J_1| - |J_0|) = |A| - r(E) + r(E - A)$. □

(37) *For any matroid $M = (E, F)$, any $R \subseteq E$, $E' = E - R$, and some M-basis B of R, let $F' = \{J \subseteq E' : J \cup B \in F\}$, then F' is the same for any choice of B; and $M' = (E', F')$ is a matroid, called the contraction of M to E'* (obtained from M by "contracting" the members of R).

Let B_1 and B_2 be M-bases of R. Suppose $J \cup B_1 \in F$, $J \subseteq E'$. Clearly $J \cup B_1$ is an M-basis of $J \cup R$. B_2 is contained in an M-basis $J_0 \cup B_1$ of $J \cup R$. Since $J_0 \subseteq J$,

$$|B_1| = |B_2|, \quad \text{and} \quad |J \cup B_1| = |J_0 \cup B_2|,$$

we have $|J_0| = |J|$ and $J_0 = J$. Thus, $J \cup B_2 \in F$.

Clearly, M' satisfies (1). We show that M' satisfies (2) simply by observing that, for any $A \subseteq E'$, and any $J \in F'$ such that $J \subseteq A$, $J \cup B$ can be extended to an M-basis $J' \cup B$ of $A \cup R$ having cardinality $r(A \cup R)$. Thus J can be extended to a $J' \in F'$, $J' \subseteq A$, having cardinality
(38) $r'(A) = r(A \cup R) - r(R)$. □

(39) *For any matroid $M = (E, F)$, and any $E' \subseteq E$, let $F' = \{J \subseteq E' : J \subset F\}$; then $M' = (E', F')$ is a matroid, called the restriction of M to E'* (obtained from M by "deleting" the members of $E - E' = S$). The rank function of this M' is simply the rank function of M restricted to subsets of E'. Obvious.

(40) For any matroid $M = (E, F)$, any $R \subseteq E$ and any $S \subseteq E - R$, $E' = E - (R \cup S)$, the matroid $M' = (E', F')$ obtained from M by contracting the members of R and deleting the members of S is called a *minor* of M.

(41) It is fairly obvious that the operations of contracting the members of R and deleting the members of S are associative and commutative. That is, M' is the same for any way of doing them.

(42) Let $M = (E, F)$ be a matroid. Let $M^* = (E, F^*)$ be the dual of M. Let $M' = (E', F')$ be obtained from M by contracting $R \subseteq E$ and deleting $S = E - R$. Then $M'^* = (E', F'^*)$, the dual of M', is the same as the matroid, say $M^{*\prime} = (E', F^{*\prime})$, obtained from M^* by deleting R and contracting S.

Proof. Let $J \subseteq E'$. $J \in F^{*\prime} \Leftrightarrow |J| = r^{*\prime}(J) = r^*(J \cup S) - r^*(S) = |J \cup S| - r(E) + r(E - (J \cup S)) - |S| + r(E) - r(E - S) \Leftrightarrow r(E - (J \cup S)) = r(E - S) \Leftrightarrow r((E' - J) \cup R) - r(R) = r(E' \cup R) - r(R) \Leftrightarrow r'(E' - J) = r'(E) \Leftrightarrow J \in F'^*$. □

(43) Let $M = (E, F)$ be a matroid; let n be an integer; and let $F^{(n)} = \{J \subseteq E : J \in F, |J| \leq n\}$; then $M^{(n)} = (E, F^{(n)})$, called the n-truncation of M, is a matroid having rank function $r^{(n)}(A) = \min[n, r(A)]$. Obvious.

(44) For any $A \subseteq E$ and any $e \in E$, e is said to M-depend on A either when $e \in A$ or when A contains some $J \in F$ such that $J \cup \{e\} \notin F$. The set of all elements in E which M-depend on A, is called the M-span or the M-closure of A.

(45) The M-closure of a set $A \subseteq E$ is the unique maximal set S such that $A \subseteq S \subseteq E$ and $r(A) = r(S)$.

Proof. Let S be any maximal set such that $A \subseteq S \subseteq E$ and $r(A) = r(S)$. Consider any $e \in E - A$. If there is some $J \in F$, $J \subseteq A$, $\{e\} \cup J \notin F$, then J is contained in some basis B of A which is also a basis of S and thus which is also a basis of $\{e\} \cup S$. Hence, $e \in S$. Conversely, if $e \in S$, then, for any basis J of A, J is also a basis for S, and so $J \cup \{e\} \notin F$. □

(46) The maximal subsets of E having given ranks are called the *spans* or *flats* or *closed sets* of M.

(47) The minimal subsets of E which are not in F are called the *circuits* of M.

(48) For any $J \in F$ and any $e \in E$, $J \cup \{e\}$ contains at most one circuit of M.

Proof. If C_1 and C_2 are two circuits contained in $J \cup \{e\}$, there is an $e_1 \in C_1 - C_2$. Then $C_1 - \{e_1\} \in F$ can be extended to a basis B of $J \cup \{e\} - \{e_1\}$ which excludes both e_1 and some element of C_2. B is also a basis of $J \cup \{e\}$, smaller than basis J of $J \cup \{e\}$, contradicting (2). □

(49) By a *matroid-algorithm* we mean an algorithm which includes as inputs a collection of arbitrary matroids. The actual application of a matroid-algorithm to given matroids depends on how the matroids are given.

A matroid $M = (E, F)$ would virtually never be given as an explicit listing of the members of F; or even of the maximal members of F (called the *bases* of M); or of the circuits of M; or of the flats of M. For moderate sizes of $|E|$ and $r(E)$, such listings would tend to be astronomical.

Matroids arise implicitly from other structures (sometimes other matroids). The following are examples; see [1] for a more general theory on the construction of matroids.

(50) Forest matroids, (E, F): where E is the edge-set of a graph G and F is comprised of the edge-sets of forests in G.

(51) Transversal matroids, (E, F): where for a given family $Q = \{Q_i\}$, $i \in I$, of sets $Q_i \subseteq E$, F is comprised of the partial transversals J of Q, i.e., sets of the form $J = \{e_i\}$, $i \in I'$, such that $e_i \in Q_i$, for some $I' \subseteq I$.

(52) Matric matroids, (E, F): where E is the set of columns in a matrix, and F is comprised of the linearly independent subsets of columns.

(53) Matroid-sums, (E, F): where (E, F_i), $i \in I$, are given matroids on E, and where F is comprised of all sets of the form $J = \bigcup J_i$, $i \in I$, $J_i \in F_i$. □

For each matroid, say $M = (E, F)$, to which a matroid-algorithm refers as an input, it is assumed that an algorithm ("subroutine") is available for,

(54) given any $J \subseteq E$, determine whether or not $J \in F$.

The efficiency of a matroid-algorithm is relative to the efficiency of these assumed subroutines.

It is obvious that there are good algorithms, relative to (54), for:

(55) given any $A \subseteq E$, determine $r(A)$ by finding a $J \subseteq A$ such that $J \in F$ and $|J|$ is maximum (indeed we have defined matroid, essentially, as a structure for which there is a certain trivial algorithm which does this);

(56) given any $A \subseteq E$, find in A, if there is one, a circuit of M;

(57) given any $A \subseteq E$, find the closure of A;

(58) given any $A \subseteq E$, determine whether or not $A \in F^*$, where $M^* = (E, F^*)$ is the dual of M;

(59) given $A \subseteq E' \subseteq E$, determine whether or not $A \in F'$, where $M' = (E', F')$ is the contraction of M to E'; and so on.

(60) Thus these operations can also be used in matroid algorithms without difficulty. Conversely, matroid-algorithms can equivalently be based on the assumption that a subroutine is available for some operation other than (54) for each matroid input if there is a good algorithm for doing (54) using that operation. For example, (57) could equivalently play the role of (54), since one can show

(61) $J \in F$ iff, for every $e \in J$, the closure of $J - e$ does not contain e.

In [2] there is a good algorithm for:

(62) given a finite collection of matroids, $M_i = (E, F_i)$, $i \in I$, and given $J \subseteq E$, determine whether or not $J \in F$, where F is defined as in (53), by finding or determining that it is not possible to find, a partition of J into a collection of sets, J_i, $i \in I$, such that $J_i \in F_i$.

Since for this F, $M = (E, F)$ is a matroid, [2] together with (55) immediately gives a good algorithm for

(63) given the matroids M_i, find sets $J_i \in F_i$, $i \in I$, such that $|\bigcup J_i|$ is maximum.

More generally, [2] together with the greedy algorithm, (24), gives a good algorithm for

(64) given $c = [c_j]$, $j \in E$, and given the matroids M_i, $i \in I$, maximize $c(J) = \sum_{j \in J} c_j$ over $J \in F$.

Let $M_a = (E, F_a)$ and $M_b = (E, F_b)$ be any two matroids on E; then [2] together with (58) gives immediately a good algorithm for:

(65) given matroids M_a and M_b, find a set $J \in F_a \cap F_b$ such that $|J|$ is maximum.

(66) Let M_a be the M_1 of (62) and let M_b^*, the dual of M_b, be the M_2 of (62), where $I = \{1, 2\}$. Use (63) and (58) to find an A of the form $A = J_1 \cup J_2$, where $J_1 \in F_a$, $J_2 \in F_b^*$, and $|A|$ is maximum. Extend J_2 to B, an M_b^*-basis of A. Let $J = A - B$. Then J maximizes $|J|$ over $J \in F_a \cap F_b$.

Proof. Clearly, $J \in F_a$. B is an M_b^*-basis of E, since otherwise it could be extended to a $B' \in F_b^*$ such that $|J_1 \cup B'| > |A|$. Thus $r_b^*(E - J) = r_b^*(E)$. Thus $J \in F_b$. If there were a $J' \in F_a \cap F_b$ larger than J, then where $B \in F_b^*$ is an M_b^*-basis of $E - J'$, also an M_b^*-basis of E, we would have $|J' \cup B| > |A|$. Thus $J \in F_a \cap F_b$ is such that $|J|$ is maximum. □

Besides finding sets $J_i \in F_i$, $i \in I$, such that $A = \bigcup J_i$ has maximum cardinality,

the algorithm in [2] for (63) finds a set $S \subseteq E$ (called S_n there), such that
(67) $A \cup S = E$ and $|A \cap S| = \sum_i r_i(S)$.
For the case (66) of (67), it is easy to show from (67) that
(68) $|J| = r_a(S) + r_b(E-S)$.
For any $J' \in F_a \cap F_b$ and any $S' \subseteq E$, we have
(69) $|J'| = |J' \cap S'| + |J' - S'| \leq r_a(S') + r_b(E-S')$.
Thus, from (68) and (69), we have the theorem:
(70) For any two matroids, $M_k = (E, F_k)$, $k = a, b$,

$$\max\{|J| : J \in F_a \cap F_b\} = \min\{r_a(S) + r_b(E-S) : S \subseteq E\}.$$

Thus, the algorithm of [2] for the problem of (66), finds a minimizing S of (70) as well as a maximizing J of (70). Clearly, for any such S and J we have that
(71) $J^a = J \cap S$ and $J^b = J \cap (E-S)$ partition J into two sets such that
(72) $\mathrm{cl}_a(J^a) \cup \mathrm{cl}_b(J^b) = E$, where $\mathrm{cl}_k(A)$ denotes M_k-closure of A.
(73) Conversely, the existence of a $J \in F_a \cap F_b$ which partitions into sets J^a and J^b satisfying (72), immediately implies Theorem (70).

(74) Given $J \in F_a \cap F_b$. We describe a direct algorithm for finding either a larger $J' \in F_a \cap F_b$ or else a partition of J into sets, J^a and J^b, which satisfy (72). It is essentially the algorithm of [2] applied to (66).
(75) Let $J_0^b = \emptyset$. Let $K_{-1} = \emptyset$.
(76) For each J_i^b, let $J_i^a = J - J_i^b$, and
(77) $K_i = E - \mathrm{cl}_a(J_i^a)$.
(78) For each K_i such that $K_{i-1} \subset K_i \subseteq \mathrm{cl}_b(J)$, let
(79) $J_{i+1}^b = \bigcup \{C : C \subseteq J, C \cup e \text{ is an } M_b\text{-circuit for some } e \in K_i\}$.
We thus generate sequences
(80) $\emptyset = J_0^b \subset J_1^b \subset \cdots \subset J_{n-1}^b \subseteq J_n^b$, $J = J_0^a \supset J_1^a \supset \cdots \supset J_{n-1}^a \supseteq J_n^a$, and $K_{-1} \subset K_0 \subset K_1 \subset \cdots \subset K_{n-1} \subseteq K_n$ such that either
(81) $K_{n-1} = K_n$, or else
(82) $K_{n-1} \subset K_n$, and $K_n - \mathrm{cl}_b(J) \neq \emptyset$.
(83) In the case of (81), we have $\mathrm{cl}_a(J_n^a) \cup \mathrm{cl}_b(J_n^b) = E$ since $E - \mathrm{cl}_a(J_n^a) = K_n = K_{n-1} \subseteq \mathrm{cl}_b(J_n^b)$, and the algorithm is done.
(84) In the case of (82):
(85) For every $e_i \in K_i - K_{i-1} = \mathrm{cl}_a(J_{i-1}^a) - \mathrm{cl}_a(J_i^a)$, $i = 1, \ldots, n$, let $\{e_i\} \cup D_i$ be the unique M_a-circuit of $\{e_i\} \cup J_{i-1}^a$. Clearly, $D_i \cap (J_{i-1}^a - J_i^a) \neq \emptyset$. That is, there is an $h_i \in J_{i-1}^b - J_i^b = J_{i-1}^a - J_i^a$ such that $h_i \in D_i$.
(86) For every $h_i \in J_i^b - J_{i-1}^b$, $i = 1, \ldots, n$, clearly by (79) there is an $e_{i-1} \in K_{i-1} - K_{i-2}$ such that $h_i \in C_{i-1}$ where $\{e_{i-1}\} \cup C_{i-1}$ is the M_b-circuit of $\{e_{i-1}\} \cup J_i^b$. ($K_{-1} = \emptyset$).
(87) Choose a single e_i for each $i = 0, \ldots, n$ and a single h_i for each $i = 1, \ldots, n$, such that
(88) $e_n \in K_n - \mathrm{cl}_b(J)$, and such that (85) and (86) hold.

(89) Clearly, the algorithm can specify an h_i for every $e_i \in K_i - K_{i-1}$ and an e_{i-1} for every $h_i \in J_i^b - J_{i-1}^b$ as the sequences (80) are being generated, or it can determine the $\{e_0, \ldots, e_n\}$ and $\{h_1, \ldots, h_n\}$ after (80).

(90) Let $J' = J \cup \{e_0, \ldots, e_n\} - \{h_1, \ldots, h_n\}$. Then $|J'| > |J|$ and $J' \in F_a \cap F_b$.

(95) To prove $J' \in F_a$, let $I_i^a = J \cup \{e_0, e_1, \ldots, e_i\} - \{h_1, h_2, \ldots, h_i\}$ for $i = 0, 1, \ldots, n$. Observe that $I_0^a \in F_a$, and that $I_i^a \in F_a$ implies $I_i^a \cup \{e_{i+1}\} - \{h_{i+1}\} = I_{i+1}^a \in F_a$ because $\{e_{i+1}\} \cup D_{i+1} \subseteq I_i^a \cup \{e_{i+1}\}$ since $\{h_1, h_2, \ldots, h_i\} \cap D_{i+1} = \emptyset$ by (85), and hence by (48), $\{e_{i+1}\} \cup D_{i+1}$ is the only M_a-circuit in $I_i^a \cup \{e_{i+1}\}$.

(96) To prove $J' \in F_b$, let $I_i^b = J \cup \{e_i, e_{i+1}, \ldots, e_n\} - \{h_{i+1}, h_{i+2}, \ldots, h_n\}$ for $i = n, n-1, \ldots, 0$. Observe that $I_n^b \in F_b$, and that $I_i^b \in F_b$ implies $I_i^b \cup \{e_{i-1}\} - \{h_i\} = I_{i-1}^b \in F_b$ because $\{e_{i-1}\} \cup C_{i-1} \subseteq I_i^b \cup \{e_{i-1}\}$ since $\{h_{i+1}, h_{i+2}, \ldots, h_n\} \cap C_{i-1} = \emptyset$ by (86), and hence by (48), $\{e_{i-1}\} \cup C_{i-1}$ is the only M_b-circuit in $I_i^b \cup \{e_{i-1}\}$.

Problems (65) and (23) are of course both special cases of the loco problem:

(97) given matroids $M_a = (E, F_a)$ and $M_b = (E, F_b)$, and given any real-valued $c = [c_j]$, $j \in E$, find a set $J \in F_a \cap F_b$, $|J| \leq m$, which maximizes $c(J)$.

(98) We show, in analogy to the Hungarian method for the optimum assignment problem, how algorithm (75)–(90) can be used as a subroutine in an algorithm for solving (97).

(99) The vector $x^J = [x_j^J : j \in E]$ of any $J \in F_a \cap F_b$, $|J| \leq m$, obviously satisfies the inequalities.

(100) For every $j \in E$, $x_j \geq 0$; and for $k = a$ or b, and for every M_k-closed set $A \subseteq E$, $\sum_{j \in A} x_j \leq r_k(A)$; and $\sum_{j \in E} x_j \leq m$. Since $c(J) = cx^J = \sum_{j \in E} c_j x_j^J$, if we maximize cx over all x satisfying (100) by the vector x^J of some $J \in F_a \cap F_b$, then by (99), this J will solve (97).

If we can do this for any rational (or integer-valued) c then by (20) we will have proved (6ii).

This is precisely what the algorithm does.

The dual of the linear program,

(101) maximize cx subject to (100) is the linear program minimize

(102) $yr = my_m + \sum r_k(A^k) y_k(A^k)$, where the summation is over $k = a, b$, and all M_k-closed sets A^k, subject to

(103) $y = [y_m, y_k(A^k) : k = a, b; A^k, M_k\text{-closed}]$ satisfying

(104) $y \geq 0$ and

(105) for every $j \in E$, $y_m + \sum \{y_k(A^k) : k = a, b; j \in A^k\} \geq c_j$.

(106) Let $t_k(y, j) \equiv \sum y_k(A^k)$, summation over the M_k-closed sets A^k which contain j. Thus (105) can be written as

(107) $t(y, j) \equiv y_m + t_a(y, j) + t_b(y, j) \geq c_j$.

The algorithm terminates with a $J \in F_a \cap F_b$, $|J| \leq m$, and a solution y of (104) and (105), such that

(108) $cx^J = yr$.

By (21), this x^J solves the primal linear program, (101).

We will ascertain (108) from the "complementary slackness" conditions,

(109) $x_j^J > 0$, i.e. $j \in J$, only if $t(y, j) = c_j$; and
(110) $y_k(A^k) > 0$ only if $|J \cap A^k| = \sum_{j \in A^k} x_j^J = r_k(A^k)$, i.e. only if
(111) $J \cap A^k$ is an M_k-basis of A^k; and
(112) $y_m > 0$ only if $|J| = m$.

Conditions (109)–(112) immediately imply that the corresponding instance of (15) holds as an equation, and thus that the corresponding instance of (16) holds as an equation, namely equation (108).

(113) At any stage of the algorithm we have a solution y^0 of (103)–(105) such that, for each $k = a, b$, $y_k^0(A) > 0$ if and only if A is an M_k-closed set A_i^k of a sequence $\emptyset \subseteq A_1^k \subset A_2^k \subset \cdots \subset A_{a(k)}^k \subseteq E$.

(114) Let $E^0 = \{j \in E : t(y^0, j) = c_j\}$. We also have a set $J^0 \in F_a \cap F_b$, $|J^0| \leq m$, such that

(115) $J^0 \subseteq E^0$, and such that condition (110) holds. That is

(116) for $k = a, b$, and $i = 1, \ldots, q(k)$, $J^0 \cap A_i^k$ is an M_k-basis of A_i^k. The set $J^0 = \emptyset$, and a y^0 which is zero-valued except for $y_m^0 = $ maximum of 0 and c_j, $j \in E$, are suitable for a starting J^0 and y^0.

Since y^0 and J^0 satisfy all the complimentary slackness conditions (109)–(112) except possibly (112), if y^0 and J^0 satisfy (112) then J^0 is an optimum solution of (97) and the algorithm stops.

In fact, where $m_0 = |J^0|$ clearly J^0 optimizes $c(J)$ subject to $J \in F_a \cap F_b$ and $|J| = m_0$.

If (112) is not satisfied then, for $i = 1, \ldots, q(k) + 1$, let

(117) $M_k^i = (A_i^k - A_{i-1}^k, F_k^i)$ be the minor obtained from M_k by deleting $E - A_i^k$ and contracting A_{i-1}^k. ($A_0^k \equiv \emptyset$ and $A_{q(k)+1}^k \equiv E$).

For each k, let $M_k' = (E, F_k')$ be the disjoint union of these matroids M_k^i.

(118) That is, $F_k' = \{J \subseteq E : J \cap (A_i^k - A_{i-1}^k) \in F_k^i \text{ for } i = 1, \ldots, q(k)\}$

Clearly, from (37)–(38), M_k' is a matroid having rank function,

(119) $r_k'(S) = \sum_{i=1}^{q(k)+1} [r_k((S \cap A_i^k) \cup A_{i-1}^k) - r_k(A_{i-1}^k)]$, where A_0^k denotes $\emptyset$ and $A_{q(k)+1}^k$ denotes E.

Let $M_k^0 = (E^0, F_k^0)$ be the restriction of M_k' to E^0. That is, $F_k^0 = \{J \subseteq E^0 : J \in F_k'\}$. The formula for rank in M_k^0 is the same as (119).

(120) It follows immediately from the definition (38) of "minor" that $J^0 \in F_k^0$.

(122) Thus we apply algorithm (75)–(90) where the J is J^0 and where for $k = a, b$ the M_k of (75)–(90) is M_k^0.

(123) If we get a J', then we replace J^0 by J', keep the same y^0, and return to (113). (Returning to (113) is the same as stopping if $|J'| = m$, and returning to (122) if $|J'| < m$).

If we fail to get a set J' as in (123), that is in case (81) of algorithm (75)–(90), then we get a partition of J^0 into two sets J_n^a and J_n^b such that

(124) $\text{cl}_a^0(J_n^a) \cup \text{cl}_b^0(j_n^b) = E^0$, where $\text{cl}_k^0(A)$ denotes the M_k^0-closure of A.

(125) In this case, let $S_k = \text{cl}_k'(J_n^k) \supseteq \text{cl}_k^0(J_n^k)$, where $\text{cl}_k'(A)$ denotes the M_k'-closure of A.

Let $\text{cl}_k(A)$ denote the M_k-closure of A. By the construction of M_k' from M_k

and the M_k-closed sets $A_1^k \subset \cdots \subset A_{q(k)}^k$, and because S_k is M_k'-closed, we easily have that

(126) $B_i^k \equiv \mathrm{cl}_k((S_k \cap A_i^k) \cup A_{i-1}^k) = (S_k \cap A_i^k) \cup A_{i-1}^k$ for $i = 1, \ldots, q(k)+1$, and

(127) $E \supseteq B_{q(k)+1}^k \supseteq A_{q(k)}^k \supseteq B_{q(k)}^k \supseteq \cdots \supseteq B_2^k \supseteq A_1^k \supseteq B_1^k \supseteq A_0^k = \emptyset$.

(128) Thus, $S_k = (B_{q(k)+1}^k - A_{q(k)}^k) \cup \cdots \cup (B_2^k - A_1^k) \cup (B_1^k - A_0^k)$, and $E - S_k = (E - B_{q(k)+1}^k) \cup (A_{q(k)}^k - B_{q(k)}^k) \cup \cdots \cup (A_1^k - B_1^k)$, where all the terms of these unions are mutually disjoint.

Clearly, for $k = a, b$, we have

(129) $r_k'(S_k) = |J_n^k|$, and so

(130) $r_a'(S_a) + r_b'(S_b) = |J^0| < m$. By (119) and (126) we have

(131) $D \equiv -m + \sum_{k=a,b} \sum_{i=1}^{q(k)+1} [r_k(B_i^k) - r_k(A_{i-1}^k)] < 0$.

From the formula (102) for yr, we see immediately that

(132) $y^1 r = y^0 r + \delta D$, where y^1 is obtained from y^0 by raising by δ, for each positive term of D, the corresponding component of y^0, and lowering by δ, for each negative term of D, the corresponding component of y^0.

More precisely (since as many as three consecutive members of the sequence (127) can be identical sets), let

(133) $y_m^1 = -\delta + y_m^0$, and for every M_k-closed set $A \subseteq E$, let

(134) $y_k^1(A) = \delta N_A^k + y_k^0(A)$, where N_A^k is the number of indices $i = 1, \ldots, q(k)+1$, such that $A = B_i^k$, minus the number of indices $i = 2, \ldots, q(k)$, such that $A = A_{i-1}^k$.

(135) Clearly by (127), $N_A^k < 0$ only if $N_A^k = -1$ and A is one of the sets A_i^k.

By (127), (128), and (134) we have

(136) for every $e \in S_k$, $t_k(y^1, j) = t_k(y^0, j) + \delta$, and

(137) for every $e \in E - S_k$, $t_k(y^1, j) = t_k(y^0, j)$.

(138) Thus, by (107) and (133), for every $j \in E$, $t(y^1, j) =$ either $t(y^0, j) + \delta$ or $t(y^0, j)$ or $t(y^0, j) - \delta$, according to whether j is in both sets S_k, one of them, or neither.

(139) Letting δ be the minimum of the numbers y_m^0, $t(y^0, j) - c_j$ for $j \in E - S_a \cup S_b$, and $y_k^0(A)$ for $N_A^k = -1$, we have that

(140) $\delta > 0$, and that

(141) y^1 is a feasible solution of (104), (105). By (131), (132), and (140), we have

(142) $y^1 r < y^0 r$.

(143) We now replace y^0 by y^1 and return to (113).

(144) Clearly, y^1 has the property required by (113).

(145) J^0 relative to y^1 has the property required by (116) of J^0 relative to y^0.

(146) Where $E^1 = \{j \in E : t(y^1, j) = c_j\}$, $J^0 \subseteq E^1$ is in keeping with requirement (115).

Thus, (139) is a legitimate instruction.

(147) In fact, by (83) and (138), the set K_n, which arose in the application of algorithm (75)–(90) to M_a^0, M_b^0, and J^0, is contained in E^1.

(148) Though it is not essential to the present algorithm, it can be shown that the structures generated by subroutine (75)–(90) in an unsuccessful attempt to augment J^0 to a larger J in $F_a^0 \cap F_b^0$, can be used as part of the generation of the like structures which either augment J^0 to a larger J in $F_a^1 \cap F_b^1$ corresponding to y^1, or else lead as we have just described to a better dual solution y^2 and matroids M_k^2. The structure developed by the successive applications of the subroutine can be used until either augmentation of J^0 or a y such that $y_m = 0$ is obtained.

(149) Though (148) leads to better bounds on the algorithm, it is simpler to bound the algorithm by observing that for integer-valued c, the vectors y^0 and y^1 of the algorithm, and hence the numbers $y^0 r$ and $y^1 r$ of the algorithm, are integer-valued. Therefore, by (142), the algorithm terminates in no more than $\max\{0, c_j : j \in E\}$ iterations of (113)–(139).

The same consideration proves the theorem:

(150) Where c is integer-valued, $\max\{c(J) : J \in F_a \cap F_b\} = \min\{yr : y$, an integer valued solution of (103)–(105)$\}$.

This is somewhat stronger than the same statement with the words "integer-valued" left out. The latter is immediately equivalent to (6ii), using basic linear programming theory, but (150) does not follow immediately from (6ii).

(151) Theorem (70) is essentially the special case of Theorem (150) where c is all ones.

Note added in proof

A simpler non-algorithmic proof of a generalisation of (150) appears in [1]. The present paper appeared as a report in 1970 though its methods, and announcements private and public of its methods, predate [1].

References

[1] J. Edmonds, Submodular functions, matroids, and certain polyhedra, Proceedings of Calgary Inter. Conf. on Combinatorial Structures and their Applications, June 1969 (Gordon and Breach, New York, 1970).
[2] J. Edmonds, Matroid partition, Math. of the Decision Sciences, Amer. Math. Soc. Lectures in Appl. Math. 11 (1968) 335–345.
[3] J. Edmonds, Optimum branchings, same as [2], 346–361.
[4] R. Rado, A theorem on independence relations, Quart. J. Math. Oxford Ser. 13 (1942) 83–89.
[5] W.T. Tutte, Menger's theorem for matroids, J. Res. Nat. Bur. Standards Sect. B 69B (1965) 49–53.
[6] H. Whitney, On the abstract properties of linear dependence, Amer. J. Math. 57 (1935) 509–533.

BOOLEAN ELEMENTS IN COMBINATORIAL OPTIMIZATION*

Peter L. HAMMER

Department of Combinatorics and Optimization, University of Waterloo, Waterloo, Ontario, Canada

Introduction

The possibility of using Boolean elements in the formulation and interpretation of combinatorial optimization problems has been first pointed out by R. Fortet [12, 13]. This approach was continued by P. Camion [5], R. Faure and Y. Malgrange [11], P.L. Hammer (Ivanescu), I. Rosenberg and S. Rudeanu [29]. A monograph [31] on this subject has appeared in 1968, and since then numerous publications have been devoted both to theoretical and to practical (algorithmic) aspects of this topic. Rudeanu's recent monograph [44] is devoted to the problems of Boolean equations.

Most of the generally available algorithms for the solution of discrete optimization problems are based either on implicit enumeration, or on linear algebra. The use of linear algebra is motivated by the excellent results it yields in the solution of (continuous) linear programming problems, and by the possibility of "relaxing" a typical discrete condition of the form $x \in \{0, 1\}$ to its continuous counterpart $0 \le x \le 1$. However, in this relaxation one risks to lose essential features of the original discrete problem. (Consider for example the system $2x - 6y \ge -5$, $2x + 6y \ge 1$, with $x, y \in \{0, 1\}$; this system obviously implies $x = 1$. If we relax $x, y \in \{0, 1\}$ to $0 \le x, y \le 1$ and examine *all* the possible surrogates of the above two inequalities, i.e. all inequalities of the form $(2 + 2\lambda)x + (-6 + 6\lambda)y \ge -5 + \lambda$ (where $\lambda \ge 0$), we see that they have the following 0–1 solutions: $(0, 0)$, $(1, 0)$, $(1, 1)$ for $0 \le \lambda \le \frac{1}{5}$; $(0, 0)$, $(0, 1)$, $(1, 0)$, $(1, 1)$ for $\frac{1}{5} \le \lambda \le 5$; and $(0, 1)$, $(1, 0)$, $(1, 1)$ for $\lambda \ge 5$. In other words, there is no surrogate of our problem implying $x = 1$.) On the other hand, the degree of implicitness of an enumeration-type algorithm depends heavily on the art of using it. The interaction of constraints being usually hard to realize (unless it is strong enough to be detected in the continuous relaxation of the problem) is bypassed and taken care of only at later steps when sufficient

---

* This survey is based on the author's contribution to the volume *Combinatorial Programming: Methods and Applications*, ed. B. Roy, Reidel Publishing Company, Dordrecht-Holland, 1975, pp. 67–92. Thanks are expressed to Reidel Publishing Company for their kind agreement to include this paper in the present volume.

variables have been fixed to arrive at conclusions from one of the particular constraints of the problem (e.g. how "implicit" is the enumeration which tells us that in every solution of the above problem $x = 1$, while y is arbitrary?). The difficulties arising in connection with discrete nonlinear problems are even greater.

The necessity of complementing rather than replacing the presently utilized methods with other ones seems obvious and Boolean algebra appears to be a likely candidate for this task. In our above discussed example, it would tell that the first inequality is equivalent to $\bar{x}y = 0$, the second to $\bar{x}\bar{y} = 0$, and the system to $\bar{x}y \vee \bar{x}\bar{y}(= \bar{x}) = 0$, i.e. to $x = 1$).

On the other hand, the role of a Boolean viewpoint in combinatorial optimization does not reduce to that of assisting the computations. Boolean procedures can be used to transform problems to simpler ones and to get a better insight into their structure. Irrelevant elements can be disposed of (in the above example the variable y was irrelevant, since our problem did not depend on it), inessential data simplified (e.g. the inequality $2x + 6y \geq 1$ can be reduced to $x + y \geq 1$). Further, some familiar problems can be given new and possibly advantageous formulations (e.g. see [48] for a new formulation of the plant-location problem). Moreover one can expect connections to be established between apparently different questions and structural results to be obtained (e.g. "almost" every 0–1 programming problem can be reduced to a covering problem in the original variables, there is a strong connection between prime implicants of threshold functions and facets of the polytope of 0–1 solutions of knapsack problems, different concepts of value in n-person characteristic function games can be viewed as linear approximations of nonlinear pseudo-Boolean functions, etc.).

The study of Boolean methods in integer programming has had its ups and downs. Much of the early enthusiasm, motivated mainly by the fact that a seemingly useless area of mathematics can find its applications, has soon turned into disillusion, when it became clear that many problems of the real world have too high dimensions to be solved by the simplistic application of the rules of Boolean manipulations alone. The optimistic attitude of today emerged when it became apparent that — while Boolean algebra cannot solve magically all integer programming problems — it can however play (in combination with other methods) a highly successful role in the solution of these problems. This possibility was amply substantiated by computational experiments.

Some of the demonstrated possibilities of using Boolean techniques in the study of discrete optimization problems are the following.

— creation of mixed algorithms, using Boolean and non-Boolean techniques;
— pre-processing of 0–1 programs (fixing some variables, simplfying some constraints, reducing coefficients, linearizing some nonlinearities, eliminating some of the redundant constraints, etc.);
— reformulation of problems (reducing, if possible, a set packing-problem to a

knapsack problem, eliminating redundant variables, reducing "monotone" constraints to set-covering relations, etc.);

— modelling some discrete optimization problems (e.g. different location problems can be viewed as unconstrained pseudo-Boolean optimization problems, etc.).

But mainly, a Boolean viewpoint can serve as a unifying tool in the examination of the combinatorial structure of discrete optimization problems. This tool allows (in principle — always) the X-raying of a problem and the discovery of its skeleton. To know when this information is dangerously costly, and when it is useful, to know when all of this X-ray, and when only a small fragment of it is needed — this might be regarded as part of the art of integer programming.

The aim of this survey is not to present a comprehensive bibliography of all pertinent developments, but rather to discuss a relatively small (and subjective) selection of possibly useful ideas which have been reported in the literature of the last few years.

1. Elements of Boolean algebra

Let $B = \{0, 1\}$. For $x \in B$ we shall denote $\bar{x} = 1 - x$ its *complement* or *negation*. We shall also write frequently $x^\alpha = x$ if $\alpha = 1$, and $x^\alpha = \bar{x}$ if $\alpha = 0$. This notation can cause no confusion, because the regular powers of $x \in B$ being all equal to x (idempotency of multiplication) we shall never use them.

For any $x, y \in B$, we shall define their *union* $x \vee y$ by $x \vee y = x + y - xy$.

Some of the most commonly utilized properties of the above defined operations are the following.

$$x \vee y = y \vee x \quad \text{(commutativity)},$$
$$x \vee (y \vee z) = (x \vee y) \vee z \quad \text{(associativity)},$$
$$x \vee x = x \quad \text{(idempotency)},$$
$$x \vee y = 0 \quad \text{if and only if} \quad x = y = 0,$$
$$x \vee 0 = x, \quad x \vee 1 = 1, \quad x \vee \bar{x} = 1,$$
$$x \vee yz = (x \vee y)(x \vee z) \quad \text{and} \quad x(y \vee z) = xy \vee xz \quad \text{(distributivities)},$$
$$x \vee xy = x \quad \text{(absorption)},$$
$$x \vee \bar{x}y = x \vee y,$$
$$\overline{xy} = \bar{x} \vee \bar{y} \quad \text{and} \quad \overline{x \vee y} = \bar{x} \cdot \bar{y} \quad \text{(De Morgan's Laws)},$$
$$\bar{\bar{x}} = x \quad \text{(double negation)},$$
$$x \leq y \quad \text{if and only if} \quad xy = x,$$
$$x \leq y \quad \text{if and only if} \quad x\bar{y} = 0,$$
$$x = y \quad \text{if and only if} \quad x\bar{y} \vee \bar{x}y = 0.$$

A function $f(x_1, \ldots, x_n)$ whose variables and values belong to B, will be called a *Boolean function*. Examples of such functions are $x \vee yz$, $x \vee yz \vee \bar{x}\bar{z}$, $(x \vee \bar{y})\overline{(y \vee xz)}$, etc. The algebraic expression of a Boolean function is not unique, e.g. the expressions $x \vee y \vee \bar{z}$ and $x \vee yz \vee \bar{x}\bar{z}$ define the same function (this can be seen either by giving to x, y, z all 2^3 possible combinations of values, or noticing that

$$x \vee yz \vee \bar{x}\bar{z} = x \vee yz \vee \bar{z} = x \vee y \vee \bar{z}).$$

A variable x, or its negation $\bar{x}$ will be called a *literal* X. A finite product of literals will be called an *elementary conjunction* $C = \prod_{j \in S} x_j^{\alpha_j}$. By convention, we shall consider sometimes also the constant 1 as being an elementary conjunction (with $S = \emptyset$). A finite union of elementary conjunctions $E = C_1 \vee C_2 \vee \cdots \vee C_m$ will be called a *disjunctive form*. It can be shown easily that every Boolean function can be expressed in a disjunctive form.

We shall say that an elementary conjunction C is *contained* in the elementary conjunction C' if every literal appearing as a factor in C is also a factor of C'. e.g. $x\bar{y}$ is contained in $x\bar{y}zu$, also in $x\bar{y}$, but is not contained in xy or in xyz.

An elementary conjunction I is said to be an *implicant* of the Boolean function $f(x_1, \ldots, x_n)$, if $I = 1$ implies $f(x_1, \ldots, x_n) = 1$. For example, $x\bar{y}$ is an implicant of $x\bar{y} \vee \bar{y}z(x \vee \bar{z})$. Also, $x\bar{y}$ is an implicant of $xz \vee \bar{y}\bar{z}$ (indeed, if $x\bar{y} = 1$, then $x = 1, y = 0$, and hence $xz \vee \bar{y}\bar{z}$ becomes $z \vee \bar{z}$ which is equal to 1).

An implicant P of a Boolean function $f(x_1, \ldots, x_n)$ is said to be a *prime implicant* if there is no other implicant P' of f contained in P. For example, $x\bar{y}$ is a prime implicant of $f = xz \vee \bar{y}\bar{z}$, but $x\bar{y}z$ is a non-prime implicant of f. If all the prime implicants of a Boolean function f are $P_1, \ldots, P_t$, then it is easy to see that $f = P_1 \vee \cdots \vee P_t$.

We shall see later that the knowledge of the prime implicants of a given Boolean function is extremely useful. A way of finding all the prime implicants is offered by the so-called *consensus* method.

Given two elementary conjunctions C and C', such that there is precisely one variable (x_0) appearing unnegated (x_0) is one of them, and negated $(\bar{x}_0)$ in the other, then the elementary conjunction obtained from the juxtaposition CC' of C and C' after deleting $x, \bar{x}$ and repetitions of literals, will be called the *consensus* of C and C'. For example, let $C = x\bar{y}\bar{z}u$ and $C' = \bar{y}zu\bar{w}$; then their consensus is $C'' = x\bar{y}u\bar{w}$.

The *consensus method* consists in applying as many times as possible the following two operations to a disjunctive form of a Boolean function.

(i) eliminate any elementary conjunction which contains another one;

(ii) add as a new elementary conjunction the consensus of two elementary conjunctions, provided this consensus does not include any of the listed (undeleted) elementary conjunctions.

All the different expressions obtained along this process represent the same Boolean function, and the elementary conjunctions appearing in the final form at the end of this (finite, but long) process are exactly the prime implicants of the given functions.

It is likely that in practical problems finding all the prime implicants of a Boolean function might require an excessive amount of computation. Therefore, in the more practical procedures described in Section 3, we shall work with implicants which are not necessarily prime, but which allow an efficient solution of many 0–1 programs. A particular way of finding them is described in [20], and numerous other alternatives are easy to describe.

A real-valued function with variables taking only the values 0, 1 will be called a *pseudo-Boolean function*; an equation (inequality) $f(x) = g(x)$ ($f(x) \leq g(x)$), where f and g are pseudo-Boolean functions, will be called a *pseudo-Boolean equation* (*inequality*).

2. The resolvent

Let $S \subseteq B^n$ be the set of solutions of the system Σ of pseudo-Boolean inequalities $f(X) \leq 0$ ($i = 1, \ldots, m$) and let $\rho(X)$ be a Boolean function which takes the value 0 iff $X \in S$. The function ρ will be called the *resolvent* of Σ, and also the resolvent of S.

Let us consider the linear inequality

$$\sum_{j=1}^{n} a_j x_j \leq a_0 \tag{1}$$

and let l be the family of all minimal covers of (1), i.e. the family of all the minimal sets $C \subseteq \{1, \ldots, n\}$ with the property

$$\sum_{j \in C} |a_j| > a_0 - \sum_{j=1}^{n} \min(0, a_j).$$

It can be seen ([19]) that the function

$$\phi(X) = \bigvee_{C \in l} \prod_{j \in C} x_j^{\alpha_j} \tag{2}$$

(where $\alpha_j = 1$ if $a_j \geq 0$ and $\alpha_j = 0$ if $a_j < 0$) is the resolvent of (1).

It has been shown in [29] (see also [31]) that every pseudo-Boolean function $f(X)$ has a polynomial expression, which is linear in each variable. Hence, every pseudo-Boolean inequality can be written in the form

$$\sum_{h=1}^{l} b_h y_h \leq b_0 \tag{3}$$

where

$$y_h = \prod_{j \in H_h} x_j \quad (h = 1, \ldots, k) \tag{4}$$

are themselves taking only the values 0 and 1. If $\psi(Y)$ is the resolvent of (3) (viewed as a linear inequality in the y_h's), then it is easy to see that the resolvent $\phi(X)$ of (3) (viewed as an inequality in the x_j's) can be obtained from $\psi(Y)$ by simply substituting (4) into it.

Further, if $\phi_i(X)$ are the resolvents of the pseudo-Boolean inequalities $f_i(X) \leq 0$ ($i = 1, \ldots, m$) then $\phi(X) = \bigvee_{i=1}^{m} \phi_i(X)$ will be the resolvent of the system Σ.

Consider for example the system consisting of $x_j \in B$ ($j = 1, \ldots, 6$) and

$$5x_1 - 4x_2 - 2x_3 - x_4 - 4x_5 + 3x_6 \leq -2, \tag{5.1}$$

$$-5x_2 + 6x_2 x_6 - 8x_1 x_3 x_4 - 4x_2 x_4 \leq -7 \tag{5.2}$$

or

$$5x_1 + 4\bar{x}_2 + 2\bar{x}_3 + \bar{x}_4 + 4\bar{x}_5 + 3x_6 \leq 9, \tag{5.1}'$$

$$5\bar{x}_2 + 6x_2 x_6 + 8\overline{x_1 x_3 x_4} + 4\overline{x_2 x_4} \leq 10. \tag{5.2}'$$

The resolvents of these inequalities are, respectively,

$$\phi_1 = x_1 \bar{x}_2 \bar{x}_3 \vee x_1 \bar{x}_2 \bar{x}_4 \vee x_1 \bar{x}_2 \bar{x}_5 \vee x_1 \bar{x}_2 x_6 \vee x_1 \bar{x}_3 \bar{x}_5 \tag{6.1}$$
$$\vee x_1 \bar{x}_4 \bar{x}_5 \vee x_1 \bar{x}_5 x_6 \vee x_1 \bar{x}_3 x_6 \vee \bar{x}_2 \bar{x}_3 \bar{x}_5 \vee \bar{x}_2 \bar{x}_5 x_6$$
$$\vee \bar{x}_2 \bar{x}_3 \bar{x}_4 x_6 \vee \bar{x}_3 \bar{x}_4 \bar{x}_5 x_6,$$

$$\phi_2 = \bar{x}_1 \bar{x}_2 \vee \bar{x}_2 \bar{x}_3 \vee \bar{x}_1 x_6 \vee \bar{x}_3 x_6 \vee \bar{x}_4, \tag{6.2}$$

while the resolvent of the system (5.1)–(5.2) is

$$\phi = \phi_1 \vee \phi_2 = \bar{x}_1 \bar{x}_2 \vee \bar{x}_1 x_6 \vee x_1 \bar{x}_3 \bar{x}_5 \vee \bar{x}_2 \bar{x}_3 \vee \bar{x}_2 \bar{x}_5$$
$$\vee \bar{x}_2 x_6 \vee \bar{x}_3 x_6 \vee \bar{x}_4 \vee \bar{x}_5 x_6, \tag{7}$$

(showing in particular that in every solution of (5.1)–(5.2), $x_4 = 1$).

3. Algorithms[1]

Due to the fact that the resolvent of a system of inequalities might involve an excessive number of (prime) implicants, practical algorithms based on the ideas outlined in the previous section can utilize only partially the information contained in it. Spielberg's minimal preferred inequalities method [44] belongs essentially to this class. Another example, APOSS (A Partial Order in the Solution Space), an algorithm given in [32] for solving linear 0–1 programs, utilizes only those minimal covers of the individual constraints which involve at most 3 elements. The corresponding implicants are combined to produce more implicants of lengths 1, 2 and 3. To every implicant of length 2 an order relation between variables is naturally associated ($x\bar{y} = 0$ means $x \leq y$, $\bar{x}y = 0$ means $\bar{x} \leq y$, etc.) If two binary relations involving the same pair of variables can be detected, then one of the variables can be eliminated ($xy = x\bar{y} = 0$ implies $x = 0$, $\bar{x}y = \bar{x}\bar{y} = 0$ implies $x = 1$, $xy = \bar{x}\bar{y} = 0$ implies $x = \bar{y}$, $x\bar{y} = \bar{x}y = 0$ implies $x = y$). When all these

---
[1] A survey on Boolean-based algorithms is given in [20].

informations are exhausted, the same binary relations are re-used as cuts in the associated linear program and finally, if no further use of the binary relations is apparent, a branching technique is applied.

Consider for example a problem involving the constraints

$$8x_1 + 7x_2 + 5x_3 + 4x_4 + 2x_5 + 2x_6 \leq 14,$$
$$4x_1 + 2x_2 + 6x_3 + 3x_4 + x_5 + 5x_6 \geq 12.$$

The minimal covers of lengths not exceeding 3 give rise to the "partial resolvents"

$$\psi_1 = x_1 x_2 \vee x_1 x_3 \vee x_1 x_4 \vee x_2 x_3 x_4,$$
$$\psi_2 = \bar{x}_1 \bar{x}_3 \vee \bar{x}_2 \bar{x}_3 \bar{x}_4 \vee \bar{x}_2 \bar{x}_4 \bar{x}_6 \vee \bar{x}_3 \bar{x}_4 \bar{x}_5 \vee \bar{x}_3 \bar{x}_6.$$

From $x_1 x_3 = \bar{x}_1 \bar{x}_3 = 0$ it follows that $x_3 = \bar{x}_1$. Substituting we get

$$\psi'_1 = x_1 x_2 \vee x_1 x_4 \vee x_2 x_4,$$
$$\psi'_2 = x_1 \bar{x}_2 \bar{x}_4 \vee \bar{x}_2 \bar{x}_4 \bar{x}_6 \vee x_1 \bar{x}_4 \bar{x}_5 \vee x_1 \bar{x}_6,$$

hence $\psi' = \psi'_1 \vee \psi'_2 = x_1 \vee x_2 x_4 \vee \bar{x}_2 \bar{x}_4 \bar{x}_6$, implying in particular $x_1 = 0$, and hence $x_3 = 1$. Substituting $x_1 = 0$, $x_3 = 1$ into our original system, we get

$$7x_2 + 4x_4 + 2x_5 + 2x_6 \leq 9, \qquad 2x_2 + 3x_4 + x_5 + 5x_6 \geq 6,$$

the partial resolvents of which are

$$\psi''_1 = x_2 x_4 \vee x_2 x_5 x_6,$$
$$\psi''_2 = \bar{x}_2 \bar{x}_6 \vee \bar{x}_4 \bar{x}_6 \vee \bar{x}_5 \bar{x}_6 \vee \bar{x}_2 \bar{x}_4 \bar{x}_5;$$

hence

$$\psi'' = \psi''_1 \vee \psi''_2 = \bar{x}_6 \vee x_2 x_4 \vee x_2 x_5 \vee \bar{x}_2 \bar{x}_4 \bar{x}_5$$

showing that $x_6 = 1$ in every feasible solution.

This algorithm has been coded on a CDC-6600 and a few hundred test problems involving up to 200 variables have been solved; the execution times (varying from 0.35 up to 65 sec.) compare favourably with those given by other methods. Recently A. Bachem and K. Räder have conducted a series of excellent computational experiments on the IBM 370/168 of the Institute of Operations Research and Econometrics of the University of Bonn.

The special case of quadratic 0–1 programs has been examined in [21, 30]. Consider a quadratic function in 0–1 variables

$$f = \sum_{j=1}^{n} c_j x_j + \sum_{i,j=1}^{n} d_{ij} x_i x_j,$$

and let us put

$$\Delta_j = c_j + \sum_{i=1}^{j-1} d_{ij} x_i + \sum_{i=j+1}^{n} d_{ji} x_i \quad (j = 1, \ldots, n)$$
$$\Delta_{jk} = \Delta_j - \Delta_k - d_{jk} x_k + d_{kj} x_j \quad (j, k = 1, \ldots, n; j < k).$$

It is easy to see that in every minimizing point of f, $\Delta_j > 0$ ($\Delta_j < 0$) implies $x_j = 0$ ($x_j = 1$), while $\Delta_{jk} > 0$ ($\Delta_{jk} < 0$) implies $x_j \leq x_k$ ($x_j \geq x_k$). These relations can be exploited exactly as in the linear case to obtain information about variables with fixed values and about equal or complementary variables. If for example,

$$f = -x_1 + 3x_2 + x_1 x_4 - 3x_1 x_3 + 2x_2 x_4 + 3x_3 x_4 - 4x_2 x_3,$$

then from $\Delta_1 = -1 - 3x_3 + x_4$ we get $x_4 = 0 \to \Delta_1 < 0 \to x_1 = 1$, and from $\Delta_4 = x_1 + 2x_2 + 3x_3$ we get $x_1 = 1 \to \Delta_4 > 0 \to x_4 = 0$, i.e. $\bar{x}_1 \bar{x}_4 = x_1 x_4 = 0$, or $x_4 = \bar{x}_1$. Replacing now x_4 by $1 - x_1$ in f gives

$$f' = -x_1 + 5x_2 + 3x_3 - 2x_1 x_2 - 6x_1 x_3 - 4x_2 x_3;$$

now $\Delta_1 < 0$, and hence $x_1 = 1$; f' becomes

$$f'' = -1 + 3x_2 - 3x_3 - 4x_2 x_3,$$

where $\Delta_3 < 0$, showing that $x_3 = 1$; finally, f'' becomes

$$f''' = -4 - x_2,$$

showing that $x_2 = 1$, and the minimum (-5) is obtained in $(1, 1, 1, 0)$.

Another device which gives some insight into the problem is the examination of a "penalty relaxation inequality". This inequality has the form $l(x) \leq b^*$, where $l(x)$ is a linear lower bound of the quadratic function $f(x)$, and b^* is an upper bound of the minimum of $f(n)$; the role of b^* can be played by the value of $f(x)$ in an arbitrary 0–1 point, while the construction of $l(x)$ (see [21]) is based on Hansen's additive penalties [34]. Such an $l(x)$ for our function is $-6 + \frac{5}{2}\bar{x}_1 + x_2 + \frac{7}{2}\bar{x}_3$, and if we take as b^* the value $f(1\,0\,1\,0) = -4$, we find that $\frac{5}{2}\bar{x}_1 + x_2 + \frac{7}{2}\bar{x}_3 \leq 2$, i.e. $x_1 = x_3 = 1$.

Of course, the examination of the Δ_j's, Δ_{ij}'s and of the penalty-relaxation inequalities does not usually solve the entire problem, but can give valuable information when coupled with a branch-and-bound type method. Since every quadratic 0–1 problem can be brought to a form where the quadratic form is positive (negative) definite (see [30]), there are possibilities of "bounding" by the use of continuous quadratic programming.

A special case of quadratic 0–1 programming has been studied in [15]. The question of maximizing a quadratic function with a single linear constraint ("quadratic knapsack problem") arose in connection with a location problem for airports in Italy, and the method suggested in [15] for its solution consists in determining linear upper bounds of the objective function and solving a sequence of associated (linear) knapsack problems.

A question which arises frequently in applications is that of minimizing an unconstrained polynomial in 0–1 variables. A method of successive elimination of variables has been given in [29] (see also [31]) for its solution. Branch-and-bound methods for the same problem have been devised in [3, 25, 33, 47, 51]; the main characteristic of these methods is the fact that branching is not performed according to single variables, but according to the 0–1 values of the nonlinear terms appearing in the polynomial. A variant of these procedures (see [25]) has been programmed on an IBM 360/50; problems with 10–30 variables, involving 10–50 nonlinear terms required between 0.48 and 239 seconds of execution time (including input-output time).

An efficient method for minimizing quotients of linear functions in 0–1 variables has been given by M. Florian and P. Robillard [41, 42] (see also [31]).

Another question which has been examined was that of constraint pairing and its application to knapsack problems. Single linear constraints can be used in a straightforward way for deriving bounds on the variables of discrete optimization problems from the examination of all the surrogate constraints associated to pairs of constraints. Different surrogates might be helpful in fixing the values (or at least improving the bounds) of different variables; it might of course happen that no surrogate constraint fixes a variable, although the system does. It was however shown in [24] that if any variables can be fixed (or their bounds improved) by using arbitrary surrogates, then the same conclusion can also be obtained from the examination of $n+2$ "special" surrogates (n of which correspond to those multipliers for which the coefficient of one of the variables in the surrogate is 0). R. Dembo [9] shows that many of the conclusions so obtainable, are also available from the "best" surrogate. A. Charnes, D. Granot and F. Granot [6] show how to extend these ideas to the case of more than two constraints.

An efficient application of this approach to knapsack problems [10]

(KP) $\quad$ maximize $\quad \sum_{j=1}^{n} a_j x_j,$

$\quad\quad\quad$ subject to $\quad \sum_{j=1}^{n} b_j x_j \leq b_0,$

$\quad\quad\quad\quad\quad\quad\quad\quad x_j \in \{0, 1\}, \quad j = 1, \ldots, n.$

associates to (KP) a pair of constraints and derives conclusions from the resulting system.

Let us assume that $(a_1/b_1) \geq \cdots \geq (a_n/b_n)$. Let $\Xi = (\xi_1, \ldots, \xi_n)$ be the optimal solution of (KP'), obtained from (KP) by replacing all the constraints $x_j \in \{0, 1\}$ by $0 \leq x_j \leq 1$ $(j = 1, \ldots, n)$. If Ξ is not an integer vector, then we have $\xi_j = 1$ $(j = 1, \ldots, t)$, $\xi_j = 0$ $(j = t+2, \ldots, n)$, and $0 < \xi_{t+1} < 1$. A very good (frequently optimal) solution is obtained by fixing $x_j^* = 1$ $(j = 1, \ldots, t)$, $x_{t+1}^* = 0$, and re-solving a new KP' for b_0 replaced by $b_0 - \sum_{j=1}^{t} b_j$, etc., until arriving to a problem with x_j fixed for $j = 1, \ldots, t^*$, and such that all the b_j's $(j = t^*+1, \ldots, n)$ are larger than the remaining b_0. Then the already fixed values x_j^* $(j = 1, \ldots, t^*)$ together with $x_j^* = 0$ $(j = t^*+1, \ldots, n)$ form a good initial solution: let $a^* = \sum_{j=1}^{n} a_j x_j^*$. If the data are integer, than any better solution will have a^*+1 as a lower bound, an upper bound to it is $\hat{a} = \sum_{j=1}^{t} a_j + [a_{t+1} \xi_{t+1}]$, (where $[\alpha]$ means the integer part of α). Hence if X^* is not an optimal solution, then any better solution satisfies

$$\sum_{j=1}^{n} a_j x_j = a^* + p + 1,$$

where p is a nonnegative integer not exceeding $\hat{a} - a^* - 1$. Pairing this with the constraint

$$\sum_{j=1}^{n} b_j x_j + s = b_0 \quad (s \geq 0)$$

usually supplies enough information to fix at least some of the variables. These informations can be supplemented by those given by the binary and ternary relations among the variables.

Consider for example the 0–1 knapsack problem of maximizing

$$15x_1 + 16x_2 + 13x_3 + 9x_4 + 17x_5 + 11x_6,$$

subject to

$$9x_1 + 10x_2 + 11x_3 + 8x_4 + 16x_5 + 11x_6 \leq 29$$

Here $\Xi = (1, 1, \frac{10}{11}, 0, 0, 0)$, $X^* = (1, 1, 0, 1, 0, 0)$, $\hat{a} = 42$, $a^* = 40$. Hence, if X^* is not optimal, then any optimal solution satisfies

$$15x_1 + 16x_2 + 13x_3 + 9x_4 + 17x_5 + 11x_6 = 41 + p \quad (0 \leq p \leq 1),$$
$$9x_1 + 10x_2 + 11x_3 + 8x_4 + 16x_5 + 11x_6 + s = 20 \quad (s \geq 0).$$

Multiplying the first equation by 11, the second one by 13 and subtracting, we get

$$48x_1 + 46x_2 - 5x_4 - 21x_5 - 22x_6 - 13s = 74 + 11p$$

or

$$48x_1 + 46x_2 + 5\bar{x}_4 + 21\bar{x}_5 + 22\bar{x}_6 + 11\bar{p} = 133 + 13s,$$

implying $x_1 = 1$, $x_2 = 1$, $x_5 = 0$, $x_6 = 0$, and the last relation reduces to $5\bar{x}_4 + 11\bar{p} = -4 + 13s$, which obviously has no nonnegative integer solutions, showing that X^* was the optimal solution of our problem.

Experiments carried out with this idea show that it is extremely useful for fixing variables in 0–1 knapsack problems. In experiments carried out on an IBM 370/145 it turned out that in randomly generated problems involving 50–10,000 variables, the average number of fixed variables was between 74% and 93% of the total number of variables, while the computing time was less than one second.

4. Equivalent forms of 0–1 programs

Let us rewrite the resolvent $\phi(X)$ of a system Σ of linear or nonlinear inequalities, in the form

$$\phi(X) = \bigvee_{t=1}^{T} \left(\prod_{j \in U_t} x_j \right) \left(\prod_{j \in V_t} \bar{x}_j \right) \tag{8}$$

where U_t, V_t $(t=1,\ldots,T)$ are disjoint subsets of $\{1,\ldots,n\}$ Then, it is easy to see that $\phi(X)=0$ iff X is a solution of the following *generalized covering problem*

$$\sum_{j\in V_t} x_j - \sum_{j\in U_t} x_j \geq 1-|U_t| \quad (t=1,\ldots,T) \tag{9}$$

Hence ([19])[2] every linear or nonlinear 0–1 programming problem is *strongly equivalent* to (i.e. has the same set of feasible solutions as) a generalized covering problem.

Consider now the problem (PI) of minimizing a pseudo-Boolean function $f(X)$ subject to Σ. Assume that f_0 is *strictly monotonic*, i.e. changing any 1 of any $X \in B^n$ to a 0 strictly decreases the value of f_0. This assumption holds for example for linear f_0's having only positive coefficients. Specializing (8) to the case where $(\prod_{j\in U_t} x_j)(\prod_{j\in V_t} \bar{x}_j)$ $(t=1,\ldots,T)$ are the prime implicants of $\phi(X)$, and assuming that $U_t=\emptyset$ for $t=1,\ldots,T_0$ and $U_t \neq \emptyset$ for $t=T_0+1,\ldots,T$, it can be shown ([22]) that (PI) is equivalent to (i.e. has the same optimal solutions as) the problem (PII) of minimizing $f_0(X)$ subject to $\phi'(X)=0$, where

$$\phi'(X) = \bigvee_{t=1}^{T_0} \prod_{j\in V_t} \bar{x}_j, \tag{8}$$

i.e. to the *covering problem*: minimize $f_0(X)$ subject to $\sum_{j\in V_t} x_j \geq 1$ $(t=1,\ldots,T_0)$.

This equivalence holds because every feasible solution of PI is a feasible solution of the covering problem, while a feasible solution of the covering problem cannot be optimal unless it is feasible for PI too.

For example minimizing $\sum_{j=1}^{6} c_j x_j$ $(c_j > 0, j=1,\ldots,6)$ subject to (5.1)–(5.2) is strongly equivalent to the generalized covering problem: minimize $\sum_{j=1}^{6} c_j x_j$ subject to $x_4=1$ and to $x_1+x_2 \geq 1$, $x_1-x_6 \geq 0$, $-x_1+x_3+x_5 \geq 1$, $x_2+x_3 \geq 1$, $x_2+x_5 \geq 1$, $x_2-x_6 \geq 0$, $x_3-x_6 \geq 0$, $x_5-x_6 \geq 0$ and is equivalent to the covering problem: minimize $\sum_{j=1}^{6} c_j x_j$ subject to $x_4=1$ and to $x_1+x_2 \geq 1$, $x_2+x_3 \geq 1$, $x_2+x_5 \geq 1$ (x_6 does not appear in any of the constraints, hence $x_6=0$ in any optimal solution).

Numerous equivalences between different forms of 0–1 programs have been described in [28].

5. Packing and knapsack problems

By a *packing problem* we shall mean a set of linear inequalities in 0–1 variables of the form $x_i+x_j \leq 1$ $((i,j) \in \Gamma)$. A linear inequality $\sum_{j=1}^{n} a_j x_j \leq a_0$ $(a_j \geq 0, j=0,1,\ldots,n)$ is equivalent to a packing problem iff all its minimal covers contain exactly two elements.

The converse problem, of characterizing those packing problems which are equivalent to a single linear inequality, has been examined in [7]. It has been

---

[2] This remark appears in a somewhat stronger form for the special case of a *single* linear pseudo-Boolean inequality in [2].

shown that the following two characterizations follow from the theory of threshold functions.

(I) PP is not 0–1 equivalent to a single linear inequality iff it is possible to find 4 distinct indices h, i, j, k such that

$$(h, i) \in \Gamma, \quad (h, j) \notin \Gamma, \quad (h, k) \notin \Gamma, \quad (i, j) \notin \Gamma, \quad (i, k) \notin \Gamma, \quad (j, k) \in \Gamma,$$

or such that

$$(h, i) \in \Gamma, \quad (h, j) \notin \Gamma, \quad (h, k) \notin \Gamma, \quad (i, j) \in \Gamma, \quad (i, k) \notin \Gamma, \quad (j, k) \in \Gamma,$$

or such that

$$(h, i) \in \Gamma, \quad (h, j) \notin \Gamma, \quad (h, k) \in \Gamma, \quad (i, j) \in \Gamma, \quad (i, k) \notin \Gamma, \quad (j, k) \in \Gamma.$$

(II) PP is 0–1 equivalent to a single linear inequality iff there exists a partitioning of $\{1, \ldots, n\}$ into two subsets N' and N'' and a permutation $(j_1, \ldots, j_r)$ of the elements of N'' such that

(i) $\forall i, j \in N', (i, j) \in \Gamma$

and

(ii) $\forall i, j \in N'', (i, j) \notin \Gamma$

(iii) $\forall j_s, j_t \in N'', (s < t), \forall i \in N', (i, j_t) \in \Gamma$ implies $(i, j_s) \in \Gamma$.

An efficient algorithm was also presented in [7] for finding such a 0–1 equivalent single linear inequality, if any, or otherwise to find a "small" system of linear inequalities equivalent to the given PP. Peled studies in a recent paper the more general question of reducing the number of linear constraints in an arbitrary system of inequalities involving only 0–1 variables.

6. Coefficient transformation

It is obvious that different linear inequalities may have the same 0–1 solutions, and it might be useful to be able to transform a given inequality to an equivalent one which has a "better" form. For example $x + y \leq 1$ seems to be a better form than $173x + 89y + 244.5$, but obviously the two inequalities have the same 0–1 solutions. This problem is studied in [4] and it is shown that the "optimal" coefficients (according to a large variety of criteria) can be determined by solving an associated linear program.

Let us consider a linear inequality

$$\sum_{j=1}^{n} a_j x_j \leq a_0, \tag{11}$$

where $a_1 \geq \cdots \geq a_n \geq 0$. A minimal cover $R \subseteq \{1, \ldots, n\}$ such that $\sum_{j \in R} a_j - a_r + a_{r'} \leq a_0$ holds for any $r \in R, r' \notin R, r < r'$, is called a *roof* of (11).

Similarly a set $C \subseteq \{1, \ldots, n\}$, maximal with the property that $\sum_{j \in C} a_j \leq a_0$, and such that $\sum_{j \in C} a_j - a_c + a_{c'} > a_0$ holds for any $c \in C, c' \notin C, c > c'$, is called a *ceiling* of (11). It is shown that every inequality

$$\sum_{j=1}^{n} B_j x_j \leq b_0 \tag{12}$$

0–1 equivalent to (11) and such that $B_1 \geq \cdots \geq B_n \geq 0$, can be obtained from a solution of the system

$$\sum_{j \in R} b_j \geq b_0 + 1 \quad \text{(for all roofs } R \text{ of (12))},$$

$$\sum_{j \in C} b_j \leq b_0 \quad \text{(for all ceilings } c \text{ of (12))},$$

$$b_1 \geq \cdots \geq b_n \geq 0,$$

by taking $B_j = \lambda b_j$ ($j = 1, \ldots, n$), $\lambda > 0$.

For example all the inequalities 0–1 equivalent to

$$7x_1 + 5x_2 + 3x_3 + 3x_4 + x_5 \leq 10$$

and having the coefficients ordered in the same way, are characterized by the system

$$b_1 + b_2 \geq b_0 + 1, \qquad b_1 + b_3 \leq b_0,$$
$$b_1 + b_4 + b_5 \geq b_0 + 1, \qquad b_2 + b_3 + b_5 \leq b_0,$$
$$b_2 + b_3 + b_4 \geq b_0 + 1,$$
$$b_1 \geq b_2 \geq b_3 \geq b_4 \geq b_5 \geq 0.$$

If the criterion is to minimize b_0, the optimal solution is $(4, 3, 2, 2, 1; 6)$, i.e. the inequality

$$4x_1 + 3x_2 + 2x_3 + 2x_4 + x_5 \leq 6.$$

Numerous problems of similar nature have been studied in threshold logic (e.g. see [35, 36]). The usefulness of such transformations for increasing the efficiency of branch-and-bound methods is pointed out in [49].

7. Polytopes in the unit cube

The convex hull $\hat{S}$ of a set S of vertices of the unit cube can be characterized by its facets. The set S is characterized by a Boolean function $\delta_S(X)$ equal to 0 for $X \in S$ and to 1 elsewhere. The question of relating the Boolean and the geometric structures of a system of inequalities in 0–1 variables arises naturally. M.A.

Pollatschek [40] seems to have been the first to examine such questions. M.W. Padberg [39] has given a procedure for producing facets of $\hat{S}$. A systematic investigation of this topic has been attempted in [23]. Some of the results of [23] overlap with those of [1, 17, 18, 37, 39, 50].

It was noticed in Section 4 that every 0–1 programming problem with a strictly monotone objective function can be reduced to a covering problem. Therefore in this section we shall mainly deal with facets of covering problems. For notational convenience we shall put $y_j = \bar{x}_j$ $(j = 1, \ldots, n)$; thus the constraints $\sum_{j \in T_i} x_j \geq 1$ $(i = 1, \ldots, h)$ of the given covering problem become

$$\sum_{j \in T_i} y_j \leq t_i - 1 \quad (i = 1, \ldots, h) \tag{13}$$

where $t_i = |T_i|$. Let S be the set of 0–1 solutions of (13), and $\hat{S}$ its convex hull. We shall assume that $\hat{S}$ is n-dimensional. It can be seen easily that $-y_j \leq 0$ is a facet of $\hat{S}$ for all $j = 1, \ldots, n$, but $y_j \leq 1$ is a facet of $\hat{S}$ iff $t_1 = 2$ implies that $j \notin T_i$.

It has been shown in [23] that the constraint (13) is a facet of $\hat{S}$ iff for any $k \notin T_j$, the intersection of all those T_j which are contained in $\{k\} \cup T_i$ is nonempty. Further, if (13) is not a facet of $\hat{S}$, a procedure was given for strengthening it to a facet (by changing the coefficients of the variables y_k ($k \notin T_i$) from 0 to certain positive values. The procedure becomes particularly efficient for an apparently special class of covering problems, the so-called *regular covering problems*, i.e. those covering problems where the feasibility of any point $(y_1^*, \ldots, y_n^*)$ (where $y_{j_1}^* = \cdots = y_{j_s}^* = 1$, the other components are 0) implies the feasibility of any point $(y_1^{**}, \ldots, y_n^{**})$ having the same number of 1 components ($y_{l_1}^{**} = \cdots = y_{l_s}^{**} = 1$, the other components are 0) when $j_1 \leq l_1, \ldots, j_s \leq l_s$. However (see [22]), a very wide class of covering problems can be brought to such a form.

The extension procedure becomes extremely simple for the case of regular covering problems and it can be shown that there is a 1–1 correspondence between those facets of $\hat{S}$ which have only 0–1 coefficients and those sets T_i which have the following two properties.

(i) if $u_i = \min \{j \mid j \in T_i\}$, $w_i = \min \{j \mid j \notin T_i, j > u_i\}$ (if any), and if $P_i \in B^n$ is the point whose 1-components are all the elements of the set

$$(\{w_i\} \cup T_i) = \{u_i\} \quad \text{(if } w_i \text{ is defined)}$$

$$T_i - \{u_i\} \quad \text{(otherwise)},$$

then $P_i \in S$;

(ii) if $v_i = \min \{j \mid j \in T_i, j \neq u_i\}$ and $R_i \in S$ is the point whose 1-components are all the elements of the set $(\{1\} \cup T_i) - \{u_i, v_i\}$, then $R_i \in S$.

The most common case of a regular covering problem corresponds to knapsack problems, when the T_i's are its minimal covers. A list of all the facets of all the knapsack problems with at most 5 variables is given in [23].

8. Pseudo-Boolean functions and game theory[3]

A *characteristic function game* (N, W) is a set of "players" $N = \{1, 2, \ldots, n\}$ and a real-valued function $W: 2^N \to R$ (called the *characteristic function*), defined for all subsets T of N. If T is a "coalition" then $W(T)$ is the "payoff" it can secure. It is clear that 2^N is mapped in a 1–1 way onto B^n by mapping a subset T of N to its *characteristic vector* X, defined by $x_k = 1$ for $k \in T$ and $x_k = 0$ for $k \notin T$. Hence as remarked by Owen ([38]) a characteristic function game is actually the same as a pseudo-Boolean function.

It is well known [31] that every pseudo-Boolean function f in n variables has a unique polynomial expression of the form

$$f(x) = \sum_{T \subseteq N} \left[a_T \prod_{k \in T} x_k \right],$$

called its *canonical form*. The corresponding characteristic function game (N, W) then satisfies

$$W(T) = \sum_{S \subseteq T} a_S, \quad T \subseteq N.$$

Shapley ([45]) has shown that this relation gives

$$a_T = \sum_{S \subseteq T} (-1)^{t-s} W(S), \quad T \subseteq N,$$

where t and s are the cardinalities of T and S, respectively. Thus the a_T's can be found from the function $f(x)$.

As an example let us consider the 3-person characteristic function game defined by Table 1. The corresponding pseudo-Boolean function on B^3 is

$$f(X) = 3x_1 x_2 \bar{x}_3 + 2x_1 \bar{x}_2 x_3 + 2\bar{x}_1 x_2 x_3 + 4x_1 x_2 x_3.$$

By replacing each $\bar{x}_j$ by $1 - x_j$ and simplifying, we obtain the canonical expression

$$f(X) = 3x_1 x_2 + 2x_1 x_3 + 2x_2 x_3 - 3x_1 x_2 x_3. \tag{14}$$

Table 1

T	$\emptyset$	$\{1\}$	$\{2\}$	$\{3\}$	$\{1,2\}$	$\{1,3\}$	$\{2,3\}$	$\{1,2,3\}$
$W(T)$	0	0	0	0	3	2	2	4

A game (N, W) is said to be *superadditive* if for any disjoint sets S, T of N, we have $W(S) + W(T) \leq W(S \cup T)$ (i.e. it always "pays" to form a larger coalition). It can be easily seen that the game in the example above is superadditive. The following result holds:

[3] See [26].

Let f be the pseudo-Boolean function corresponding to the game W. Then the following are equivalent

(a) W is superadditive,
(b) $X+Y \leqslant 1$ implies $f(X)+f(Y) \leqslant f(X+Y)$ for all $X, Y \in B^n$,
(c) $XY=0$ implies $f(X)+f(Y) \leqslant f(X+Y)$ for all $X, Y \in B^n$,
(d) $f(XY)+f(X\bar{Y}) \leqslant f(X)$ for all $X, Y \in B^n$ (here $\bar{Y} = \underline{1} - Y$, $\underline{1} = (1, \ldots, 1)$).

The goal of n-person characteristic function game theory is to find a "solution", i.e. a value for each player based upon the coalitions he may join. If a game (N, W) satisfies $W(\emptyset) = 0$, then, as Shapley [45] mentions, we may regard a solution as an inessential game (N, Z) which "approximates" (N, W) by some method and which assigns a value $Z(\{j\})$ to each player j. In [26] we have discussed a few specific solutions in terms of pseudo-Boolean functions. Since such a function defines a game we can speak about the core and the Shapley value of a function. Throughout this section let f be a pseudo-Boolean function with $f(0) = 0$. A *core element* of f is a linear pseudo-Boolean function $h(X)$ satisfying $h(X) \geqslant f(X)$ for all $X \in B^n$ and $h(1) = f(1)$. We shall also say that the vector h of coefficients of $h(X)$ is a *core element* of f. The polyhedron of all the core elements of f is called the *core* of f. It may be empty for some f. In [26] we construct another polyhedron (the *selectope*) and show that it contains the core of f.

Consider the canonical form of the pseudo-Boolean function f. We denote $Y^+ = \{T \subseteq N : a_T > 0\}$, $Y^- = \{T \subseteq N : a_T < 0\}$, $Y = Y^+ \cup Y^-$. The *incidence graph* of f is a directed bipartite graph $G = (Y, N; E)$ in which an edge $e \in E$ is directed from $T \in Y^+$ to $j \in N$ if $j \in T$, and an edge $e \in E$ is directed from $j \in N$ to $T \in Y^-$ if $j \in T$. For any node $T \in Y$, $I(T)$ denotes the set of edges $e \in E$ incident with T. For any node $j \in N$, $I^+(j)$ denotes the set of edges $e \in E$ directed to j, $I^-(J)$ denotes the set of edges $e \in E$ directed away from j and $I(j) = I^+(j) \cup I^-(j)$. For each edge $e \in E$, $T(e)$ denotes its end in Y and $j(e)$ its end in N. The edge $e \in E$ corresponds to the occurrence of the variable $x_{j(e)}$ in the term $a_{T(e)} \cdot \prod_{j \in T} x_j$ of f. Fig. 1 illustrates the incidence graph of our pseudo-Boolean function (14) with the values a_T displayed next to the nodes T.

A *selector* of f is a vector $s = (e_T, T \in Y)$, such that $e_T \in I(T), \forall T \in Y$. The

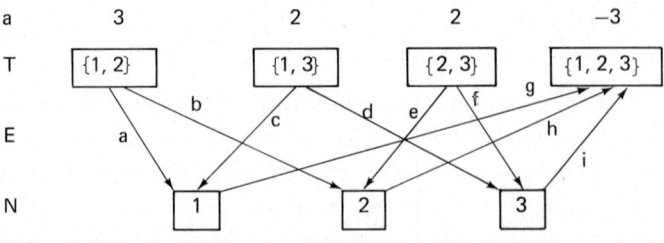

Fig. 1.

Table 2

	Selectors				Selections		
	$e_{\{1,2\}}$	$e_{\{1,3\}}$	$e_{\{2,3\}}$	$e_{\{1,2,3\}}$	h_1	h_2	h_3
1	a	c	e	g	2	2	0
2	a	c	e	h	5	−1	0
3	a	c	e	i	5	2	−3
4	a	c	f	g	2	0	2
5	a	c	f	h	5	−3	2
6	a	c	f	i	5	0	−1
7	a	d	e	g	0	2	2
8	a	d	e	h	3	−1	2
9	a	d	e	i	3	2	−1
10	a	d	f	g	0	0	4
11	a	d	f	h	3	−3	4
12	a	d	f	i	3	0	1
13	b	c	e	g	−1	5	0
14	b	c	e	h	2	2	0
15	b	c	e	i	2	5	−3
16	b	c	f	g	−1	3	2
17	b	c	f	h	2	0	2
18	b	c	f	i	2	3	−1
19	b	d	e	g	−3	5	2
20	b	d	e	h	0	2	2
21	b	d	e	i	0	5	−1
22	b	d	f	g	−3	3	4
23	b	d	f	h	0	0	4
24	b	d	f	i	0	3	1

corresponding *selection* of f is the vector $h(s) = (h_j, j \in N)$, where

$$h_j = \sum_{e_T \in J(j)} a_T .$$

By S we denote the set of all selectors of f. In our example there are $\sum_{T \in Y} |T| = 2 \cdot 2 \cdot 2 \cdot 3 = 24$ selectors, which are listed in Table 2 along with the corresponding selections (of which only 20 are distinct).

Selectors have been introduced in [27], where selections are called "linear factors", and where it is shown that if Y^- is empty then $f(X)$ is the minimum of all the linear pseudo-Boolean functions $\sum_{j \in N} h_j x_j$, where h is a selection of f. This concept has been generalized by I. Rosenberg in [43].

The *selectope* of f is the convex hull of all the selections of f. We give below a characterization of the selectope of f. Let $h = (h_j, j \in N)$ be any n-vector. A *flow* for h in G is a non-negative vector $z = (z_e, e \in E)$ satisfying the node equations

$$\sum_{e \in I(Y)} z_e |a_T|, \quad T \in Y$$

$$\sum_{e \in I^+(j)} z_e - \sum_{e \in I^-(j)} z_e = h_j, \quad j \in N.$$

Then

(1) The selectope of f is the set of those n-vectors h for which there exists a flow in G.

(2) The selectope of f contains the core of f, equality holding if and only if all the nonlinear terms in (1) have nonnegative coefficients.

We remark that from here we obtain an efficient partial test for a nonnegative vector h satisfying $h(1) = f(1)$ to be a core element of an unlinear pseudo-Boolean function f. Apply the maximal flow algorithm to G. If the value of this flow is less than $\sum_{T \in Y^+} a_T$, h cannot be a core element of f. However if the value is $\sum_{t \in Y^+} a_T$, we do not have any conclusion. (For example, of all the 20 selections of the pseudo-Boolean function f in (14), only $(2, 2, 0)$ is a core element of f.) It would be of interest to refine the test for that case.

A vector $Y \in B^n$ is said to be a *carrier* of a pseudo-Boolean function f on B^n if $f(X) = f(XY)$ for all $X \in B^n$. The product of carriers of f is a carrier of f, hence the product Y^* of all the carriers of f is the unique minimal carrier of f, and f effectively depends on x_j if and only if $y_j^* = 1$.

A mapping $\pi : B^n \to B^n$ is an *automorphism* if it is one-one and onto, and also conserves the operations $\vee, \cdot$ and $-$, i.e.

$$\pi(X \vee Y) = \pi(X) \vee \pi(Y), \qquad \pi(XY) = \pi(X)\pi(Y), \qquad \pi(\bar{X}) = \overline{\pi(X)}.$$

For convenience we shall write πX for $\pi(X)$. For any automorphism π on B^n and for any function f on B^n we define the function πf by $\pi f(X) = f(\pi^{-1} X)$ or equivalently by $\pi f(\pi X) = f(X)$. It can be seen that if π is an automorphism of B^n and X is a unit vector of B^n, then πX is a unit vector of B^n. Hence π permutes the unit vectors of B^n and permits us to view π as a permutation of the variables $j \in N$ themselves. For $j \in N$, $k = \pi(j)$ is defined so that if X is the unit vector with $x_j = 1$, then πX is the unit vector with $(\pi X)_k = 1$. Thus $(\pi X)_k = x_{\pi^{-1}k}$ for all $X \in B^n$.

We can now state the axiomatic definition of the Shapley value ([45]). Let F be the set of all pseudo-Boolean functions f on B^n such that $f(0) = 0$. A *Shapley value* is a mapping $\eta : F \to R^n$ satisfying the following axioms.

Axiom 1. *For each automorphism π of B^n and for each $f \in F$,*

$$\eta_{\pi(k)}[\pi f] = \eta_k[f], \quad k = 1, \ldots, n.$$

Axiom 2. *For each $f \in F$ and for each carrier Y of f,*

$$\sum_{k=1}^{n} \eta_k[f] y_k = f(1).$$

(*in particular then* $\sum_{k=1}^{n} \eta_k[f] = f(1)$).

Axiom 3. *For each $f, g \in F$*

$$\eta[f + g] = \eta[f] + \eta[g].$$

The following theorem is due to Shapley ([45]).
There exists a unique Shapley value, and it is given by the formula

$$\eta_k[f] = \sum_{\substack{T \subseteq N \\ k \in T}} \frac{a_T}{|T|}.$$

As an illustration, the Shapley value of (14) is

$$\eta[f] = (\tfrac{3}{2} + \tfrac{2}{2} - \tfrac{3}{3}, \tfrac{3}{2} + \tfrac{2}{2} - \tfrac{3}{3}, \tfrac{3}{2} + \tfrac{2}{2} - \tfrac{3}{3})$$
$$= (\tfrac{3}{2}, \tfrac{3}{2}, 1).$$

Let f be a pseudo-Boolean function with $f(0) = 0$. Then the Shapley value of f is the arithemetic mean, over all the selectors of f, of the corresponding selections, i.e.

$$\frac{1}{|S|} \sum_{s \in S} h(s).$$

Also the following result holds: Let $f(0) = 0$. If $a_T \geq 0$ for all $T \subseteq N$ then f is superadditive and every selection of f as well as the Shapley value of f are core elements of f.

References

[1] E. Balas, Facets of the knapsack problem, Math. Programming 8 (1975) 146–164.
[2] E. Balas and R. Jeroslow, Canonical cuts on the unit hypercube, SIAM J. Appl. Math. 23 (1972) 61–66.
[3] G. Berman, A branch-and-bound method for the maximization of a pseudo-Boolean function. Technical Report, Faculty of Mathematics, University of Waterloo.
[4] G. Bradley, P.L. Hammer and L.A. Wolsey, Coefficient reduction in 0–1 programming, Math. Programming 7 (1974) 263–282.
[5] P. Camion, Une methode de resolution par l'algèbre de Boole des problèmes combinatoires ou interviennent des entiers, Cahiers Centre Études Recherche Opér. 2 (1960) 234–289.
[6] A. Charnes, D. Granot and F. Granot, On improving bounds for variables in linear integer programs by surrogate constraints, INFOR-Canad. J. Operational Res. 13 (1975) 260–269.
[7] V. Chvátal and P.L. Hammer, Aggregation of inequalities in integer programming, Annals of Discrete Mathematics 1 (1977) 145–162.
[8] A.W. Colijn, Pseudo-Boolean programming. Ph.D. Thesis, University of Waterloo, Department of Combinatorics and Optimization, July 1973.
[9] R. Dembo, A note on constraint pairing in integer linear programming. University of Waterloo, Department of Management Science, Working Paper, April 1974.
[10] R. Dembo and P.L. Hammer, A reduction algorithm for knapsack problems. University of Waterloo, Combinatorics and Optimization Research Report CORR 74–11, April 1974.
[11] R. Faure and Y. Malgrange, Une méthode booléenne pour la résolution des programmes linéaires en nombres entiers, Gestion, Numéro Spécial, April 1963.
[12] R. Fortet, L'algèbre de Boole et ses applications en recherche opérationelle. Cahiers Centre Étude Recherche Opér. 1 (4) (1959) 5–36.
[13] R. Fortet, Applications de l'algèbre de Boole en recherche opérationnelle. Rev. Francaise Recherche Opér., 4 (1960) 17–26.
[14] D. Gale, A theorem on flows in networks, Pacific J. Math. 7 (1957) 1073–1082.

[15] G. Gallo, P.L. Hammer and B. Simeone, Quadratic knapsack problems in: M. Padberg, ed., Combinatorial Optimization, Math. Programming Studies (North-Holland, Amsterdam, forthcoming).

[16] A.M. Geoffrion, Lagrangean relaxation for integer programming, University of California at Los Angeles, Western Management Science Institute, Working Paper No. 195, December 1973.

[17] F. Glover, Unit-coefficient inequalities for zero–one programming, University of Colorado, Management Science Report Series, No. 73-7, July 1973.

[18] F. Glover, Polyhedral annexations in mixed integer programming. University of Colorado, Management Science Report Series, No. 73-9, August 1973.

[19] F. Granot and P.L. Hammer, On the use of Boolean functions in 0–1 programming, Methods of Operations Research 12 (1972) 154–184.

[20] P.L. Hammer, Boolean procedures for bivalent programming. in: P.L. Hammer and G. Zoutendijk, eds., Mathematical Programming in Theory and Applications (North-Holland, Amsterdam, 1974).

[21] P.L. Hammer and P. Hansen, Quadratic 0–1 programming. University of Montreal, CRM 186, 1972.

[22] P.L. Hammer, E.L. Johnson and U.N. Peled, Regular 0–1 programs, Cahiers Centre Études Recherche Opér. 16 (1974) 267–276.

[23] P.L. Hammer, E.L. Johnson and U.N. Peled, Facets of regular 0–1 polytopes, Math. Programming. 8 (1975) 179–206.

[24] P.L. Hammer, M.W. Padberg and U.N. Peled, Constraint pairing in integer programming, INFOR-Canad. J. of Operational Res. 13 (1975) 68–81.

[25] P.L. Hammer and U.N. Peled, On the maximization of a pseudo-Boolean function, J. Assoc. Comput. Mach. 19 (1972) 265–282.

[26] P.L. Hammer, U.N. Peled and S. Sorensen, Pseudo-Boolean functions and game theory I. Core elements and Shapley value. Cahiers Centre Études Recherche Opér. 19 (1977) 159–176.

[27] P.L. Hammer and I.G. Rosenberg, Linear decomposition of a positive group-boolean function, in: L. Collatz and W. Wettering, eds., Numerische Methoden bei Optimierung, Vol. II (Birkhauser Verlag, Basel-Stuttgart, 1974) 51–62.

[28] P.L. Hammer and I.G. Rosenberg, Equivalent forms of zero-one programs, in: S. Zaremba, ed. Applications of Number Theory to Numerical Analysis (Academic Press, New York and London, 1972) 453–463.

[29] P.L. Hammer, I.G. Rosenberg and S. Rudeanu, On the minimization of pseudo-Boolean functions, Stud. Cerc. Matem. 14 (3) (1963) 359–364. (In Romanian).

[30] P.L. Hammer and A.A. Rubin, Some remarks on quadratic programming with 0–1 variables. Rev. Francaise Informat. Recherche Opérationelle 4 (1970) V-3, 67–79.

[31] P.L. Hammer and S. Rudeanu, Boolean Methods in Operations Research and Related Areas (Springer-Verlag, Berlin/Heidelberg/New York, 1968; French edition: Dunod, Paris, 1970).

[32] P.L. Hammer and Sang Nguyen, APOSS-A partial order in the solution space of bivalent programs, in: N. Christofides, ed., Combinatorial Optimization (J. Wiley, New York, forthcoming).

[33] P. Hansen, Un algorithme SEP pour les programmes pseudo-Booléens nonlinéaries, Cahiers Centre Études Recherche Opér. 11 (1969) 26–44.

[34] P. Hansen, Pénalités additives pour les programmes en variables zéro-un. C.R. Acad. Sci. Paris 273 (1971) 175–177.

[35] S.T. Hu, Threshold Logic (University of California Press, Berkeley and Los Angeles, 1965).

[36] S. Muroga, Threshold Logic and its Applications (Wiley-Interscience, New York, 1971).

[37] G.L. Nemhauser and L.E. Trotter, Jr., Properties of vertex packing and independence systems polyhedra, Math. Programming 6 (1974) 48–61.

[38] G. Owen, Multilinear extensions of games, Management Sci. 18 (5) (January 1972) 64–79.

[39] M.W. Padberg, A note on zero–one programming. University of Waterloo, Combinatorics and Optimization Research Report CORR No. 73–5, March 1973.

[40] M.A. Pollatschek, Algorithms on finite weighted graphs. Ph.D. Thesis, Technion-Isreal Institute of Technology, Faculty of Industrial and Management Engineering, 1970. (In Hebrew, with English synopsis).

[41] P. Robillard, (0, 1) hyperbolic programming problems, Naval Res. Logist. Quart. 18 (1) (1971) 47–57.

[42] P. Robillard and M. Florian, Note concernant la programmation hypérbolique en variables bivalentes. Université de Montréal Département d'Informatiques, Publication 41, June 1971.

[43] I.G. Rosenberg, Decomposition of a positive group-Boolean function into linear factors, Université de Montréal, Centre de recherches mathématiques, Publication 345, October 1973.

[44] S. Rudeanu, Boolean Functions and Equations (North-Holland Amsterdam, 1974).

[45] L.S. Shapley, A value for n-person games, in: H.W. Kuhn and A.W. Tucker, eds., Contributions to the Theory of Games, Vol. II, Annals of Mathematics Studies, No. 28 (Princeton University Press, Princeton, NJ, 1953) 307–317.

[46] K. Spielberg, Minimal preferred variable reduction methods in zero–one programming. IBM Philadelphia Scientific Center Report No. 320–3013, July 1972, Revised, 1973.

[47] H.A. Taha, A Balasian-based algorithm for zero–one polynomial programming, Management Sci. 18 (1972) 328–343.

[48] A. Warszawski, Pseudo-Boolean solutions to multidimensional location problems, Operations Res. 22 (1974) 1081–1085.

[49] H.P. Williams, Experiments in the formulation of integer programming problems. University of Sussex, March 1973.

[50] L.A. Wolsey, Faces for linear inequalities in 0–1 variables, Math. Programming 8 (1975) 165–177.

[51] Y. Yoshida, Y. Inagaki and T. Fukumura, Algorithms of pseudo-Boolean programming based on the branch and bound method, J. Inst. Elec. and Commun. Engs. Japan, 50 (1967) 277–285.

THE ROLE OF UNIMODULARITY IN APPLYING LINEAR INEQUALITIES TO COMBINATORIAL THEOREMS

A.J. HOFFMAN

IBM T.J. Watson Research Center, Yorktown Heights, N.Y. 10598, U.S.A.

In this lecture we are concerned with the use of linear programming duality to prove combinatorial theorems, and the rôle played in such investigations by unimodular matrices and related concepts.

1. Introduction

In this lecture we are concerned with the use of linear programming duality to prove combinatorial theorems, and the rôle played in such investigations by unimodular matrices and related concepts. To illustrate the order of ideas we shall be considering, we begin with the theorem that first attracted us to this topic twenty-six years ago, when we were struggling, under George Dantzig's patient instruction, to understand the duality theorem of linear programming. We were deep in the penumbra of shadow prices and other ghostly ideas and sought an environment where dual variables could be exposed to sunlight. We found it in the following classical theorem of Egervary-Koenig:

If M is $(0, 1)$ matrix, the largest cardinality of a set of 1's in M, no two in the same row or column, is the smallest cardinality of a set of rows and columns containing all the 1's of M.

A proof via linear programming can be constructed as follows. Consider the primal problem

$$x_{ij} \geq 0, \quad \text{all } i \text{ and } j$$

$$\sum_j x_{ij} \leq 1, \quad \text{all } i$$

$$\sum_i x_{ij} \leq 1, \quad \text{all } j \tag{1.1}$$

$$\text{maximize} \sum_{i,j} m_{ij} x_{ij}.$$

The dual problem is

$$u_i \geq 0, \quad \text{all } 1; \quad v_j \geq 0, \quad \text{all } j$$

$$u_i + v_j \geq m_{ij}, \quad \text{all } i \text{ and } j \tag{1.2}$$

$$\text{minimize} \sum_i u_i + \sum_j v_j.$$

If we knew that the maximum in (1.1) is attained at an (x_{ij}), in which all coordinates are integers, then we would know that the primal problem would have for its value the largest cardinality of a set of 1's in M, no two in the same row or column. The reason is that a nonnegative integral X having $\sum_j x_{ij} \leq 1$ for all i must have all coordinates in row i of X equalling 0, except for at most one nonzero which would have to be 1. Similarly, every column of X would be all 0 with at most one 1.

Similarly, if we knew that the minimum in (1.2) is attained at integral u_i and v_j, since each is nonnegative, and our minimization problem involves satisfying $u_i + v_j \geq 0$ or 1, each u_i and v_j would be 0 or 1. Thus, the value of the dual problem would be the smallest cardinality of a set of rows and columns including all 1's of M. Then the duality theorem of linear programming yields the Egervary-Koenig theorem.

What remains is to prove the assertion that optima in (1.1) and (1.2) occur at integral points. One proof is to use the observation that Dantzig's method for solving transportation problems and their duals only uses additions and subtractions. Since the entries are all 0 or 1, this method will produce integral answers.

But we prefer the proof which uses the fact that optima of linear functions on polyhedra occur at vertices of the polyhedra. At a vertex, a system of independent type planes intersect at a point, and the coordinates of the point can (in principle) be calculated by Cramer's rule. In the case of (1.1), the system of independent hyperplanes has the form

$$x_{ij} = 0 \quad \text{for certain } i \text{ and } j,$$

$$\sum_j x_{ij} = 1 \quad \text{for certain } i,$$

$$\sum_i x_{ij} = 1 \quad \text{for certain } j.$$

But one can see that a nonsingular system of that type has determinant ± 1. Therefore, all the corresponding x_{ij} are integers. And a similar analysis shows the same for (1.2).

What we have used is that a (0, 1) matrix in which the rows can be partitioned into two parts such that every column contains exactly two 1's one in each part, is *totally unimodular*, i.e., has the property that every square submatrix of every order has determinant $0, \pm 1$.

In Section 2, we will briefly survey some facts about totally unimodular matrices (useful sufficient conditions, not so useful necessary and sufficient conditions), and also present a sort of characterization of combinatiorial theorems which depend on total unimodularity. Apart from the necessary sufficient conditions for total unimodularity, the rest of this material is essentially twenty years old (although its corollaries are often discovered by the uninitiated). But a very remarkable application to combinatorics of total unimodularity due to Baranyai [1] deserves special mention and commendation.

The "modern" period, although foreshadowed by Ford and Fulkerson's marvelous first proof in [7] of the max flow min cut theorem (unfortunately omitted from their book [6]) begins with Edmonds's work on the matching polytope [3]. In this work and in subsequent developments leads by Edmonds and Fulkerson, very interesting instances of combinatorial extremal theorems (max of one class = min of another) were found even where the matrix of inequalities was not totally unimodular. In some of these cases, the Cramer's rule argument still has a rôle to play, which we now explain (but in others, most notably the perfect graph theorem, it seems at this time irrelevant).

The most common situation is as follows: we have a pair of linear programming problems

$$Ax \leq b, \quad x \geq 0, \quad \max(c, x) \tag{1.3}$$

$$y'A \geq c', \quad y \geq 0, \quad \min(b, y) \tag{1.4}$$

where A, b and c are integral, and the duality theorem will be combinatorially interesting if there are integral optima to (1.3) and (1.4). Typically in one problem (say (1.3)), A and b are fixed and we are considering all possible c. We shall find in the cases we shall discuss that *every* vertex of $\{x \mid Ax \leq b, x \geq 0\}$ is integral. And, typically though not invariably, we shall prove this by showing that, for each c, *among* the optimal solutions to (1.4) is an integral one.

Next, we will raise the question why these vertices are integral. Consider $\{x \mid Ax \leq b, x \geq 0\}$, and a vertex x of this polyhedron. Let B be the submatrix of A formed by the rows i of A such that $(Ax)_i = b_i$ and columns j of A such that $x_j > 0$. Since x is a vertex, rank B is the number of its columns, and the nonzero coordinates of x are given by the unique solution of $Bz = b$, where b is the part of b corresponding to rows of B. If the g.c.d. of all full rank minors of B is 1, z is integral by a well known theorem [23]. In this case, we will say the polyhedron is *locally unimodular at x*. It may be that something stronger holds. For example, if at least one of the full rank minors of B has determinant ± 1, we will say the polyhedron is *locally strongly unimodular at x*. Perhaps B is totally unimodular; then we will say the polyhedron is *locally totally unimodular at x*. Of course, similar definitions will be given with respect to optimal integral vertices of the dual problem.

These viewpoints will be described with respect to matching polytopes in Section 3, flow polyhedra in Section 4, matroid polyhedra in Section 5, cut set polyhedra in Section 6, lattice polyhedra in Section 7 (all these terms will be described later).

Finally, in Section 8, we discuss briefly balanced and perfect matrices, which are classes of (0, 1) matrices permitting combinatorial applications of duality that are not as strong, but in the spirit of, total unimodularity.

This discussion does not exhaust the relevance of linear programming duality to polyhedral combinatorics (see [22] for example), but covers most of what we can currently claim as related to unimodularity.

2. Total unimodularity

We repeat the definition that a matrix A is totally unimodular if every square submatrix of A has determinant $0, 1, -1$. This implies, of course, that every entry of A is $0, 1, -1$. The most frequently invoked sufficient condition for a matrix to be totally unimodular arises from trees. Let T be a tree on $m+1$ vertices so that $|E(T)| = m$. Choose arbitrarily an orientation of $E(T)$. Let (i, j) be any ordered pair of vertices of T, and $C^{(i,j)}$ be the column vector with m components (corresponding to $E(T)$) in which the kth entry is $+1$ if the unique path in T from i to j includes edge k oriented with the path, -1 if oriented opposite to the path, 0 if edge k is not in the path. For an example, see Fig. 1.

	(1,2)	(1,3)	(1,4)	(1,5)	(2,3)	(2,4)	(2,5)	(3,4)	(3,5)	(4,5)
(1,3)	1	1	1	1	0	0	0	0	0	0
(3,2)	1	0	0	0	-1	-1	-1	0	0	0
(3,4)	0	0	1	1	0	1	+1	1	1	0
(5,4)	0	0	0	-1	0	0	-1	0	-1	-1

Fig. 1

The best-known necessary and sufficient condition for a $(0, 1, -1)$ A to be totally unimodular is that every square Eulerian submatrix B of A (B has all row and column sums even) have the property that the sum of the entries of B be divisible by 4 (alternatively, B is singular). There is a considerable literature on conditions for total unimodularity, featuring work of Ghouila-Houri, Camion, Truemper, Tamir, Padberg, Chandresekaran, Gondran, Commoner, Heller, Tompkins, Edmonds, Hoffman and Kruskal, and the reader is directed to surveys by Padberg [24] and by Truemper [26].

The use of total unimodularity in extremal combinatorial theorems was illustrated in Section 1. There are other uses in which the application is stated most directly in terms of existences of solutions (e.g., in Philip Hall's theorem on system of district representatives). In this connection, the following result [13] (see also [2] and [12]) is a characterization of such cases. Let A be a $(0, 1, -1)$ matrix, $a \leq b$, $l \leq u$ real vectors, and

$$P = P(A, a, b, l, u) \equiv \{x \mid a \leq Ax \leq b, l \leq x \leq u\}.$$

It is easy to verify that, if P is nonempty, the following condition holds (2.1) for every pair of totally unimodular vectors w and v such that $w'A = v'$,

$$\sum_{i|w_i=-1} a_i + \sum_{j|v_j=+1} l_j \leq \sum_{i|w_i=+1} b_i + \sum_{j|v_j=-1} u_j.$$

We consider the question of when (2.1) is a sufficient condition for P to be non-empty, or even to contain an integral point. (The reason is that, in Hall's theorem and its relatives, it turns out that (2.1) is the "obviously necessary" condition for the existence of a combinatorial pattern satisfying desired conditions). And the answer is that the following statements about A are equivalent:

(2.2) A is totally unimodular,
(2.3) For all $a \leq b$, $l \leq u$ (2.1) is a sufficient condition for P to be non-empty
(2.4) if a, b, l, u are integral and P is non-empty, P contains an integral point.

3. Matching polytopes

There are various versions of the matching polytope of Edmonds, and we choose to discuss that version where we understand at least part of its relation to local unimodularity. Let G be a graph with nodes $1, \ldots, n$, and edges $a, b, c, \ldots$. Let $d_1, d_2, \ldots, d_n$ be given integers. We use the notation $i \in e$ to mean that i is an endpoint of edge e, and $e \in S$ to mean that both ends of edge e are contained in the subset S of $V(G)$. Consider the linear programming problem:

$$x_e \geq 0, \quad \text{all } e$$

$$\sum_{e \ni i} x_e \leq d_1, \quad \text{all } i$$

$$\sum_{i \in S} x_e \leq \left\lfloor \tfrac{1}{2} \sum_{i \in S} d_1 \right\rfloor \quad AS \subset V(G), \quad \text{with } |S| \geq 2$$

$$\text{maximize } \sum C_e x_e$$

Here $\{C_e\}$ are nonnegative integers.

Theorem 3.1 [3]. *The linear programming problem and its dual have optimal integral solutions.*

Edmonds' original proof was based on an algorithm for solving primal and dual (and the algorithm is probably more significant than the theorem). Since the $\{C_e\}$ were arbitrary nonnegative integers, it follows that every vertex of (3.1) is integral. It turns out [15] that the polyhedron (3.1) is locally strongly unimodular at every vertex. This is proved by induction (along with an alternate proof of the integrality of the vertices), based on the fact that a matrix with $n+1$ rows and n columns in which the first n rows are the node-edge incidence matrix of a connected graph with n nodes and edges with one odd cycle, and the $(n+1)$th row consists entirely of 1's has the property that, deleting any row other than the last, yields a square matrix of determinant ± 1. On the other hand, we do not know any unimodularity property of integral optima for the dual. A proof can be found combining ideas of [16] and [21].

A curiosity is that the fact that each vertex of (3.1) is locally strongly unimodular (and that there is an optimal integral dual) uses superfluous inequalities. If $\sum_{i \in S} d_i$ is an even integer, then

$$\sum_{e \in S} x_e \leq \left\lfloor \tfrac{1}{2} \sum_{i \in S} d_i \right\rfloor$$

becomes

$$\sum_{e \in S} x_e \leq \tfrac{1}{2} \sum_{i \in S} d_i,$$

which follows from adding the inequalities

$$\sum_{i \in S} \sum_{e \ni i} x_e = \tfrac{1}{2} \sum_{i \in S} di$$

So this inequality need not be adjoined to specify (3.1). But if it is not adjoined, there is trouble. Consider the case where G is a triangle, and each $d_i = 2$.

In connection with matching, there is a related result [8] worth mentioning since it promulgates a method that seems not to have been exploited elsewhere. Let G be a graph with the property that any pair of odd cycles C_1 and C_2 contain respective nodes x_1 and x_2 such that $x_1 = x_2$ or x_1 is adjacent to x_2. Let A be the node edge incidence matrix of G, and let

$$B \begin{bmatrix} A & 0 \\ I & I \end{bmatrix},$$

and consider $Bz = f$, for some integral f.

Theorem 3.2 [8]. *If $Bz = f$ has a nonnegative solution and an integral solution, then it has a nonnegative integral solution.*

One can use this result to prove the Erdös-Gallai characterization the possible degree sequences of a graph. The reason is that it is not difficult to specify the conditions on f permitting solutions of the two types.

4. Flow polyhedra

We shall follow [14], which is much more in the spirit of [7] than subsequent proofs of the max flow min cut theorem. Let U be a set, $\mathscr{S} = \{S_1, \ldots, S_n\}$ a family of subsets of U, each of which is simply ordered. However there may be elements u and v in both S_1 and S_2, where $u < v$ in S_1 and $v < u$ in S_2. What we do require of these orderings is a "switch" property:

(4.1) for every S_1, u, S_2, where $u \in S_1 \cap S_2$, there exists a number $f(S_1, u, S_2)$ such that

$$S_{f(S_1,u,S_2)} \subset \{v \mid v <_{S_1} u\} \cup \{u\} \cup \{v \mid u <_{S_2} v\}.$$

What we have in mind is that U consists, say, of edges, and the S's are paths from source to sink. Then (4.1) is satisfied. But there are other models which could also satisfy (4.1). Further, we do not have to have all paths from source to sink, only enough paths to satisfy (4.1). Henceforth, we will call the S's paths.

We also assume a nonnegative integral function r defined on paths such that

$$r(f(S_i, u, S_j)) + r(f(S_j, u, S_i)) \geq r(S_i) + r(S_j) \qquad (4.2)$$

for all triples u, S_1, S_j such that $u \in S_i \cap S_j$.

Let A be the $(0, 1)$ matrix whose rows correspond to paths, columns to elements of U, with $a_{iu} = 1$ if $u \in S_i$, 0 otherwise. Let c be a nonnegative integral vector.

Theorem 4.1 [14]. *The linear programming problem*

$$x \geq 0, \quad Ax \geq r, \quad \text{minimize } (c, x) \qquad (4.3)$$

and its dual have optimal integral solutions. Further every vertex of (4.3) *is locally unimodular.*

The last sentence is not proved in [14], and is left here as an exercise. Concerning the dual, we know nothing about unimodularity. The theorem is proved by observing that there is a solution to the dual which can be gleaned from a solution to a related transportation problem.

Observe that, if each r_i is 1, we have the max flow min cut theorem.

There is an interesting connection between the case $r_i = 1$ of flow polyhedra and the case $r_i = 1$ of lattice polyhedra. These will be defined later in Section 7, but we give enough here to establish the connection, in order to point out that (4.1) is not just an artificial abstraction of the concept of source-sink paths needed for the proofs in [7] or [14], but has a natural justification. Let U be a set, and $\mathcal{L} = \{L_1, L_2, \ldots\}$ a clutter of subsets on which is imposed a lattice structure (*not* the usual lattice of subsets) satisfying

$$u \in L_1 \cap L_3, L_1 < L_2 < L_3 \Rightarrow u \in L_2, \qquad (4.4)$$

and

$$u \in (L_1 \wedge L_2) \cup (L_1 \vee L_2) \Rightarrow u \in L_1 \cup L_2. \qquad (4.5)$$

Let $\mathcal{S} = \{S_1, S_2, \ldots\}$ be the clutter of subsets of U, each of which has a non-empty intersection with each $L_i \in \mathcal{L}$ and is minimal with respect to this property. Then each S_i can be simply ordered in such a way that (4.1) holds. Conversely, let $\mathcal{S} = \{S_1, S_2, \ldots\}$ be a clutter of simply ordered sets satisfying (4.1), then the analogous "blocking" clutter, which is $\mathcal{L}$ (by [19]), can be given a lattice structure so that (4.4) and (4.5) hold. And if we move in this way $\mathcal{L} \to \mathcal{S} \to \text{``}\mathcal{L}\text{''} \to \text{``}\mathcal{S}\text{''}$, the lattice structure on "$\mathcal{L}$" is the same as that of $\mathcal{L}$, and likewise the ordering on the sets of "$\mathcal{S}$" is the same as S.

5. Matroid polyhedra

These were introduced by Edmonds [4]. Let r be a nonnegative integral function on the subsets of U which is submodular, i.e.,

$$r(S \cup T) + r(S \cap T) \leq r(S) + r(T). \qquad (5.1)$$

Consider the linear programming problem

$$x_u \geq 0, \quad \text{all } u$$

$$\sum_{u \in S} x_u \leq f(S), \quad \text{all } S \subset U \tag{5.2}$$

maximize (c, x).

Theorem 5.1 [4]. *Every vertex of* (5.1) *is integral and locally totally unimodular (indeed, triangular). If c is a nonnegative integral vector, there is always an optimum integral solution to the dual program, which is locally totally unimodular (indeed, traingular).*

Next, let r_1 and r_2 be two nonnegative integral functions, each satisfying (5.1), and consider the problem

$$x_u \geq 0,$$

$$\sum_{u \in S} x_u \leq \min(r_1(S), r_2(S)), \tag{5.3}$$

maximize (c, x).

Theorem 5.2 [4]. *Every vertex of* (5.3) *is integral, and locally strongly unimodular. If c is a nonnegative integral vector, there is always an optimum integral solution to the dual problem which is locally totally unimodular (indeed, consists of two triangular matrices on disjoint rows).*

Generalization of Theorem 5.2 as well as indications of its proof will be given in Sections 6 and 7. Interesting applications of the theorem and of algorithms for solving the problems have been made by Edmonds, Fulkerson, Iri and others. The idea behind the non-algorithmic proofs will be explained in Section 6.

6. Cut-set polyhedra

We follow here the discussion in Johnson [20] and Edmonds and Giles [5].

Let G be a graph, directed or undirected. If $S \subset V(G)$, $e \in w(S)$ means that (in the directed case) the initial node of edge e is in S, the terminal node is not or (in the undirected case) one end of e is in S, the other end is not. Let v be a distinguished node of $V(G)$, and let $\mathcal{F}$ be the set of all $S \subset V(G)$ such that $v \in S$. Let r be a nonnegative integral function defined on $\mathcal{F}$ which is supermodular, i.e.,

$$r(S \cup T) + r(S \cap T) \geq r(S) + r(T). \tag{6.1}$$

Theorem 6.1 [20]. *The linear program*

$$x_e \geq 0, \quad \text{all } e$$

$$\sum_{e \in w(S)} x_e \geq r(S), \quad \text{all } S \in \mathcal{F} \tag{6.2}$$

minimize $\sum c_e x_e$,

where $\{c_e\}$ are nonnegative and integral has optimal primal and dual solutions which are integral. Every vertex of (6.2) is integral.

One can prove that (6.2) is locally strongly unimodular at each vertex, and that, among the optimal vertices of each dual there is one which is locally totally unimodular.

Next, consider the case where G is a directed graph, and $\mathcal{F}$ is a family of subsets of $V(G)$ such that $S, T \in \mathcal{F}$, $S \cap T \neq \emptyset$ and $S \cup T \neq V(G)$ imply $S \cap T$, $S \cup T \in \mathcal{F}$. Let r be an integral supermodular function defined on $\mathcal{F}$.

Theorem 6.2 [5]. *The linear program*

$$b_{ij} \geq x_{ij} \geq a_{ij}, \quad \text{all edges } (i, j)$$

$$\sum_{\substack{i \in S \\ i \notin S}} x_{ij} - \sum_{\substack{i \notin S \\ j \in S}} \geq r(S), \quad \text{all } S \in \mathcal{F} \tag{6.3}$$

$$\text{minimize} \sum_{i \in j} c_{ij} x_{ij},$$

where a_{ij}, b_{ij}, c_{ij} are integers has optimal primal and dual solutions which are integral. Every vertex of (6.3) is integral and locally unimodular. Among the optimal solutions to the dual there is one which is integral and locally totally unimodular.

The proof of these assertions is based on the observation that, among the optimal dual solutions there is one in which the positive variables corresponding to $S \in \mathcal{F}$ have the property that the associated S's form a cross-free subfamily $\mathcal{F}' \subset \mathcal{F}$. This means that

$$S, T \in \mathcal{F}' \Rightarrow S \subset T, T \subset S, \quad \text{or } S \cap T = \emptyset, \quad \text{or } S \cup T = V(G). \tag{6.4}$$

Then it is proved that the corresponding submatrix arises from a tree as described in the beginning of Section 2. The proof that every vertex of (6.3) is locally unimodular we leave as an exercise.

Theorem 6.2 is shown in [5] to include as special cases Theorems 5.1 and 5.2, the circulation theorem for networks [12], and various results on directed cuts of acyclic graphs such as the Lucchesi-Younger theorem.

7. Lattice polyhedra

This material is taken from [17] and [18]. Let L be a partially ordered set with commutative binary operations $\wedge$ and $\vee$ (not necessarily g.l.b. and l.u.b.) such that $a \leq b \Rightarrow a \wedge b = a$, $a \vee b = b$; $a \wedge b \leq a, b$, $a \vee b \geq a, b$.

Let U be a set and $f: L \to 2^U$ satisfy

$$a < b < c \Rightarrow f(a) \cap f(c) \subset f(b) \tag{7.1}$$

and

$$f(a \vee b) \cup f(a \wedge b) \subset f(a) \cup f(b) \tag{7.2}$$

or

$$f(a \vee b) \cup f(a \wedge b) \supset f(a) \cup f(b) \tag{7.2'}$$

and

$$f(a \vee b) \cap f(a \wedge B) \supset f(a) \cap f(b).$$

Let r be a nonnegative integral function on L.

Theorem 7.1 [17] *If r is supermodular, and (7.1) and (7.2) hold, the linear program*

$$d_u \geq x_u \geq 0, \quad \text{all } u \in U$$

$$\sum_{u \in f(a)} x_u \geq r(a), \quad \text{all } a \in L \tag{7.3}$$

$$\text{minimize } \sum c_u x_u$$

for c, d integral has optimal primal and dual solutions which are integral. Every vertex of (7.3) is locally strongly unimodular. Among the optimal vertices to the dual there is one which is totally unimodular.

Analogous statements hold for the linear program

$$d_u \geq x_u \geq 0,$$

$$\sum_{u \in f(a)} x_u \leq r(a)$$

$$\text{maximize } \sum c_u d_u,$$

where (7.1) and (7.2)' hold, and r is submodular.

The proof follows from showing there is always an optimal solution to the dual in which positive variables associated with elements of L have the property that the related elements of L form a chain. From (7.1) and the tree discussion in Section 2, we know the corresponding matrix is totally unimodular.

This theorem includes Section 5 and theorems of Greene and Kleitman [10], [11] on partitions of a partially ordered set as special cases. It also includes Theorem 6.1 and, with some additional hypotheses, can be made to include Theorem 6.2.

8. Perfect and balanced matrices

Here we follow mostly the discussion in Padberg [24]. A $(0, 1)$ matrix is balanced if it contains no square submatrix of odd order with every row and column sum 2. This concept was introduced by Berge. Its significance lies in the fact that, for every square submatrix B of A, every vertex of $\{x \mid Bx = 1, x \geq 0\}$ is a $(0, 1)$ vector. In fact [9] this property characterizes balanced matrices. It is clear that, in the same manner that total unimodularity yields extremal combinatorial theorems, balanced matrices will also yield such theorems (where right hand side and objectives are $(0, 1, \infty)$ vectors). But we do not know of cases of independent interest arising from balanced matrices that do not come either from the totally unimodular matrices or from perfect matrices.

A perfect matrix is a $(0, 1)$ matrix A with the property that every vertex of $P(A)\{x \mid Ax \leq 1, x \geq 0\}$ is a $(0, 1)$ vector. This concept is due to Fulkerson, who proposed exploiting it to prove Berge's conjecture, if G is a graph such that its chromatic number $X(G)$ and the size of its largest clique $K(G)$ are equal, and also $X(H) = K(H)$ for all induced subgraphs $H \subset G$, then the same properties hold for the complementary graph $\mathbf{G}$. This program was carried out by Lovász [21]. The first step was to show that, if A is the clique-node incidence matrix of G, where $X(A) = K(H)$ for all $H \subset G$, then every vertex of $P(A)$ is a $(0, 1)$ vector. Next, just using the integrality of the vertices of $P(A)$, and induction on $\sum c_j$, where $c = (c_1, \ldots, c_u)$ is a nonnegative integral vector, the linear program $\max(c, x)$, $x \in P(A)$ has a dual with at least one optimal integral selection. In particular, considering the case where each c_j is 0 or 1, we have $X(H) = K(H)$ for all $H \subset \mathbf{G}$.

Lovász subsequently derived a proof of Berge's conjecture which did not use linear programming.

Padberg [25] proved that a $(0, 1)$ matrix with m rows A is perfect if and only if A contains no collection of k columns containing a nonsingular $k \times k$ submatrix B with constant row and column sums, and with every other row of these k columns the duplicate of a row of B or having row seem smaller than that of B.

References

[1] Z. Baranyai, On the factorization of the complete uniform hypergraph. Proc. Erdos Colloquium, Keszthely.
[2] P. Camion, Modules unimodulaires, J. Combinatorial Theory 4 (1968) 301–362.
[3] J. Edmonds, Path, trees and flowers, Canad. J. Math. 17 (1965) 449–467.
[4] J. Edmonds, Submodular functions, matroids and certain polyhedra, in Combinatorial Structures and their Applications (Gordon and Breach, London, 1970) 69–87.
[5] J. Edmonds and R. Giles, A min-max relation for submodular functions on graphs, CORE discussion paper 7615 (1976).
[6] L.R. Ford, Jr. and D.R. Fulkerson, Flows in networks (Princeton University Press, NJ, 1966).
[7] L.R. Ford, Jr., Maximal flow through a network, Canad. J. Math. 8 (1956) 399–404.
[8] D.R. Fulkerson, A.J. Hoffman and M.H. McAndrew, Some properties of graphs with multiple edges, Canad. J. Math. 17 (1965) 166–177.

[9] D.R. Fulkerson, A.J. Hoffman and R. Oppenheim, On balanced matrices, Math. Programming Study 1 (1974) 120–133.
[10] C. Greene, Some partitions associated with a partially ordered set, J. Combinatorial Theory Ser. A (1976) 69–79.
[11] C. Greene and D.J. Kleitman, The structure of Sperner k-families, J. Combinatorial Theory Ser. A (1976) 41–68.
[12] A.J. Hoffman, Some recent applications of the theory of linear inequalities to extremal combinatorial analysis, in Proc. Symp. Appl. Math. 10, Combinatorial Analysis (Providence, RI, 1960) 113–127.
[13] A.J. Hoffman, Total unimodularity and combinatorial theorems, Linear Algebra and Appl. 13 (1976) 103–108.
[14] A.J. Hoffman, A generalization of max flow-min cut, Math. Programming 6 (1974) 352–359.
[15] A.J. Hoffman and R. Oppenheim, Local unimodularity in the matching polytope, Ann. Discrete Math. 2 (1978) 201–209.
[16] A.J. Hoffman and J.B. Kruskal, Integral Boundary Points of Convex Polyhedra, in: Ann. Math. Study 38 (Princeton University Press, NJ, 1956) 223–241.
[17] A.J. Hoffman and D. Schwartz, On lattice polyhedra, in: Proc. Conf. on Combinatorial Mathematics, Keszthely (to appear).
[18] A.J. Hoffman, On lattice polyhedra II: Construction and examples, Math. Programming Study 8 (1978).
[19] A.J. Hoffman, On lattice polyhedra III: Blockers and anti-blockers of lattice clutters, Math. Programming Study B (1978).
[20] E.L. Johnson, On cut-set integer polyhedra, Cahiers Centre Eludes Recherche Opér. 17 (1975) 235–251.
[21] L. Lovász, Normal hypergraphs and the perfect graph conjecture, Discrete Math. 2 (1972) 253–267.
[22] L. Lovász, Certain duality principles in integer programming, Ann. Discrete Math. 1 (1977) 363–374.
[23] C.C. MacDuffee, The Theory of Matrices (Chelsea Publishing Co., New York, 1946).
[24] M.W. Padberg, Characterization of totally unimodular, balanced and perfect matrices, in: Combinatorial Programming: Methods and Applications (D. Reidel Publishing Company, Dordrechet, 1975) 275–284.
[25] M.W. Padberg, Perfect zero-one matrices, Math. Programming 6 (1974).
[26] K. Truemper, Algebraic characterization of totally unimodular matrices (to appear).

APPROXIMATIVE ALGORITHMS FOR DISCRETE OPTIMIZATION PROBLEMS

Bernhard KORTE

Institut für Ökonometrie und Operations Research, Nassestrasse 2, D-5300 Bonn, W. Germany

1. Introduction

The general *Discrete Optimization Problem* is defined over an arbitrary finite set S (i.e. $|S|<\infty$) on which a mapping into the real numbers is defined

$$c: S \to \mathbb{R},$$

thus the optimization problem can be specified as

$$\max\{c(s) \mid s \in S\} \quad \text{or} \quad \min\{c(s) \mid s \in S\}. \tag{1}$$

S will be called the set of feasible solutions or the *feasible set*. In order to provide problem (1) with more algebraic structure we assume for Combinatorial Optimization Problems that the set of feasible solutions is a subset of the power set of a given finite set, i.e. $S \subseteq \mathcal{P}(E)$ with $|E|<\infty$. This enables us to define intersections and unions of elements of S and to identify these elements by a $(0, 1)$-incidence vector of length $|E|$. In the case of combinatorial optimization problems we consider only separable objective functions, i.e. functions $c: E \to \mathbb{R}$ in a canonical way:

$$c(s) = \sum_{e \in s} c(e).$$

Most well known combinatorial optimization problems as well as the classical integer programming problem can be stated in the general form (1) with additional assumptions.

In addition to (1) we will state the *Discrete Decision Problem*: Given a set S and a subset

$$P := \{s \in S \mid s \text{ has some "well defined property"}\},$$

the decision problem can be stated as:

Decide whether $P \neq \emptyset$. \hfill (2)

Again we might assume S to be a subset of the power-set of some finite set. The difference between (1) and (2) can be demonstrated by the following example: Let $G = (V, E)$ be a complete undirected and loopless graph and c a function assigning weights to the edges:

$$c: E \to \mathbb{R}$$

The maximum weighted Hamiltonian cycle (*Travelling Salesman Tour*) can be determined by

$$\max\{c(t) \mid t \subseteq E \text{ and } t \text{ is a Hamiltonian cycle}\},$$

whereas the decision problem whether a given arbitrary undirected graph $G = (V, E)$ has a Hamiltonian cycle is of type (2) with

$$P := \{t \subseteq E \mid t \text{ is a Hamiltonian cycle}\}.$$

Most of the problems we will deal with in this paper are of type (1). One advantage of decision problems (or "yes–no-problems") of type (2) is that they are canonically related to *Language Recognition Problems* on *Turing Machines* in that the set P corresponds to the set of languages accepted by the Turing machine.

Although we do not deal with *computational complexity* in this paper, it might be worthwhile to state the main result of this discipline here very briefly since it directly implies the need for and the recent development of approximative algorithms. For a painstaking and solid introduction to the field of formal languages and Turing machines the reader is referred to the textbooks of Hopcroft and Ullman [1969] and Aho, Hopcroft and Ullmann [1974] whereas the results of computational complexity as a consequence of the famous theorem of Cook [1971] are summarized in Karp [1972], Karp [1975] and Aho, Hopcroft and Ullmann [1974], Chapter 10.

Before we go on, let us make some general remarks on the manner and the level of this survey paper:

Although most of this paper deals with algorithms we will not define an algorithm precisely. And this is true for almost all papers on algorithms. The reason for this is that there is only one way of a precise definition of an algorithm or a computational procedure, namely by a Turing machine. All different approaches in computer science for defining an algorithm so far have turned out to be equivalent to the Turing machine definition. There is a lot of mathematical tools developed in centuries for describing *analytical phenomena* in mathematics, but up to this day there are only very few precise instruments (like the Turing machine) for defining and describing *algorithmic approaches.* For instance an analytic definition of a function is trivial, but a precise algorithmic definition of a function in general is very difficult, long, and unesthetical. We may hope that further research in the fields of theory of algorithms and automata will provide us with a somewhat more efficient and precise algorithmic apparatus in mathematics, since all applications of mathematics are algorithmically oriented. For the time being we can use only the Turing machine approach for a precise description. But it is well known that this approach if exactly applied is very longish, non-elegant, and not very familiar to analytically trained mathematicians. For this reason many authors try to avoid a description completely based on the notation of a Turing machine. We have to confess that we will follow this line. Practically it means that we will give a precise definition of the Turing machine and some of its derivatives, but almost all other arguments, proofs, and deductions concerning algorithms are based on the Turing machine approach but they do not go down to the complete Turing machine notation. This policy seems to us most convenient for this survey. The diligent reader may (as a very time consuming exercise) try to rewrite all deductions in an exact Turing machine notation.

For the purpose of these general remarks a *language* is considered to be a well defined set of elements called *strings* or *words* which are tupels or vectors with elements of a given finite *alphabet*; e.g. the set P of the decision problem (2) can be considered a language over the alphabet $\{0, 1\}$ where the strings are incidence vectors of the elements of P.

A (deterministic) *Turing machine* with k tapes can be formalized as a tuple $(Q, T, I, \delta, b, q_0, q_f)$ with

Q	a finite set of states of the Turing machine,
T	a set (alphabet) of tape symbols,
$I \subseteq T$	the set of input symbols,
$b \in T \setminus I$	a special symbol (blank),
$q_0 \in Q$	the initial state,
$q_f \in Q$	the final (or accepting) state,
δ	the next move function

The essential part of this definition is the mapping

$$\delta : Q \times T^k \to Q \times \{T \times \{L, R, S\}\}^k \tag{3}$$

where the set $\{L, R, S\}$ indicates that the tape head of the kth tape will be moved left (L), right (R), or remain stationary (S). That is, for some $(k+1)$-tuple consisting of a state and k tape symbols δ gives a new state and k pairs, each pair consisting of a new tape symbol and a direction for the tape head. The next move function describes the algorithmic or procedural aspects of a Turing machine. The Turing machine starts applying δ to q_0 and it halts if q_f appears in the image of δ. One specified tape of the machine will be called the *input tape* on which prior to the first application of δ the *input string* must be placed.

A *non-deterministic Turing machine* can be defined in exactly the same way as a 7-tuple with (3) except that δ is not a mapping, but a *correspondence* (*set-valued* or *point-to-set-mapping*). That means with each movement (application of δ) the Turing machine is reproducing itself yielding finitely many successors. For each state and k-tuple of tape symbols there is a finite set of choices for the next move. The "flow chart" of a deterministic Turing machine can be considered as a straight line whereas we can compare the procedural flow of a non-deterministic Turing machine with a rooted tree of a tree-search or backtracking algorithm. A deterministic Turing machine is closely related to a random access memory (RAM)-machine or a real world computer, while a non-deterministic Turing machine has no hardware realisation.

A deterministic or non-deterministic Turing machine can be extended by an *oracle* to a so called *query machine*. This machine has an additional tape, the oracle or query tape and three or more additional states of which one is the so called query state and two or more are the oracle output states (in the case of two one may call them yes-or-no-states). The action of the query machine is identical to that of a Turing machine except when it enters the query state. Then the next operation of the machine is determined by the oracle. In the case of two additional output states the oracle itself can be defined as a finite set X which will place the query machine into its yes-state if the current string on the oracle tape is an element of X; otherwise the oracle places the machine into the no-state. In the case of more than two oracle output states the oracle is a finite collection of decision sets. In other words the Turing machine is well-defined by the next-move mapping δ except where δ is applied to the query state. In this case the next state is not defined by the mapping δ but by the outcome of the oracle i.e. by the fact that the current string of the oracle tape belongs to the oracle set X or not. In this respect the oracle can be considered as another formulation of non-determinism. And in fact in his famous article Cook [1971] describes a non-deterministic Turing machine as a simple Turing machine with an oracle which is described by the range of the set valued mapping δ. Although we need the oracle set X for a formal definition of a query machine, the elements of X are indeed hidden for the normal actions of the Turing machine. If we look at the algorithmic aspects of a Turing machine we may compare it with a well-coded algorithm with the addition of a special subroutine. This subroutine is equivalent to the oracle. Whenever the subroutine is called by the main program its input will be an element x, and the output of the oracle will be yes or no depending on whether $x \in X$ or not. The set X itself and the way how the subroutine decides on the membership of X is absolutely unknown to the main program.

In Hausmann and Korte [1977a] we give a somewhat modified definition of an oracle as a mapping on the query tape itself, but for the present introduction we consider the above given description more useful (cf. Baker, Gill, and Solovay [1975]).

As it is common in the theory of computational complexity we now define two important classes of languages, but first let us mention that we understand by *time-complexity* $T(n)$ of a Turing machine for an input string of length n the maximum number of moves (applications of δ) made by the machine in processing this input. We do not deal here with problems of encoding the input, so in general we assume binary encoding; for a string l we denote its length by $|l|$.

Definition 1.1. $\mathscr{P} := \{L \mid L$ language for which a deterministic Turing machine M and a polynomial $p(\cdot)$ exist such that for all strings l, M has the time-complexity $p(|l|)$ and L will be accepted by $M\}$.

Analogously:

Definition 1.2. $\mathcal{NP} := \{L \mid L$ language for which a non-deterministic Turing machine and a polynomial $p(\cdot)$ exist such that for $l \in L$, M has time complexity $p(|l|)$ and L will be accepted by $M\}$.

We note that the above definitions differ only by "deterministic" and "non-deterministic" respectively. Of course, $\mathcal{P} \subseteq \mathcal{NP}$. But the question whether the opposite inclusion holds is still unsolved although some very eminent researchers have attacked it. The common opinion is that there is no good hope to prove $\mathcal{P} = \mathcal{NP}$, but also for proving the contrary there is no substantial idea or guide-line so far.

Definition 1.3. A language $L_0 \in \mathcal{NP}$ is called *NP-complete* if the following condition is satisfied: Given a deterministic Turing machine of time complexity $T(n)$ which recognizes L_0. Then we can determine for *any* language $L \in \mathcal{NP}$ a deterministic Turing machine with time complexity $T(p_L(n))$ which recognizes L where p_L is a polynomial depending on L.

This essential definition characterizes the subset of NP-complete languages as the crucial one since any member of $\mathcal{NP}$ will be recognized with a time complexity which is except for a polynomial the same as that for an NP-complete language. Using the idea of query machines we can state an equivalent definition:

Definition 1.3a. A language $L_0 \in \mathcal{NP}$ is called *NP-complete* if *any* language $L \in \mathcal{NP}$ can be recognized by a deterministic Turing machine with polynomial time-complexity using L_0 as an oracle set.

We shall note that this definition is due to Cook [1971] and if the conditions of Definition 1.3 or 1.3a resp. are fulfilled we say L is Cook-*reducible* to L_0. Karp [1972] used a slightly modified definition which is known in the literature as Karp-reducibility and which is based on polynomial transformability of languages.

Cook [1971] was the first to characterize an NP-complete language:

Theorem 1.4 (Cook [1971]). *The language B of all satisfiable Boolean expressions (or all tautologies) is NP-complete.*

A Boolean expression of length n is a string of n Boolean or (0, 1)-variables which can be negated or not and arbitrarily connected by logical "and" and "or". Such an expression is called satisfiable if there exists a special assignment of "zeros" and "ones" to the variables which makes the expression equal "one". A tautology is a Boolean expression that has the value 1 for all possible assignments of values 0–1 to its variables (always true). Beside the original proof of this famous and fundamental theorem which uses oracle techniques the reader may find an easily understandable proof in Aho, Hopcroft and Ullmann [1974], p. 379ff.

As an immediate corollary of Theorem 1.4 we state:

Corollary 1.5. $\mathcal{P} = \mathcal{NP} \Leftrightarrow B \in \mathcal{P}$.

So far we have used the precise notations "Turing Machine" and "language". As mentioned above let us now turn to a relaxation. Instead of the term Turing machine we use in the following the unspecified notation "algorithm". And instead of "a language is recognized by a Turing machine" we say "a decision problem (2) is solved by an algorithm". The reader may easily convince himself that the set P in the definition of (2) directly corresponds to a recognizable language (as a set of strings). A Turing machine with polynomial time complexity may now be called a polynomial algorithm, i.e. an algorithm whose number of steps is a polynomial in the input length. Edmonds [1965] was the first to call such algorithms "good" ones.

In all known instances, optimization problems of type (1) can be reduced to polynomially many language recognition problems. This is possible because there exists an a priori bound $U(S, c)$, such that

$$c(s) \leq U(S, c) \quad \forall s \in S$$

and

$$U(S, c) \leq 2^{p \text{ (length of instance } (S,c))}.$$

We can then answer the optimization problem by using *binary search* on the interval $[0, U(S, c)]$.

With the introduced relaxations we may now to a certain extent identify $\mathcal{P}$ with the class of problems (1) or (2) which can be solved by a polynomial algorithm, whereas $\mathcal{NP}$ is the class of discrete optimization or decision problems which can be solved by an enumerative or backtracking search algorithm where the depth of the search tree is a polynomial in the dimension (input length) of the problem. A search tree of polynomial depth in general has exponentially many nodes. Thus the number of steps of an algorithm for problems in $\mathcal{NP}$ is exponential in the dimension of the problem.

After these additional interpretations the reader can imagine that the definitions of $\mathcal{P}$ and $\mathcal{NP}$ are not superfluous but substantial: For problems in $\mathcal{P}$ we have good algorithms, for problems in $\mathcal{NP}$ we have only inefficient enumerative algorithms. What about NP-completeness? An interpretation of Theorem 1.4 yields that any problem in $\mathcal{NP}$ (i.e. any discrete optimization or decision problem) could be solved by a good algorithm if for some NP-complete problem like the satisfiability problem there is a good algorithm. Since the general satisfiability problem has no structure except that there are n Boolean variables which arbitrarily form a logical expression, there is absolutely no idea how a good algorithm for this problem could look like. A practical comment about the degree of the polynomial for problems in $\mathcal{P}$ should be added: With very sophisticated tricks like splitting, binary search etc. one can push down the degree of the polynomial close to one. We do not know of a practical problem in $\mathcal{P}$ which does not have an algorithm whose number of steps is bounded by a polynomial of degree 4 or even less. A natural question to be raised has been answered by extensive research during the last years, namely: Are there many other problems in $\mathcal{NP}$ which are NP-complete? The answer is, that almost all problems in $\mathcal{NP}$ for which we do not know a polynomial algorithm are NP-complete and therefore have the same degree of difficulty as the satisfiability problem. In Karp [1972], Karp [1975], Aho, Hopcroft and Ullmann [1974], Sahni [1974], Rinnooy Kan [1976], and Lawler [1976] the reader may find an extensive list of NP-complete problems. All classical but hard combinatorial or graph optimization problems are NP-complete: the vertex packing, the clique problem, the Hamiltonian circuit problem, the symmetric travelling salesman problem, the Steiner tree problem and the chromatic number problem on an undirected graph as well as the asymmetric travelling salesman problem, the acyclic subgraph problem on a directed graph, many scheduling problems, the 3-dimensional assignment problem, the general set covering problem, the discrete multicommodity by flow problem. It can easily be shown that the knapsack problem and the general integer programming problem are NP-complete, too.

There are only two substantial problems in $\mathcal{NP}$ for which so far we neither know a polynomial algorithm nor whether they are NP-complete: In the category of decision problems the *graph isomorphism problem* that is the problem to decide whether two given graphs are isomorphic or not and in the category of optimization problems the classical *linear programming problem*. Although they have been seriously attacked by numerous researchers both problems to our understanding remain the key-problems in this area.

We should mention that some slightly relaxed derivatives of these key-problems are known to be NP-complete. To decide whether a given graph is isomorphic to some subgraph of another graph is NP-complete, and recently Jeroslow [1976] has shown that the parametric linear programming problem is NP-complete, too.

For completeness let us mention some problems which are known to be in $\mathcal{P}$: deciding whether a graph is planar or not; various connectivity problems in directed and undirected graphs; isomorphism of trees; various shortest path and distances problems; network flow problems, matching problems and optimization problems over intersections of no more than two matroids; colouring, clique and stable set problems in special graphs like comparability graphs (Golumbic [1976]) and chordal graphs (Giavril [1972]).

2. Definitions and performance measures of approximative algorithms

We have learned in the preceding section that there is no good hope for being able to design polynomial algorithms for optimally solving any NP-complete problem. Therefore we have to look for polynomial algorithms which do not guarantee the optimal solution of a discrete optimization problem but which have in a somewhat well defined way a guaranteed performance. In the past such *"suboptimal"* algorithms were designed especially by practicians for very special practical problems and they were called *"heuristics"*. There is no exact definition of what heuristic means, it is a more or less well defined algorithm which does not necessarily yield the optimal solution but a feasible one which cannot be improved by the algorithm itself. In the following we will use the notation *"approximative algorithm"* since it is analogous to some similar methods in non-discrete numerical analysis and it enables us to state a precise definition. Under a *problem class* of optimization problems we will understand in the following a set or collection of problems of type (1) which have some similarities in the definition of the feasible set and the objective function, such as the set of all travelling salesman problems or all (0–1)-knapsack problems. We have to admit that the definition of a problem class is merely intuitive by the attributes of its elements. A problem class of decision problems can be defined similarly.

Definition 2.1. An approximative algorithm A for a given problem class is a well-defined procedure which for any problem instance halts after finitely many steps with a feasible solution $s_A \in S$ and with its objective value $c(s_A)$.

Definition 2.2. Let A be an approximative algorithm for a given problem class and $c(s_A)$ the objective value of the (approximative) solution found by the algorithm. We say A has a *guaranteed performance* or a *worst case bound* k if for any problem instance
  (1) $k \leq c(s_A)/c(s_0) \leq 1$ (in the case of a maximization problem), or
  (2) $1 \leq c(s_A)/c(s_0) \leq k$ (in the case of a minimization problem), or in general
  (3) $|(c(s_0) - c(s_A))/c(s_0)| \leq r$
where (3) indicates the relative error for both types of problems with $r = 1 - k$ for the maximization and $r = k - 1$ for the minimization problem.

As outlined in the definition, the quotients (1), (2) or (3) measure the worst case behaviour of an algorithm for a given problem class. Now one may raise the question whether such a pessimistic measure is appropriate to judge the performance of an algorithm. But one can look at this measure also from a positive point of view: With k or r we have a standard for a guaranteed performance: for example if we have an algorithm which yields for the general knapsack problem a value k, then for any application of this algorithm with arbitrary coefficients specifying the knapsack problem we get an approximative solution for which the value of the objective function is not less than k-times the optimal value.

As will be outlined later, there are algorithms for which k is a universal constant. For other algorithms k depends on the dimension (length of the input) of the problem or on some structural parameter characterizing the feasible set. In some cases this relation implies that k tends to zero if the dimension goes to infinity. Of course, the best performance guarantee will be obtained if k is a universal constant close to one.

Beside the worst-case analysis one might be interested in an *average performance* measure of algorithms, but in this case the term "average" has to be specified exactly. For a given class of problems one has to define an average problem or a meaningful distribution function for the input parameters in order to derive average results. This has been done up to now for sorting and merging algorithms (cf. Aho, Hopcropft and Ullmann [1974] or Knuth [1973]). In these simple cases the assumption about the distribution of the input data is straightforward. The data to be sorted or merged can be assumed to be equally distributed. In more complicated cases an average behaviour analysis is extremely complicated. Recently there are first promising approaches by Karp [1976] for graph optimization problems using the theory of random graphs introduced by Erdös and Renyi. Another recent approach is due to Rabin [1976].

We think that it is more appropriate to judge an algorithm for practical applications on the basis of a well defined measure of average behaviour. However, probabilistic analysis of algorithms is extremely hard and worst-case analysis may at least help in structuring and classifying types of algorithms.

As an additional example for demonstrating the difference between worst-case and average behaviour on the one hand but also for the extreme difficulties in measuring the average behaviour on the other hand we may name the well known *simplex algorithm* for linear programming. Special examples of Klee and Minty [1972] and Zadeh [1973] have demonstrated that in the worst case the number of steps (pivots) of this algorithm grows exponentially with the dimension of the problem, since in these examples all vertices of the corresponding polyhedron have to be checked. However, there is a tremendous number of practical linear programming problems which were solved very efficiently by the simplex algorithm. For all these practical problems the number of pivots needed to find the optimal vertex was bounded by a polynomial of the problem dimension having a very low degree (less than 3). It is still an open question how to describe the "average structure" of practical problems and how to derive exact bounds for the average behaviour. In the case of the simplex algorithm we have immense statistical experience (of a period of more than 30 years) but we cannot yet verify the evident quality of the average behaviour.

Before giving some more definitions let us turn to some examples of very easy approximative algorithms (which can be considered as folklore in the literature) with a performance guarantee bounded by a constant.

(a) *Maximum weighted acyclic subgraph problem.* Consider the complete directed loopless graph $G = (V, E)$ with $V = \{1, \ldots, n\}$, $E = V \times V \setminus \{(i, i) \mid i \in V\}$, and weights assigned to the arcs: $c : E \to \mathbb{R}^+$. Find that acyclic subgraph, i.e. a subgraph without circuits, which has maximum weight. Let $(i, j) \in E$ denote an arbitrary arc, π a permutation of $(1, \ldots, n)$,

$$T_\pi := \{(\pi(i), \pi(i+1)) \mid i = 1, \ldots, n-1\}$$

a set of $n-1$ consecutive arcs,

$$T_{\pi^{-1}} : \{(\pi(i+1), \pi(i)) \mid i = 1, \ldots, n-1\}$$

the "inverse" of T_π, and for a subset $R \subseteq E$ we write $\bar{R}$ for its *transitive closure*. Since the maximum weighted acyclic subgraph contains $\binom{n}{2}$ arcs we need to consider only the sets $\bar{T}_\pi$ and $\bar{T}_{\pi^{-1}}$.

We state a very simple algorithm with a guaranteed performance of $\frac{1}{2}$ and which needs only $O(n^2)$ iterations.

We use an ALGOL-like notation:

procedure ACYCLIC;
begin
   let π be an arbitrary permutation;
   evaluate $c(\bar{T}_\pi)$ and $c(\bar{T}_{\pi^{-1}})$;
   if $c(\bar{T}_\pi) \geq c(\bar{T}_{\pi^{-1}})$ **then** $s_A := \bar{T}_\pi$;
   else $s_A := \bar{T}_{\pi^{-1}}$;
   output $s_A, c(s_A)$;
end ACYCLIC.

Since the sum of all weights on E is constant and $c(\bar{T}_\pi) + c(\bar{T}_{\pi^{-1}}) = c(E)$ it is evident that

$$\frac{c(s_A)}{c(s_0)} \geq \frac{c(s_A)}{c(E)} \geq \frac{1}{2}.$$

(b) *Maximum weighted Hamiltonian circuit in a directed graph* (asymmetric travelling salesman maximization problem). Again we have $G = (V, E)$, $V = (1, \ldots, n)$ $E := V \times V \setminus \{(i, i) \, i \in V\}$, $c : E \to \mathbb{R}^+$. Find a Hamiltonian circuit with maximum weight. Here again we propose an approximative algorithm with guaranteed performance $\frac{1}{2}$ which requires polynomially many steps: The procedural notation will be somewhat sloppy:

procedure MAXHAM;
begin
  (i) solve the maximum weighted matching problem for the canonically corresponding bipartite graph (the solution corresponds to a set of arcs s in the graph G);
  (ii) from any circuit z_i in s with length less than n, remove the arc with the smallest weight;

(iii) complete the remaining set of arcs in s with arbitrary other arcs to a Hamiltonian circuit H;
$$s_A := H;$$
$$c(s_A) := c(H)$$
output $s_A, c(s_A)$;
end MAXHAM.

Let $\alpha_i := \min\{c(e): e \in z_i\}$ for all i. Then the following shows that the performance guarantee is $k = \frac{1}{2}$:

$$c(s_A) \geq \sum (c(z_i) - \alpha_i),$$

$$c(s_0) \leq c(s) = \sum c(z_i),$$

hence since $|z_i| \geq 2$

$$\frac{c(s_A)}{c(s_0)} \geq 1 - \frac{\sum \alpha_i}{\sum c(z_i)} \geq 1 - \frac{\sum \alpha_i}{\sum |z_i| \alpha_i} \geq 1 - \frac{\sum \alpha_i}{2 \sum \alpha_i} = \frac{1}{2}.$$

For the maximum weighted Hamiltonian cycle problem in an undirected graph (symmetric travelling salesman max-problem) $G = (V, E)$ we can slightly modify the above procedure:
(i)* instead of the bipartite matching use the 2-matching algorithm for G,
(ii)* and (iii)* replace "circuit" and "arc" in (ii) and (iii) by "cycle" and "edge".
Since in this case $|z_i| \geq 3$, it follows by similar inequalities as above

$$\frac{c(s_A)}{c(s_0)} \geq \frac{2}{3}.$$

(c) *General integer knapsack problem.*

$$\max\{c'x \mid ax \leq b, x \in \mathbb{Z}_n^+\}. \tag{4}$$

The algorithm which we will design for this problem is a slight modification of the *greedy algorithm* which will be treated extensively in the last chapter. By *weight density* of an element i we define the quotient c_i/a_i.

procedure KNAPSACK;
begin
let $\{1, \ldots, n\}$ be a renumbering of the elements according to non-increasing weight densities, i.e.

$$\frac{c_i}{a_i} \geq \frac{c_{i+1}}{a_{i+1}};$$

```
for i = 1 to n do
begin
    x_i := [b/a_i];
    b := b - a_i x_i;
end;
s_A := (x_1, ..., x_n);
output s_A, c(s_A);
end KNAPSACK.
```

The guaranteed performance is given by

Proposition 2.3. *For all problem instances of the general knapsack problem (i.e. for all c, a, b and n) we have*

$$\frac{c(s_A)}{c(s_0)} \geq \tfrac{1}{2}.$$

Proof. Since we have non-increasing weight density, it is clear that

$$c(s_0) \leq \frac{c_1}{a_1} \cdot b$$

Moreover it is evident that

$$c(s_A) \geq c_1 \left[\frac{b}{a_1}\right].$$

This implies

$$\frac{c(s_A)}{c(s_0)} \geq \frac{[b/a_1]}{b/a_1} \geq \tfrac{1}{2}. \qquad \square$$

We should mention that this easily derived result is only true for the general knapsack but not for the binary problem for which we will derive a different worst-case-bound later on.

The purpose of these examples was to demonstrate the underlying structures of some easy heuristics, especially the last example gives the reader a first impression of the basic idea for approximative algorithms.

3. Greedy type approximative algorithms

In this section we discuss a very fundamental approximative algorithm which can be applied to almost all discrete optimization problems of type (1.1), the *greedy algorithm*. As a matter of fact the greedy approach is a basic tool in many approximative or heuristic algorithms for combinatorial optimization problems. Worst-case bounds for the greedy will enable us to judge the performance of many other approximative algorithms. Furthermore we will prove that the greedy algorithm "dominates" a very broad class of other general approximative algorithms. It turns out that we have to consider the greedy algorithm as the key- and master-algorithm in approximative combinatorial optimization.

There are different historical sources for the greedy approach which was sometimes called *myopic*. One of the first implicit applications of the greedy approach was done by Kruskal [1956] for the spanning tree problem. The extension to general matroid problems is due to Rado [1957] and later independently to Edmonds [1971], Gale [1968] and Welsh [1968].

Let us restate the general discrete optimization problem: Let $E = \{1, \ldots, n\}$ be a finite set, $S \subseteq \mathcal{P}(E)$ the set of feasible solutions, $c : E \to \mathbb{R}$ the objective function, then we have the following two optimization problems:

$$\max\{c(S) \mid s \in S\} \tag{1a}$$

or

$$\min\{c(S) \mid s \in S\}. \tag{1b}$$

We now formulate the greedy algorithm for problem (1a):

procedure GREEDYMAX;
begin
   let $(e_1, \ldots, e_n)$ be an ordering of $E = \{1, \ldots, n\}$ with $c(e_i) \geq c(e_{i+1})$;
   $s_G = \emptyset$;
   for $i = 1$ **to** n **do**
      if $s_G \cup \{e_i\} \in S$ **then** $s_G := s_G \cup \{e_i\}$;
   output $s_G, c(s_G)$;
end GREEDYMAX

For the minimization problem (1b) an analogous algorithm "GREEDYMIN" may be formulated where the only alteration is taking $(e_1, \ldots, e_n)$ to be an ordering of E with non-decreasing weights.

The greedy algorithm for the maximization (minimization) problem starts with the empty set and adds in each step to a known partial solution that of the remaining elements of E which has the largest (smallest) weight such that the obtained new set remains feasible.

Not counting the sorting and the decision whether $S_G \cup \{e_i\}$ is feasible or not the greedy algorithm is of linear time complexity in the dimension n of the problem and it halts with a feasible solution which generally is not the optimum for (1a) or (1b). In our ALGOL-like procedural notation we did not code the decision on feasibility. So we look at the greedy as an *oracle* algorithm, where the decision whether a subset of E is feasible or not is completely left to the oracle.

In order to give some performance guarantees or worst case bounds we have only to modify our problems (1a) or (1b) slightly, since one can imagine that for an absolutely arbitrary feasible set $S\subseteq\mathcal{P}(E)$ the straightforward strategy of the greedy may not guarantee some substantial bound. But we should clearly state here that this minor modification does not lead to a substantial loss of the generality in the formulation of discrete optimization problems.

From now on we assume that S is an *independence system*:

Definition 3.1. Let E be a finite set and $S\subseteq\mathcal{P}(E)$ a system of subsets of E with the *monotonicity property*

$$s_1 \subseteq s_2 \in S \Rightarrow s_1 \in S. \tag{1}$$

Then (E, S) is called an *independence system* and a set $s \in S$ is called *independent*.

If we would restrict c to be a mapping into the non-negative reals, i.e. $e: E \to \mathbb{R}^+$ then for any maximization problem over an arbitrary feasible set $S \subseteq \mathcal{P}(E)$ we may without loss of generality replace S by its *hereditary closure* $S^* = S \cup \{s \mid \exists s': s \subseteq s' \in S\}$. Of course, (E, S^*) is an independence system, and the reader may easily verify that by replacing S by S^* we do not change the problem, i.e. optimal value and set of optimum solutions are identical for both problems. The assumption that c maps E into the non-negative reals is very often naturally given for real world applications, for example costs, weights, utilities and distances are usually non-negative. But even if this additional property is not guaranteed by the nature of the problem we can modify most practical problems in order to obtain non-negative weights for all elements of E. For instance if the travelling salesman problem has negative weights on some arcs of the graph one may add some sufficient constant value to all arcs adjacent to the negative weight arc at some given node. Since all Hamiltonian cycles are of equal cardinality this constant value changes the objective value only by a constant but does not change the optimal Hamiltonian cycle. If the acyclic subgraph problem has some negative weighted arc one may add to it and the opposite arc a constant which again does not change the optimum. For many optimization problems one may do such modifications without loss of generality. For example, in the maximum weighted-clique problem, any negative-weighted vertex and all edges incident to it can be omitted from the graph.

For minimization problems over independence systems with $c: E \to \mathbb{R}^+$ we have to restrict feasibility not to independent sets but to maximal independent sets (cf. Definition 3.2) otherwise the problem would be trivial.

We will give some more definitions and notations for independence systems:

Definition 3.2. Let (E, S) be an independence system. A set $s \in S$ is called *maximal independent* (or *basis*) $\Leftrightarrow \forall e_i \in E \setminus s : s \cup \{e_i\} \notin S$. All sets $c \in \mathcal{P}(E) \setminus S$ are called *dependent*. A set $c \in \mathcal{P}(E) \setminus S$ is called *minimal dependent* (or *circuit*): $\Leftrightarrow$ all proper subsets of c are independent.

An independent set which has the largest cardinality among all elements of S is called *maximum independent*. The following definition is very substantial for all performance measures of algorithms for optimization problems on independence systems:

Definition 3.3. Let (E, S) be an independence system. For each subset $F \subseteq E$ we define two fundamental measures, the *lower rank* and the *upper rank* as the cardinality of the smallest and the largest maximal independent subset of F, respectively:

(2) upper rank: $\text{ur}(F) := \max\{|s| \mid s \in S, s \subseteq F\}$,
(3) lower rank: $\text{lr}(F) := \min\{|s| \mid s \in S, s \subseteq F, \forall e \in F \setminus s : s \cup \{e\} \notin S\}$.

This definition enables us to give a short definition of a matroid:

Definition 3.4. An independence system (E, S) is called a *matroid*:

$$\Leftrightarrow \forall F \subseteq E : \text{lr}(F) = \text{ur}(F).$$

The minimum $\min\{\text{lr}(F)/\text{ur}(F) \mid F \subseteq E\}$ over all subsets $F \subseteq E$ can be used as a measure of the worst-case behaviour of the greedy algorithm for maximization over independence systems, as the following important theorem demonstrates:

Theorem 3.5. *Let (E, S) be an arbitrary independence system, s_G the solution of the algorithm* GREEDYMAX, *s_0 the optimum solution of* (1a). *Then for any weight function $c : E \to \mathbb{R}^+$*

$$\min_{F \subseteq E} \frac{\text{lr}(F)}{\text{ur}(F)} \leq \frac{c(s_G)}{c(s_0)} \leq 1. \tag{4}$$

We should mention that (4) was first conjectured by Nemhauser. Proofs were given by Jenkyns [1976], and Korte and Hausmann [1976] independently. Since we consider the proof of Korte and Hausmann [1976] elegant and extremely short we will repeat it here.

Proof. Setting $E_i := \{e_1, \ldots, e_i\}$, $c(e_{n+1}) = 0$, we get by a suitable summation

$$c(s_G) = \sum_{i=1}^{n} |s_G \cap E_i| (c(e_i) - c(e_{i+1})), \tag{5}$$

$$c(s_0) = \sum_{i=1}^{n} |s_0 \cap E_i| (c(e_i) - c(e_{i+1})). \tag{6}$$

Since $s_0 \cap E_i \in S$ we have

$$|s_0 \cap E_i| \leq \text{ur}(E_i). \tag{7}$$

Due to the procedure of GREEDYMAX, $s_G \cap E_i$ is a maximal independent subset of E_i, hence

$$|s_G \cap E_i| \geq \text{lr}(E_i). \tag{8}$$

(7) and (8) inserted into (6) and (5) yield for the quotient:

$$\frac{c(s_G)}{c(s_0)} \geq \min_i \frac{\mathrm{lr}\,(E_i)}{\mathrm{ur}\,(E_i)} \qquad (9)$$

$$\geq \min_{F \subseteq E} \frac{\mathrm{lr}\,(F)}{\mathrm{ur}\,(F)}. \qquad (10)$$
□

Remark. Although (10) proves the general theorem, (9) yields a tighter bound in the case of special applications with a fixed objective function.

Edmonds [1971], Welsh [1976] gave special proofs for the following theorem which is now an immediate corollary of Theorem 3.4.

Theorem 3.6. *The greedy algorithm* GREEDYMAX *yields the optimum solution for* (1a) *for any objective function* $c: E \to \mathbb{R}^+$ *iff the independence system* (E, S) *is a matroid.*

Proof. "if". Direct application of Definition 3.4 to formula (4).
"only if". Cf. Theorem 3.7. □

Two questions are suggested by Theorem 3.4:
 (a) Is the bound given by (4) tight, and
 (b) are there some more natural bounds for

$$\min_{F \subseteq E} \frac{\mathrm{lr}\,(F)}{\mathrm{ur}\,(F)}$$

and how can one derive these bounds?

These questions and some other ones were extensively answered in Korte and Hausmann [1976] and Jenkyns [1976]. We will summarize the answers here.

Theorem 3.7. *The worst case bound* (4) *of Theorem 3.4 is sharp in the following sense: For every independence system* (E, S) *there exists a function* $c: E \to \mathbb{R}^+$ *such that*

$$\frac{c(s_G)}{c(s_0)} = \min_{F \subseteq E} \frac{\mathrm{lr}\,(F)}{\mathrm{ur}\,(F)}. \qquad (11)$$

Proof. (cf. Korte and Hausmann [1976]). Let $F_0 \subseteq F$ be the set for which the rank quotient attains its minimum, i.e.

$$\frac{\mathrm{lr}\,(F_0)}{\mathrm{ur}\,(F_0)} = \lim_{F \subseteq E} \frac{\mathrm{lr}\,(F)}{\mathrm{ur}\,(F)}.$$

Then we define

$$c(e) = \begin{cases} 1 & e \in F_0, \\ 0 & e \notin F_0. \end{cases}$$

With this objective function one can easily demonstrate the equality in (11). □

Before we will give a general lower bound for the minimum of the rank quotient we will calculate its exact value for some special examples. In some cases the value of the rank quotient is a constant, but in other cases it will depend on the dimension of the problem and unfortunately it will tend to zero when the dimension goes to infinity.

(a) *Stable Set Problem (Vertex Packing Problem) in a graph.* Let $G = (V, E)$ be a graph containing an induced subgraph isomorphic to $K_{1,m}$: Let (V, S) be the independence system for the stable set problem in this graph; i.e. S contains all subsets of vertices the elements of which are not linked by an edge. Then

$$\lim_{m \to \infty} \min_{F \subseteq V} \frac{\mathrm{lr}\,(F)}{\mathrm{ur}\,(F)} = 0. \tag{12}$$

(b) *Acyclic Subgraph Problem.* Let $G = (V, E)$ be a directed graph with $E := V \times V \setminus \{(i, i) \mid i \in V\}$ and

$$S := \{s \subseteq E \mid s \text{ contains no circuit}\}.$$

Of course, (E, S) is an independence system. Then we will show that

$$\lim_{|E| \to \infty} \min_{F \subseteq E} \frac{\mathrm{lr}\,(F)}{\mathrm{ur}\,(F)} = 0 \tag{13}$$

Let $n := |V|$ and $V = \{v_1, \ldots, v_n\}$.

$$F_1 := \{(v_i, v_{i+1}) \mid 1 \leq i \leq n-1\},$$
$$F_2 := \{(v_i, v_j) \mid 1 \leq j < i \leq n\},$$
$$F_0 := F_1 \cup F_2.$$

We have

$$\mathrm{lr}\,(F_0) = |F_1| = n - 1,$$
$$\mathrm{ur}\,(F_0) = |F_2| = \binom{n}{2}.$$

Thus

$$\min_{F \subseteq E} \frac{\mathrm{lr}\,(F)}{\mathrm{ur}\,(F)} \leq \frac{n-1}{\binom{n}{2}} = \frac{2}{n} \to 0. \qquad \square$$

(c) *Matching.*

Theorem 3.8. *Let (E, S) be the independence system of the matchings in an undirected graph $G = (V, E)$. In the trivial cases that every connected component of*

G has less than 4 vertices

$$\min_{F \subseteq E} \frac{\mathrm{lr}\,(F)}{\mathrm{ur}\,(F)} = 1.$$

In all other cases we have

$$\min_{F \subseteq E} \frac{\mathrm{lr}\,(F)}{\mathrm{ur}\,(F)} = \tfrac{1}{2}. \tag{14}$$

For a proof cf. Korte and Hausmann [1976].

We should note that for practical purposes the bound (14) for the greedy algorithm does not mean very much since Edmonds proposed a polynomially bounded algorithm which is optimal for the matching problem.

(d) *Symmetric travelling salesman problem.*

Theorem 3.9. *Let $G = (V, E)$ be a complete undirected graph without loops. Let S be the set of all subsets of Hamiltonian cycles in G. Then for the independence system (E, S) we have*

$$\tfrac{1}{2} \leq \min_{F \subseteq E} \frac{\mathrm{lr}\,(F)}{\mathrm{ur}\,(F)} \leq \tfrac{1}{2} + \tfrac{3}{2} \frac{1}{|V|}. \tag{15}$$

Again, a proof can be found in Korte and Hausmann [1976].

Remark. The bound (15) for the worst case behaviour of the greedy algorithm is worse than the bound for the special algorithm mentioned in Section 2. That algorithm was based on the matching algorithm and had a guaranteed performance of $\tfrac{2}{3}$.

So far we have given four examples for the value of the minimum of the rank quotient for special independence systems. It is natural to ask whether there is a general lower bound on this quotient. Since any independence system can be interpreted as an intersection of finitely many matroids, the following theorem gives an answer to this question.

Theorem 3.10. *Let (E, S^i), $i = 1, \ldots, k$, be matroids on the same set E. Let*

$$S := \bigcap_{i=1}^{k} S^i := \{s \mid s \in S^i, i = 1, \ldots, k\}.$$

Clearly, (E, S) is an independence system for which we have

$$\min_{F \subseteq E} \frac{\mathrm{lr}\,(F)}{\mathrm{ur}\,(F)} \geq \frac{1}{k}. \tag{16}$$

Proof. Let $F \subseteq E$ be any subset and s_1, s_2 maximal independent subsets of F. We have to show $|s_1|/|s_2| \geq 1/k$. For $i = 1, \ldots, k$ and $j = 1, 2$, let s_j^i be a maximal

S^i-independent subset of $s_1 \cup s_2$ containing s_j. If there were an element $e \in s_2 \setminus s_1$ with
$$e \in \bigcap_{i=1}^{k} s_1^i \setminus s_1,$$
then
$$s_1 \cup \{e\} \subseteq \bigcap_{i=1}^{k} s_1^i \in S,$$
a contradiction to the maximality of s_1. Hence each $e \in s_2 \setminus s_1$ can be an element of $s_1^i \setminus s_1$ for at most $k-1$ indices i. It follows
$$\sum_{i=1}^{k} |s_1^i| - k|s_1| = \sum_{i=1}^{k} |s_1^i \setminus s_1| \leq (k-1)|s_2 \setminus s_1| \leq (k-1)|s_2|.$$
By the definition of a matroid, we have
$$|s_1^i| = |s_2^i| \quad \text{for any } i, 1 \leq i \leq k.$$
Hence the above inequality implies
$$|s_2| \leq \left(\sum_{i=1}^{k} |s_2^i| - k|s_2| \right) + |s_2|$$
$$= \sum_{i=1}^{k} |s_1^i| - (k-1)|s_2| \leq k|s_1|.$$
The theorem follows. □

Remark. This is a shortened version of a proof to be found in Korte and Hausmann [1976], shortened due to a hint of T. A. Jenkyns in a private communication.

The question whether the bound of (16) is tight can be answered in two different directions:

(a) For each $k \geq 1$ there exists an independence system (E, S) which is an intersection of k matroids and which cannot be represented as an intersection of less than k matroids for which (16) is tight, i.e.
$$\min_{F \subseteq E} \frac{\text{lr}(F)}{\text{ur}(F)} = \frac{1}{k}.$$

(b) On the other hand there are examples of independence systems for which the bound in (16) is not tight. Again, the interested reader may find proofs in Korte and Hausmann [1976].

So far we have dealt only with maximization problems over independence systems. What about minimization problems? The next theorem gives the answer:

Theorem 3.11. *Let (E, S) be an independence system, $c : E \to \mathbb{R}^+$ an objective function and let us consider the minimization problem*
$$\min \{c(s) \mid s \text{ basis of } S\}$$

Then there is no upper bound for the worst-case behaviour of the algorithm GREEDYMIN, i.e. the algorithm can behave arbitrarily bad.

Proof. For any $M>0$ we have to construct a special independence system (E, S) and an objective function c such that

$$\frac{c(s_G)}{c(s_0)} > M.$$

Let us consider the graph $K_{1,2}$

$$\circ\!\!-\!\!-\!\!\circ\!\!-\!\!-\!\!\circ$$
$$v_1 \quad v_2 \quad v_3$$

and let (V, S) be the independence system of the stable set problem in this graph, i.e.

$$S = \{\{v_1\}, \{v_2\}, \{v_3\}, \{v_1, v_3\}\}.$$

Let

$$c(v) = \begin{cases} 1 & \text{for } v = v_1, \\ 2 & \text{for } v = v_2, \\ 2M > 2 & \text{for } v = v_3. \end{cases}$$

Then

$$c(s_G) = 2M+1 \quad \text{and} \quad c(s_0) = 2. \qquad \square$$

With the above mentioned results we have completely described the behaviour of the algorithms GREEDYMAX and GREEDYMIN. Although this is a considerable step towards answering the question how approximative algorithms behave for general optimization problems of type (1a) and (1b) the open question remains: are there other general classes of algorithms (not very special algorithms for special problems, like the examples in Section 2) which have a better behaviour than the greedy? The following part of this chapter will deal with this question. The answer will be NO for a very general class of algorithms to be specified. But to verify this answer we have to outline some theorems.

A natural way of improving the guaranteed performance of a greedy-type algorithm could be to combine the greedy approach with an enumerative pre-step (like several methods for the knapsack-problem which we will discuss later on). In this case we first completely enumerate all possible subsets of E of cardinality $\leq k$ ($k \leq |E|$ fixed) and check their feasibility. Afterwards we increase the set s^* with

$$c(s^*) = \max \{c(s) \mid |s| \leq k\},$$

using the greedy algorithm. One might think intuitively that this modification should have a better guaranteed performance than the simple greedy algorithm since we invest more computational effort in the pre-step. Of course if $k = |E|$ then this procedure results in a complete enumeration which guarantees optimality. Another modification of this approach is allowing that not only at a pre-step but at each step of the greedy at most k elements can be added. These k elements will be chosen among all possible candidates by complete enumeration. Again, one may assume that this algorithm should have a better guaranteed performance since at any step it involves more computational effort in order not to be as myopic as the greedy algorithm. We were able to prove (Hausmann and Korte [1977c]) with similar techniques as for the above mentioned theorems that these modifications do not have a better worst-case bound than the one for the simple greedy algorithm for general independence systems. We do not sketch the proofs here because the following more general theorem implies these results.

At the beginning of this section we stated that one may consider the greedy algorithm as an *oracle* algorithm as defined in Section 1. The greedy algorithm is precisely coded except for the "oracle" which checks feasibility (or independence) of a set. Let us now consider some general oracle algorithms for optimization problems over independence systems which may have the same oracle as the greedy algorithm. But apart from this they are completely arbitrary. That means the non-oracle part of the algorithm (which was for the greedy mainly a single loop) may have any possible structure and the way how the oracle is called can be done in any possible way. It is obvious that this is a very broad and extensive definition of a class of algorithms. Dobkin [1976] has called algorithms of this general type *search tree algorithm*. Since we consider any call of the oracle during the process of the algorithm as a conditional branching statement, whereas the condition for this branching is checked in the oracle-subroutine, the "flow-chart" of a general algorithm as outlined above is indeed a search tree. For the time being we restrict ourselves only to that type of an oracle which checks whether a set is independent or not. Beside this the algorithm may have any possible structure. Later on we will also relax this restriction insofar as other decisions are left to the oracle.

The reader will agree that this general type of oracle algorithms covers almost all heuristic approaches for combinatorial optimization. For the purpose of classification again we state that we admit any structure for the (main) program in the following discussion. So all well-known heuristic tricks and tools (like exchange, delete, permutate etc.) are covered herewith.

Let us now state a main theorem which shows that all arbitrary (search tree) algorithms which have polynomial time complexity and use an oracle that checks independence have a performance guarantee not better than that of the simple greedy algorithm as demonstrated by Theorem 3.4.

Theorem 3.12 (Hausmann and Korte [1977a]). *Let A be an algorithm for maximization over independence systems (E, S) which accepts as an input the set E, an objective function $c: E \to \mathbb{R}^+$ and an oracle which checks for a given subset $s \subseteq E$ whether $s \in S$. Let s_A denote the solution of this algorithm and s_0 the optimal solution. If for any input (where the independence system is not a matroid) this* algorithm has a better guaranteed performance than the greedy, i.e.

$$\frac{c(s_A)}{c(s_0)} > \min_{F \subseteq E} \frac{\mathrm{lr}(F)}{\mathrm{ur}(F)}, \tag{17}$$

then A can require $O(2^{|E|})$ calls on its oracle, i.e. the time complexity of A is exponential.

This theorem states that there exists no oracle algorithm which calls an oracle like the one described above at most polynomially many times and which has a globally better performance guarantee.

Proof. A formal proof of this theorem can be found in Hausmann and Korte [1977a]. Therefore we will here stress the intuitive main ideas. We consider a class of problems (1a) where E with $|E| = n$ is fixed and where $c(e) = 1$ for any $e \in E$. Let A be an algorithm as described in the theorem and suppose that it requires only a polynomial number (thus less than $\binom{n}{m}$ with $m = n/2$) of calls on its oracle R_S. We can think of A as a flow chart which can branch arbitrarily often, but as E and c are fixed, all branchings are due to calls on R_S. Now we construct a special path through the flow chart by choosing one of the two possible results YES or NO (i.e. a subset belongs to S or not) for any call on R_S. Let R_S be called for $s \subseteq E$, then we choose the YES-path iff $|s| < m$. Next we show that the constructed path can be realized by an independence system (E, S) which does not satisfy the inequality (17) thus yielding the desired contradiction. Let $s_1, \ldots, s_{\binom{n}{m}-1}$ be the first $\binom{n}{m} - 1$ sets for which A following the constructed path calls R_S. If on this path A calls R_S for only $r < \binom{n}{m} - 1$ sets and afterwards yields the output s_A then let $s_i := s_A$, $r + 1 \leq i \leq \binom{n}{m} - 1$. Obviously there exists a subset $s' \subseteq E$ of cardinality m which is different from $s_1, \ldots, s_{\binom{n}{m}-1}$. Now let S' be the system of all subsets $s \subseteq E$ with $|s| < m$ plus the single subset s'. It is easy to see that (E, S') is an independence system and that A applied to an oracle $R_{S'}$ follows exactly the path constructed above, stops after $r < \binom{n}{m} - 1$ calls on R_S and produces the output $s_A = s_{r+1} \neq s'$. Hence

$$\frac{c(s_A)}{c(s_0)} = \frac{|s_{r+1}|}{|s'|} \leq \frac{(m-1)}{m} = \min_{F \subseteq E} \frac{\mathrm{lr}(F)}{\mathrm{ur}(F)},$$

a contradiction to (3). □

This theorem again illuminates the central role of the greedy algorithm in combinatorial optimization. We can even generalize the preceding theorem allowing the use of oracles which can do more than simply decide whether a subset is independent.

Theorem 3.13. *Theorem 3.11 remains valid even if we admit the use of an oracle*

which checks whether
   (a) a given set is a member of S
   (b) a given set is a basis of the independence system (E, S),
   (c) a given set is a circuit of the independence system (E, S),
   (d) the incidence vectors of two given sets are adjacent vertices of the convex hull of all incidence vectors of sets in S.

Proof. Straightforward extension of the proof of Theorem 3.12 to be found in Hausmann and Korte [1977a]. □

Theorem 3.13 states that even if we know substantially more structure of the problem, like a complete characterization of the adjacency structure (cf. Hausmann and Korte [1976]) we would not be able to design a more efficient algorithm for maximization problems over independence systems.

We recall that in Theorem 3.11 we have stated that the greedy behaves arbitrarily bad for minimization problems over the bases of independence systems. The same result is true for the broad classes of algorithms defined in Theorems 3.12 and 3.13.

Theorem 3.14 (Hausmann and Korte [1977b]). *Let A be an algorithm for the minimization problem over an independence system (E, S), $\min\{c(s) \mid s$ basis of $(E, S)\}$ which accepts as an input the set E, an objective function $c: E \to \mathbb{R}^+$ and an oracle which checks for a given set $s \subseteq E$ whether it is a basis of (E, S) or not. Let s_A denote the solution of this algorithm and s_0 an optimal solution. If*

$$\frac{c(s_A)}{c(s_0)} \leq \alpha \tag{18}$$

for some finite bound α, then A requires $O(2^{|E|})$ calls on its oracle, i.e. A has an exponential time complexity.

Proof. Suppose A is an algorithm with the above properties that calls the oracle R less than $\binom{n-1}{m} - 1$ times where $m = \frac{1}{2}(n-1)$. Then for some $n \geq 5$ we define the objective function $c: E \to \mathbb{R}^+$ by $c(1) := \alpha m$ and $c(e) := 1$ for $e \neq 1$. As in the proof of Theorem 3.12 we construct a special path through the flow chart of A choosing the YES-path iff the oracle is called for a subset $s \subseteq E$ with $1 \in s$ and $|s| = m$. Since R is called for less than $\binom{n-1}{m} - 1$ subsets there is a subset s' with $1 \notin s'$ and $|s'| = m$ such that, on the constructed path, the oracle is not called for s' and s' is not the output set. Let $B' := \{s \subseteq E \mid s = s' \text{ or } (1 \in s, |s| = m)\}$ and let (E, S') be the corresponding independence system where B' is its set of bases. Then it is easy to see that A applied to an oracle $R_{B'}$ follows just the constructed path and yields the output $s_A \in B' \setminus \{s'\}$. Hence

$$c(s_A) = \alpha m + m - 1 > \alpha m = \alpha \cdot c(s') = \alpha \cdot c(s_0),$$

a contradiction to (18). □

The above mentioned theorems indicate that the greedy algorithm is a central approach for optimization problems over general independence systems. To our understanding further research should be done to extend the results especially of Theorems 3.12 as to allow some other oracles.

Before concluding this chapter let us discuss briefly another greedy approach which does not deal with optimization of *linear* objective functions over general independence systems but with maximization of submodular functions over cardinality restricted subsets of a given set E. This approach was first derived by Cornuejols, Fisher and Nemhauser [1976] from a practical problem, the uncapacitated location problem. It was further generalized by Nemhauser, Wolsey, and Fisher [1976] and by Fisher, Nemhauer, and Wolsey [1976]. Finally an oracle oriented approach by Nemhauser and Wolsey [1976] states also that the simple greedy algorithm for this problem structure dominates (in the worst-case sense) a large class of algorithms which were called *black-box* algorithms. Here we will present the main results of this approach without proofs which can be found in the above mentioned references.

The general problem formulation can be given in the following way: Again we have a finite set E. The set of feasible solutions is the uniform matroid

$$S := \{s \mid s \in \mathcal{P}(E), |s| \leq k\}, \quad k \text{ fixed},$$

whereas the objective function does not have the separability property which we have assumed for optimization over general independence systems, but is *submodular*, i.e.

$$c(s_1) + c(s_2) \geq c(s_1 \cup s_2) + c(s_1 \cap s_2) \tag{19}$$

and nondecreasing, i.e.

$$s_1 \subseteq s_2 \quad \text{implies} \quad c(s_1) \leq c(s_2). \tag{20}$$

There are some different but equivalent characterizations of submodularity of which the reader may find a complete list in Nemhauser, Wolsey and Fisher [1976]. The maximization of a submodular, nondecreasing function over cardinality restricted subsets of E can then be written as

$$\max \{c(s) \mid s \subseteq E, |s| \leq k\}. \tag{21}$$

Compared to the general optimization problem (of a separable objective function) over an independence system to our understanding the application of (21) to real-world, graph theoretical or other problems is somewhat restricted. Although both general problem classes are equivalent in the sense that they both contain NP-complete problems which therefore can be transformed into each other by a polynomial algorithm, it does not make much sense to transform first an optimization problem over an independence system via an appropriate submodular function into a maximization problem of type (21) and then solve it with a greedy approach.

Cornuejols, Fisher, Nemhauser, and Wolsey [1976] have proved that the greedy algorithm for problem (21) has a remarkable performance guarantee which is a universal constant and does not depend on the problem size and the objective function. But this is strongly depending on the structure of (21). Thus if one would polynomially transform optimization problems over indepence systems into problems of type (21), solve it approximatively and then retransform it, one can not guarantee the same performance since via the retransformation one will change the objective function which means that nominator and denominator of the performance measure can change.

Beside the uncapacitated location problem which was generalized to (21) one may find some other applications of (21) in Nemhauser, Wolsey, and Fisher [1976].

After the authors have proved a good performance guarantee for the greedy algorithm they generalize their algorithm in order to obtain even better performance guarantees. They classify their method as *q-enumeration plus greedy* which means that all subsets $s_q \subseteq E$ of cardinality q where q is a fixed integer $< k$ are used as a basis for the greedy algorithm and that the output s_A is the best of all $\binom{n}{q}$ greedy solutions s. In an Algol-like formulation the algorithm is:

procedure q-GREEDYSUBMOD:
begin
  $s_A = \emptyset$;
  for all subsets $s_q \subseteq E$ with $|s_q| = q$ **do**
    begin
      $s = s_q$; LEFT $= E \setminus s_q$;
      while ($|s| < k$) **do**
        begin
          find $j^* \in$ LEFT maximizing $c(s \cup \{j\})$, $j \in$ LEFT;
          $s = s \cup \{j^*\}$;
          LEFT = LEFT $\setminus \{j^*\}$;
        end;
      if $c(s) > c(s_A)$ **then** $s_A = s$;
    end;
  output s_A and $c(s_A)$;
end q-GREEDYSUBMOD:

Obviously q-GREEDYSUBMOD requires at most

$$\binom{n}{q}((n-q)+(n-q-1)+\cdots+(n-k+1)) = O(n^{p+1})$$

function evaluations. The main theorem is

Theorem 3.15 (Nemhauser, Wolsey, and Fisher [1976]). *Let E be a finite set, k and q integers with $0 \leq q < k \leq |E|$, and $c: \mathcal{P}(E) \to \mathbb{R}^+$ a submodular nondecreasing function with $c(\emptyset) = 0$. Let s_A be the approximate solution obtained by q-GREEDYSUBMOD and s_0 an optimal solution for (21). Then the following holds:*

$$\frac{c(s_A)}{c(s_0)} \geq 1 - \left(\frac{k-q}{k}\right) \cdot \left(\frac{k-q-1}{k-q}\right)^{k-q}. \tag{22}$$

Remarks. (1) The formulation in Nemhauser, Wolsey, and Fisher [1976] is slightly more general, as the assumption that c is nondecreasing and satisfies $c(\emptyset) = 0$ is replaced by a weaker assumption. In that case one gets not a bound for $c(s_A)/c(s_0)$ but for another closely related performance measure.

(2) If $q = 0$ q-GREEDYSUBMOD is the usual greedy algorithm. Then the performance measure in (22) tends to $(e-1)/e \approx 0.63$ as k goes to infinity (e denotes the base of the natural logarithm).

In the algorithmic description of q-GREEDYSUBMOD we assumed that for any subset $s \subseteq E$ the value $c(s)$ can be computed somehow. As c is a general submodular nondecreasing function we leave this computation to an *oracle* R_c. Thus q-GREEDYSUBMOD can be considered as an oracle algorithm; any function evaluation corresponds to a call on R_c. An important result of Nemhauser and Wolsey [1977] states that q-GREEDYSUBMOD is in a sense an optimal oracle algorithm for the problem (21):

Theorem 3.16 (Nemhauser and Wolsey [1977]). *Any oracle algorithm for problem (21) having a globally better performance guarantee than (22) requires at least $O(n^{q+2})$ function evaluations, i.e. calls on its oracle.*

4. Approximative algorithms for the knapsack problem

In the preceding sections we have extensively treated optimization problems over general independence systems. As we have outlined the greedy algorithm yields worst-case bounds which cannot be globally strengthened by any other polynomial approximative algorithm. This is not so if we treat optimization problems over restricted independence systems such as the matching, the branching, and the shortest path problem which can be solved by well-known polynomial algorithms and therefore can be transformed into problems of the class $\mathcal{P}$. But even for some $\mathcal{NP}$-hard problems, e.g. the travelling salesman, the acyclic subgraph, and the knapsack problem, we can do better than the greedy; following Karp [1975a] we call an optimization problem $\mathcal{NP}$-*hard* if the existence of a polynomial algorithm for its solution implies $\mathcal{P} = \mathcal{NP}$. In the following we shall essentially deal with the 0–1-knapsack problem as on the one hand this covers a wide range of applications and on the other hand some "good" polynomial approximative algorithms have been designed for this problem.

The problem is

$$\max \{c(s) \mid s \subseteq E, a(s) \leq b\} \tag{1}$$

where $E = \{1, \ldots, n\}$ and $c, a : \mathcal{P}(E) \to \mathbb{R}^+$ with

$$c(s) = \sum_{i \in s} c_i \quad \text{and} \quad a(s) = \sum_{i \in s} a_i, \quad s \subseteq E,$$

c_i and a_i denote the "profit" and "weight", respectively, of element $i \in E$ and $b > 0$ the "size" of the knapsack.

Let us now give some more detailed definitions which are due to Garey and Johnson [1976c] and Sahni [1976b]. The aim is to define for every relative error ε an approximative algorithm which guarantees (3) of Section 2.

Definition 4.1. An *approximation scheme* A is an algorithm for a given problem class such that for any problem instance of this class and for any given relative error $\varepsilon > 0$ the approximation scheme yields a feasible solution s_A such that

$$\left| \frac{c(s_0) - c(s_A)}{c(s_0)} \right| \leq \varepsilon. \tag{2}$$

Definition 4.2. An approximation scheme is called a *polynomial approximation scheme* if for every fixed $\varepsilon > 0$ it has a polynomial complexity.

Definition 4.3. An approximation scheme is called a *fully polynomial approximation scheme* if it has a complexity which is polynomial in both the input size and $1/\varepsilon$.

The procedures ACYCLIC, MAXHAM and KNAPSACK described in Section 2 for the acyclic subgraph, the travelling salesman, and the knapsack problem are polynomial but evidently do not satisfy the definition of an approximation scheme, because they yield only fixed worst case bounds and do not allow immediate modifications which for any given relative error $\varepsilon > 0$ render suitable approximate solutions.

On the other hand, complete enumeration of all feasible solutions of a given problem can be considered as an approximation scheme, it even yields an optimal solution but is clearly not polynomial. As we will show a proper combination of both, a straightforward approximative algorithm with good time bounds like the greedy and a partial enumeration, can result in a polynomial approximative algorithm.

It may seem surprising at first view, but even for some problems which are known to be $\mathcal{NP}$-hard polynomial approximation schemes do exist, and even more surprisingly one has found fully polynomial approximation schemes for some special $\mathcal{NP}$-hard problems.

As a first example, we will now look at the *subset sum problem*, which is a special case of the 0–1 knapsack problem in (1) where the weights and the profits are identical for each element. Hence the problem is the following:

$$\max\{c(s) \mid s \subseteq E, c(s) \leq b\}. \tag{3}$$

For this problem which is known to be $\mathcal{NP}$-hard Johnson [1974b] gave a polynomial approximation scheme: Let for a positive integer q

$$s_q := \{i \in E \mid c_i > b/(q+1)\}$$

and let s_q^* be an optimal solution of

$$\max\{c(s) \mid s \subseteq s_q, c(s) \leq b\}.$$

Now we define algorithm A_q for problem (3) to operate as follows:
(1) determine s_q^*, e.g. by complete enumeration;
(2) apply the greedy algorithm starting with the partial solution s_q^* for (3) yielding the output s_{A_q}.

As one may easily check, step (1) of algorithm A_q requires at most the enumeration of

$$\sum_{j=1}^{q} \binom{n}{j} < qn^q \text{ sets } s \subseteq s_q \quad \text{because of } |s_q| \leq q.$$

Theorem 4.4 (Johnson [1974b])

$$\left| \frac{c(s_0) - c(s_{A_q})}{c(s_0)} \right| \leq \frac{1}{q+1}$$

where s_0 is an optimal solution for (3). The time complexity for A_q is $O(qn^q)$.

A polynomial approximation scheme for the general 0–1 knapsack problem (1) is due to Sahni [1975]. Let $\bar{s}$ be a solution to (1) obtained by applying the greedy algorithm starting with a partial solution s where the elements of $E \setminus s$ are assumed to be ordered according to nonincreasing "profit densities" a_i/s_i (cf. procedure KNAPSACK in Section 2). Then Sahni's algorithm A_q for problem (1) determines an approximate solution s_{A_q} by completely enumerating all possible feasible solutions to

$$\max\{c(\bar{s}) \mid |s| \leq q, a(s) \leq b\}$$

where q is a fixed positive integer. Note that E contains $\sum_{j=1}^{q} \binom{n}{j} < qn^q$ subsets of cardinality $\leq q$.

Theorem 4.5 (Sahni [1975])

$$\left|\frac{c(s_0)-c(s_{A_q})}{c(s_0)}\right| < \frac{1}{q+1}$$

where s_0 is an optimal solution to (1). The time complexity for A_q is $O(qn^{q+1})$.

Now we will describe the main ideas leading to the fully polynomial approximation scheme of Ibarra and Kim [1975] for the 0–1-knapsack problem which is polynomial in n and $1/\varepsilon$. We think that this very interesting approach can be described best as a sort of implicit enumeration (or branch and bound) method (cf. Garfinkel and Nemhauser [1972] for the terminology used below). In an implicit enumeration algorithm for the *exact* solution of the knapsack problem, the number of *live* (i.e. non-fathomed) candidate problems in general grows exponentially with n therefore yielding an exponential time (and space) complexity. The basic idea of the implicit enumeration variant introduced by Ibarra and Kim is to use slightly changed, "coarsened" profits $\tilde{c}_i$ instead of c_i which allows to bound the number of live vertices and thus the time and space complexity by a polynomial in n. If the difference between the original profits c_i and the changed profits $\tilde{c}_i$ is small enough, the optimal solution for the $\tilde{c}$-problem will be an approximate solution of the original c-problem yielding a relative error $\leq \varepsilon$.

We will first describe the implicit enumeration method for the original profits and after that show why the profits should be coarsened. Every implicit enumeration method has three basic features: separation, branching (or backtracking), and fathoming. For the Ibarra and Kim approach, separation and branching can be shortly classified as a breadth-first search in the enumeration tree: The elements $i \in E$ are processed in a certain order; for each new element i every live candidate problem is separated into two new candidate problems putting i into or leaving it out of the knapsack. Then before processing the next element, all new candidate problems are inspected and – if possible – fathomed. Obviously for every live candidate problem CP there is a corresponding partial solution $s(\text{CP})$, namely the set of all elements which have been put into the knapsack up to reaching CP.

Now there are two fathoming criteria. The first criterion says that a candidate problem CP is fathomed if the corresponding partial solution has a weight $a(s(\text{CP}))$ exceeding the knapsack size b, which is the usual feasibility check of implicit enumeration methods. The second criterion checks whether the partial solutions assigned to two candidate problems CP and CP' yield the same profit $c(s(\text{CP})) = c(s(\text{CP}'))$ but different weights, say $a(s(\text{CP})) > a(s(\text{CP}'))$. Then CP is fathomed because for every feasible solution of CP there exists a feasible solution of CP' yielding the same profit.

This second fathoming criterion implies that the number of the live candidate problems is always bounded by the number of different profits $c(s)$ where s varies through all feasible solutions. Hence in an implementation of this implicit enumeration method, the live candidate problems (more exactly the partial solutions, profits, and weights assigned to them) can be stored in the rows of an array TABLE where every row of TABLE corresponds to a distinct value $c(s)$, s feasible. Fathoming a candidate problem CP using the second criterion can then simply be implemented as replacement of the row of TABLE corresponding to CP by the row corresponding to CP' with $c(s(\text{CP})) = c(s(\text{CP}'))$ and $a(s(\text{CP})) > a(s(\text{CP}'))$.

But since in general the number of different values $c(s)$, s feasible, grows exponentially in n, the (exact) algorithm described above would require an exponential size of TABLE and thus an exponential time and space complexity. Hence the crucial point in the algorithm is to replace the profits c_i by "coarser" profits $\tilde{c}_i$ which are on the one hand sufficiently close to the c_i such that an optimal solution for the $\tilde{c}_i$ is an approximate solution for the c_i yielding a relative error $\leq \varepsilon$ and which are on the other hand sufficiently "coarse" in the sense that the number of different values $\tilde{c}(s)$, s feasible, grows only polynomially in n and $1/\varepsilon$. This coarsening can be realized in several ways. Ibarra and Kim use a certain "rounding" or "*digit truncation*" and prove

Theorem 4.6 (Ibarra and Kim [1975]). *There exists a fully polynomial approximation scheme for the 0–1-knapsack problem which has time complexity* $O(n \log n + n/\varepsilon^2)$, *space complexity* $O(n + 1/\varepsilon^3)$, *and which produces an approximate solution s satisfying* $(c(s_0) - c(s))/c(s_0) \leq \varepsilon$.

Besides digit truncation Sahni [1976b] considers two different methods of reducing the set of possible objective values by coarsening techniques. In the *interval partitioning* method the interval $[0, p]$ where $p := \max\{c(s) \mid s \text{ feasible}\}$ is partitioned into subintervals each of size $p\varepsilon/n$ and no distinction is made between values $c(s)$ belonging to the same subinterval. *Separation* is a technique very similar to interval partitioning. Here the interval $[0, p]$ is not regularly divided into subintervals but the subintervals are constructed such that as many values $c(s)$ as possible fall into the same subinterval.

We think that the basic idea of the Ibarra and Kim algorithm – implicit enumeration using a fathoming criterion for the coarsened objective function such that the number of live candidate problems is bounded by a polynomial – seems to be so general and attractive that it might be fruitful for many other discrete optimization problems. Ibarra and Kim themselves apply their idea also to the *integer knapsack problem* which differs from the 0–1-knapsack problem in that several copies of each element $i \in E$ can be put into the knapsack. For this problem they prove:

Theorem 4.7 (Ibarra and Kim [1975]). *There exists a fully polynomial approximation scheme for the integer knapsack problem which has time complexity*

$O(n/\varepsilon^3)$, space complexity $O(n+1/\varepsilon^3)$, and which produces an approximate solution s satisfying $(c(s_0)-c(s))/c(s_0) \leq \varepsilon$.

Another class of problems where Ibarra and Kim's idea applies is the wide field of *scheduling* and *sequencing* problems. Referring to Sahni [1976b] the general procedure applies to
- job sequencing with deadlines (Sahni [1976b]);
- minimizing weighted mean finish time (Horowitz and Sahni [1976], Sahni [1976a]);
- finding an optimal SPT (= shortest processing time first) schedule (Sahni [1976a]);
- finding minimum finish time schedules on identical, uniform and nonidentical machines (Horowitz and Sahni [1976], Sahni [1976a]).

We will show now that there are many NP-hard problems for which there does not exist any polynomial approximation scheme, not even any polynomial approximative algorithm with a performance guarantee bounded by a constant – unless $\mathcal{P} = \mathcal{NP}$. Hence the search for such polynomial approximative algorithms is as hard as the search for polynomial exact algorithms.

Sahni and Gonzales [1976] give many examples of problems having such nasty features, e.g. cycle cover problem, travelling salesman min-problem, knapsack-min problem, multicommodity network flow problem, quadratic assignment min-problem. Garey and Johnson [1976b] derive similar results for the graph colouring problem. To sketch the main ideas for proving these results we want to treat in detail the travelling salesman min-problem.

Let $G = (V, E)$ be a complete undirected graph with $|V| = n$ and $E = \{\{i, j\} \mid i, j \in V, i \neq j\}$. Let c_{ij} be nonnegative weights assigned to the edges $\{i, j\} \in E$. Then the symmetric travelling salesman min-problem is to find a Hamiltonian cycle of G with minimum weight.

Theorem 4.8 (Sahni and Gonzales [1976]). *Unless $\mathcal{P} = \mathcal{NP}$ there does not exist any polynomial approximation algorithm with a constant performance guarantee k for the symmetric travelling salesman min-problem.*

Proof. Suppose A is such a polynomial algorithm with a constant performance bound k, i.e. $1 \leq c(s_A)/c(s_0) \leq k$. We will show that then the well-known $\mathcal{NP}$-complete problem of deciding whether a graph is Hamiltonian can also be solved by a polynomial algorithm, which would imply $\mathcal{P} = \mathcal{NP}$. Let $\tilde{G} = (V, \tilde{E})$ be any (undirected) graph, $|V| = n$, and let $G = (V, E)$ be the corresponding complete graph on V. Consider the symmetric travelling salesman min-problem for G and the weights

$$c_{ij} = \begin{cases} 1 & \text{if } \{i, j\} \in \tilde{E}, \\ kn & \text{otherwise.} \end{cases}$$

Let c^0 be the value of the optimal solution and c^A the value of the solution yielded by A. Obviously the following implications hold

$\tilde{G}$ Hamiltonian $\Rightarrow c^0 = n,$

$\tilde{G}$ not Hamiltonian $\Rightarrow c^0 \geq kn + n - 1,$

hence

$c^A \leq kn \Rightarrow c^0 \leq kn < kn + n - 1 \Rightarrow \tilde{G}$ Hamiltonian,

$c^A > kn \Rightarrow c^0 \geq c^A/k > n \quad \Rightarrow \tilde{G}$ not Hamiltonian.

Therefore the polynomial algorithm which calculates c^A using A and then checks whether $c^A \leq kn$ can be used for deciding whether $\tilde{G}$ is Hamiltonian. □

Any discrete optimization problem can be formulated as a maximum or minimum problem. Clearly both formulations are completely equivalent, in particular the minimum formulation is $\mathcal{NP}$-hard iff the maximum formulation is $\mathcal{NP}$-hard. This is not true for the corresponding approximative problems: In Section 2 we showed that the polynomial approximative algorithm MAXHAM for the symmetric travelling salesman max-problem has a constant performance guarantee whereas Theorem 4.8 implies that the minimum formulation of the same problem does not allow such an algorithm unless $\mathcal{NP} = \mathcal{P}$. To understand this paradox we have to recall that the transformation of a maximum problem $\max\{c(s) \mid s \subseteq E\}$ with $c \geq 0$ into a minimum problem also with a nonnegative objective function involves the addition of a constant to all the weights. Let e.g. $\alpha = \max\{c_e \mid e \in E\}$ and $\tilde{c}_e = \alpha - c_e \geq 0$, $e \in E$. Then $\min\{\tilde{c}(s) \mid s \subseteq E\}$ is an equivalent minimum problem. Let s_0 and s_A be an optimal and an approximate solution of these problems, respectively. As the cardinalities $|s_0|$ and $|s_A|$ in general depend on n, also the quotient

$$\frac{\tilde{c}(s_A)}{\tilde{c}(s_0)} = \frac{\alpha |s_A| - c(s_A)}{\alpha |s_0| - c(s_0)}$$

can depend on n, even if the quotient $c(s_A)/c(s_0)$ is bounded by a constant.

In spite of this explanation the situation remains strange enough and seems to indicate that the quotient $c(s_A)/c(s_0)$ which is used by most authors might not be an appropriate tool for fully characterizing the worst-case behaviour of approximative algorithms.

References

Aho, A. V., Hopcroft, J. E. and J. D. Ullman [1974]. The Design and Analysis of Computer Algorithms (Addison Wesley, Reading, MA, 1974).

Baker, T., J. Gill and R. Solovay [1975]. Relativizations of the P = ?NP Question. SIAM J. Comput. 4 (4) (1975) 431–442.

Brucker, P. [1976]. Die Komplexität von Scheduling Problemen, in: Proceedings in Operations Research 5 (Würzburg, 1976) 357–368.

Brucker, P. [1976]. Approximative Verfahren zur Lösung von diskreten Optimierungsproblemen, Universität Oldenburg (1976, working paper).
Brucker, P., Garey, M. R., Johnson, D. S. [1976]. Scheduling Equal-Length Tasks under Treelike Precedence Constraints to Minimize Maximum Lateness (unpublished working paper).
Chandra, A. K., Wong, C. K. [1975]. Worst-Case Analysis of a Placement Algorithm Related to Storage Allocation. SIAM J. Comput. 4 (3) (September 1975).
Chang, L., Korsh, J. F. [1976]. Canonical Coin Changing and Greedy Solutions. J. Assoc. Comput. Mach. 23 (3) (July 1976) 418–422.
Christofides, N. [1976]. Worst-Case Analysis of a New Heuristic for the Traveling Salesman Problem. Working Paper. Carnegie-Mellon University, Pittsburgh, PA 15213 (Feb. 1976).
Cook, S. A. [1971]. The Complexity of Theorem-Proving Procedures, in: Proceedings of 3rd Annual ACM Symposium on Theory of Computing (1971) 151–158.
Corneil, D. G. [1974]. The Analysis of Graph Theoretical Algorithms in: Proc. 5th S.-E. Conference Combinatorics, Graph Theory, and Computing (1974) 3–38.
Cornuejols, G., Fisher, M. L. and G. L. Nemhauser [1976]. An Analysis of Heuristics and Relaxations for the Uncapacitated Location Problem. Discussion paper 7602 CORE Louvain (Jan. 1976, to appear in Management Sci.).
Dobkin, D. P. [1976]. A Nonlinear Lower Bound on Linear Search Tree Programs for Solving Knapsack Problems. J. Comput. System Sci. 13 (1976) 69–73.
Edmonds, J. [1965]. Paths, Trees, and Flowers. Canad. J. Math. 17 (1965) 449–467.
Edmonds, J. [1971]. Matroids and the Greedy Algorithm. Math. Programming 1 (1971) 127–136.
Edmonds, J. [1975]. Matroid Intersection (Unpublished Working paper).
Even, J., Tarjan, R. E. [1976]. A Combinatorial Problem which is Complete in Polynomial Space. J. Assoc. Comput. Mach. 23 (4) (October 1976) 710–719.
Fisher, M. L., Nemhauser, G. L. and L. A. Wolsey [1976]. An Analysis of Approximations for Maximizing Submodular Set Functions II. Discussion paper 7629, CORE Louvain (December 1976).
Gale, D. [1968]. Optimal Assignments in an Ordered Set: an Application of Matroid Theorie. J. Combinatorial Theory (1968) 176–180.
Garey, M. R., Graham, R. L. [1975]. Bounds for Multiprocessor Scheduling with Resource Constraints. SIAM J. Comput. 4 (2) (June 1975) 187–200.
Garey, M. R., Graham, R. L., Johnson, D. S., Yao, A. Ch.-Ch. [1976]. Resource Constrained Scheduling as Generalized Bin Packing. J. Combinatorial Theory (A) 21 (1976) 257–298.
Garey, M. R., Johnson, D. S. [1974]. "Strong" NP-Completeness Results: Motivation, Examples and Implications (Bell Laboratories, Murray Hill, NJ 07974).
Garey, M. R. Johnson, D. S. [1975]. Complexity Results for Multiprocessor Scheduling under Resource Constraints. SIAM J. Comput. 4 (4) (December 1975) 397–411.
Garey, M. R., Johnson, D. S. [1976a]. Scheduling Tasks with Nonuniform Deadlines on Two Processors. J. Assoc. Comput. Mach. 23 (3) (July 1976) 461–467.
Garey, M. R., Johnson, D. S. [1976b]. The Complexity of Near-Optimal Graph Coloring. J. Assoc. Comput. Mach. 23 (1) (January 1976) 43–49.
Garey, M. R. and D. S. Johnson [1976c]. Approximation Algorithms for Combinatorial Problems: An Annotated Bibliography, in: J. F. Traub, ed., [1976a] 41–52.
Garey, M. R., Johnson, D. S., Stockmeyer, L. [1976]. Some Simplified NP-Complete Graph Problems. Theoret. Comput. Sci. 1 (1976) 237–267.
Garfinkel, R. S. and Nemhauser, G. L. [1972]. Integer Programming (Wiley, London, 1972).
Gavril, F. [1972]. Algorithms for Minimum Coloring, Maximum Cliques and Maximum Cliques and Maximum Independent Sets of a Chordal Graph. SIAM J. Comput. 1 (1972) 180–187.
Golumbic, M. C. [1976]. The Complexity of Comparability Graph Recognition and Coloring, in: J. F. Traub, ed., [1976a] 459.
Gonzales, Th., Ibarra, O. H., Sahni, S. [1977]. Bounds for LPT Schedules on uniform Processors. SIAM J. Comput. 6 (1) (March 1977) 155–166.
Gonzales, T., Sahni, S. [1976]. Open Shop Scheduling to Minimize Finish Time. J. Assoc. Comput. Mach. 23 (4) (October 1976) 665–679.
Graham, R. L. [1966]. Bounds for Certain Multiprocessing Anomalies. Bell System Tech. J. (November 1966) 1563–1581.

Graham, R. L. [1969]. Bounds on Multiprocessing Timing Anomalies. SIAM J. Appl. Math. 17 (2) (March 1969) 416–429.

Hausmann, D. and B. Korte [1976]. Colouring Criteria for Adjacency on (0–1)-Polytopes, Math. Programming Study 8 (1978) 106–127.

Hausmann, D. and B. Korte [1977a]. Lower Bounds on the Worst-Case Complexity of Some Oracle Algorithms. Working Paper No. 7757–OR, Institut für Ökonometrie und Operations Research, University of Bonn (April 1977), to appear in Discrete Math.

Hausmann, D. and B. Korte [1977b]. Worst-Case-Behaviour of Polynomial Bounded Algorithms for Independence Systems. Working Paper no. 7760–OR, Institut für Ökonometrie und Operations Research, University of Bonn (May 1977) to appear in: Zeitschrift für Angewandte Mathematik und Mechanik (ZAMM).

Hausmann, D. and B. Korte [1977c]. K-Greedy Algorithms for Independence Systems. Working paper no. 7761–OR, Institut für Ökonometrie und Operations Research, University of Bonn (June 1977), to appear in Z. Operations Res.

Hirschberg, D. S., Wong, C. K. [1976]. A Polynomial-Time Algorithm for the Knapsack Problem with Two Variables. J. Assoc. Comput. Mach. 23 (1) (January 1976) 147–154.

Holland, J. H. [1973]. Genetic Algorithm and the Optimal Allocation of Trials, SIAM J. Comput. 2 (2) (June 1973) 88–105.

Hopcroft, J. E. and J. D. Ullman [1969]. Formal Languages and Their Relation to Automata (Addison-Wesley, Reading, MA, 1969).

Horowitz, E., Sahni, S. [1976]. Exact and Approximate Algorithms for Scheduling Nonidentical Processors, J. Assoc. Comput. Mach. 23 (2) (April 1976) 317–327.

Hwang, F. K. [1976]. On Steiner Minimal Trees with Rectilinear Distance. SIAM J. Appl. Math. 30 (1) (January 1976).

Ibarra, O.H., Kim, Ch.E. [1975]. Fast Approximation Algorithms for the Knapsack and Sum of Subset Problems. J. Assoc. Comput. Mach. 22 (4) (October 1975) 463–468.

Ibarra, O.H., Kim, Ch.E. [1977]. Heuristic Algorithms for Scheduling Independent Tasks on Nonidentical Processors. J. Assoc. Comput. Mach. 24 (2) (April 1977) 280–289.

Jenkyns, Th.A., [1976]. The Efficacy of the "Greedy" Algorithm. Proc. 7th S–E Conf. Combinatorics, Graph Theory, and Computing (1976) 341–350.

Jenkyns, Th.A. [1977]. The Greedy Travelling Salesman Problem. Working paper, Brock University (April 1977).

Jeroslow, R. [1976]. Bracketing Discrete Problems by two Problems of Linear Optimization. Working Paper, Carnegie Mellon Univ., Pittsburgh (October 1976), in: Operations Research Verfahren XXV, Methods of Operations Research (1977) 205–216.

Johnson, D.S. [1974a]. Fast Algorithms for Bin Packing. J. Comput. System Sci. 8 (1974) 272–314.

Johnson, D.S. [1974b]. Approximation Algorithms for Combinatorial Problems. J. Comput. System Sci. 9 (1974) 256–278.

Johnson, D.S. [1974c]. Worst Case Behaviour of Graph Colouring Algorithms. Proc. 5th S–E Conf. Combinatorics, Graph Theory, and Computing (1974) 513–527.

Johnson, D.S., Demers, A., Ullman, J.D., Garey, M.R., Graham, R.L. [1974]. Worst-Case Performance Bounds for Simple One-Dimensional Packing Algorithms. SIAM J. Comput. 3 (4) (December 1974) 299–325.

Kafura, D.G., Shen, V.Y. [1977]. Task Scheduling on a Multiprocessor System with Independent Memories. SIAM J. Comput. 6 (1) (March 1977) 167–187.

Karp, R.M. [1972]. Reducibility among Combinatorial Problems, in: Miller, R.E. and J.W. Thatcher, eds., [1972].

Karp, R.M. [1975]. On the Computational Complexity of Combinatorial Problems. Networks 5 (1975) 45–68.

Karp, R.M. [1975a]. The Fast Approximative Solution of Hard Combinatorial Problems, In: Proceedings of the 6th S.E. Conference on Combinatorics, Graph Theory, and Computing (1975).

Karp, R.M. [1976]. The Probabilistic Analysis of Some Combinatorial Search Algorithms, in: J.F. Traub, ed., [1976a] 1–19.

Karp, R.M., McKellar, A.C., Wong, G.K. [1975]. Near-Optimal Solutions to a 2-Dimensional Placement Problem. SIAM J. Comput. 4 (3) (September 1975) 271–286.

Klee, V. and G.J. Minty [1972]. How Good is the Simplex Algorithm? in: Shisha, O., ed., Inequalities III (Academic Press, New York 1972) 159–175.
Kleitman, D.J. [1975]. Algorithms. Advances in Math. 16 (1975) 233–245.
Knuth, D.E. [1973]. The Art of Computer Programming: Sorting and Searching. (Addison-Wesley, Reading, MA, 1973).
Korte, B. and D. Hausmann [1976]. An Analysis of the Greedy Heuristic for Independence Systems. Annals of Discrete Math. 2 (1978) 65–74.
Kruskal, J.B. [1956]. On the Shortest Spanning Subtree of a Graph and the Travelling Salesman Problem. Proc. Amer. Math. Soc. 7 (1956) 48–50.
Lawler, E.W. [1976]. Introduction to the Theory of Algorithms, in: Yeh, R.T., ed., Applied Computation Theory [1976] 61–81.
Lenstra, J.K., Rinnooy Kan, A.H.G. [1976]. A Note on the Expected Performance of Branch-and-Bound Algorithms. Afdeling Mathematische Besliskunde BW 63/76, December, 2e Boerhaavestraat 49, Amsterdam.
Lins, S. [1975]. The Use of Matroid Theory in Optimal Multi-Assignment over Totally Ordered Sets. Departemento de Matemática, Universidade Federal de Pernambuco, Cidade Universitária, Recife-Brasil (1975).
Loecks, J. [1976]. Algorithmentheorie. Springer, Berlin-Heidelberg-New York (1976).
Miller, R.E. and J.W. Thatcher, eds. [1972]. Complexity of Computer Computations (Plenum Press, New York, 1972).
Nemhauser, G.L. and L.A. Wolsey [1976]. Best Algorithms for Approximating the Maximum of a Submodular Set Function. Discussion paper 7636, CORE Louvain (December 1976).
Nemhauser, G.L., Wolsey, L.A. and M.L. Fisher [1976]. An Analysis of Approximations for Maximizing Submodular Set Functions. Discussion paper 7618, CORE Louvain (July 1976).
Papadimitriou, Ch.H., [1976]. On the Complexity of Edge Traversing. J. Assoc. Comput. Mach. 23 (3) (July 1976) 544–554.
Papadimitriou, Ch.H., Steiglitz, K. [1977]. On the Complexity of Local Search for the Traveling Salesman Problem, SIAM J. Comput. 6 (1) (March 1977) 76–83.
Petersohn, U., Voss, K., Weber, K.H. [1974]. Genetische Adaption – ein stochastisches Suchverfahren für diskrete Optimierungsprobleme. Math. Operationsforsch. Statist. 5 (1+74) (Heft 7/8) 555–571.
Rabin, M.O. [1976]. Probabilistic Algorithms, in: J.F. Traub, ed., [1976a] 21–39.
Rado, R. [1957]. Note on independence functions. Proc. London Math. Soc. 7 (1957) 300–320.
Recski, A. [1976]. Matroids and n-Ports. 9th Internat. Symposium on Math. Programming, Budapest.
Recski, A. [1976]. Matroids and Independent State Variables. Research Institute for Telecommunication H-1026 Budapest, Hungary.
Rinnooy Kan, A.H.G. [1976]. Machine Scheduling Problems. Doctoral dissertation, Amsterdam (1976).
Rosenkrantz, D.J., Stearns, R.E., Lewis, P.M. [1974]. Approximate Algorithms for the Traveling Salesperson Problem. Proceedings of 15th IEEE Symposium on Switching and Automata Theory (1974) 33–42.
Sahni, S. [1974]. Computationally Related Problems. SIAM J. Comput. 3 (1974) 262–279.
Sahni, S. [1975]. Approximative Algorithms for the 0/1 Knapsack Problem. J. Assoc. Comput. Mach. 22 (1) (January 1975) 115–124.
Sahni, S. [1976a]. Algorithms for Scheduling Independent Tasks. J. Assoc. Comput. Mach. 23 (1) (January 1976) 116–127.
Sahni, S. [1976b]. General Techniques for Combinatorial Approximation. Technical Report 76-6, Department of Computer Science, University of Minnesota (June 1976).
Sahni, S. and Gonzales, T. [1976]. P-Complete Approximation Problems. J. Assoc. Comput. Mach. 23 (1976) 555–565.
Schwefel, H.P. [1975]. Evolutionsstrategien und numerische Optimierung. Diss. TU, Berlin (1975).
Traub, J.F., ed. [1976a]. Algorithms and Complexity. New Directions and Recent Results (Academic Press, New York, San Francisco, London, 1976).
Traub, J.F., ed. [1976b]. Analytic Computational Complexity (Academic Press, New York, San Francisco, London, 1976).

Zadeh, N. [1973]. A Bad Network Problem for the Simplex Method and other Minimum Cost Flow Algorithms. Math. Programming 5 (1973) 255–266.
Yeh, R.T., ed. [1976]. Applied Computation Theory. Analysis, Design, Modeling (Prentice Hall, Engelwood Cliffs, NJ, 1976).
Welsh, D.J.A. [1968]. Kruskal's Theorem for Matroids. Proceedings Cambridge Philos. Soc. 64 (1968) 3–4.
Welsh, D.J.A. [1976]. Matroid Theory (Academic Press, London, New York, San Francisco, 1976).

COMPUTATIONAL COMPLEXITY OF DISCRETE OPTIMIZATION PROBLEMS

J.K. LENSTRA

Mathematisch Centrum, Amsterdam, The Netherlands

A.H.G. RINNOOY KAN

Erasmus University, Rotterdam, The Netherlands

Recent developments in the theory of computational complexity as applied to combinatorial problems have revealed the existence of a large class of so-called *NP-complete* problems, either *all* or *none* of which are solvable in polynomial time. Since many infamous combinatorial problems have been proved to be NP-complete, the latter alternative seems far more likely. In that sense, NP-completeness of a problem justifies the use of enumerative optimization methods and of approximation algorithms. In this paper we give an informal introduction to the theory of NP-completeness and derive some fundamental results, in the hope of stimulating further use of this valuable analytical tool.

1. Introduction

After a wave of initial optimism, integer programming soon proved to be much harder than linear programming. As integer programming formulations were found for more and more discrete optimization problems, it also became obvious that such formulations yielded little computational benefit. To this day, general integer programming problems of more than miniature size remain computationally intractable.

For some specially structured problems, however, highly efficient algorithms have been developed. Network flow and matching provide well-known examples of problems that are *easy* in the sense that they are solvable by a *good* algorithm — a term coined by J. Edmonds [11] to indicate an algorithm whose running time is bounded by a polynomial function of problem size. This notion is not only theoretically convenient, but is also supported by overwhelming practical evidence that polynomial-time algorithms can indeed solve large problem instances very efficiently; the polynomial involved is usually of low degree. For example, in a network on v vertices a maximum flow can be determined in $O(v^3)$ time [10, 37, 14] and a maximum weight matching can be found in $O(v^3)$ time [18, 41].

It is commonly conjectured that no good algorithm exists for the general integer programming problem. A similar conjecture holds with respect to many other combinatorial problems that are *notorious for their computational intractibility* [33], such a graph coloring, set covering, traveling salesman and job shop scheduling problems. Typically, all optimization methods that have been proposed so far for these problems are of an enumerative nature. They involve some type of backtrack search in a tree whose depth is bounded by a polynomial function of problem size. In the worst case, those algorithms require superpolynomial (*e.g.*, exponential) time.

For the time being, we shall loosely denote the class of all problems solvable in polynomial time by $\mathcal{P}$ and the class of all problems solvable by polynomial-depth backtrack search by $\mathcal{NP}$. It is obvious that $\mathcal{P} \subset \mathcal{NP}$.

The battle against hard combinatorial problems dragged on until S.A. Cook [8] and R.M. Karp [34] showed the way to *peace with honor* [16]. They exhibited the existence within $\mathcal{NP}$ of a large class of so-called NP-complete problems [39] that are equivalent in the following sense:

none of them is known to belong to $\mathcal{P}$;

if one of them belongs to $\mathcal{P}$, then all problems in $\mathcal{NP}$ belong to $\mathcal{P}$, which would imply that $\mathcal{P} = \mathcal{NP}$.

NP-completeness of a problem is generally accepted as strong evidence against the existence of a good algorithm and consequently as a justification for the use of enumerative optimization methods such as branch-and-bound or of approximation algorithms. By way of examples, even restricted versions of all hard problems mentioned above are NP-complete.

NP-completeness theory has proved to be an extremely fruitful research area. The computational complexity of many types of combinatorial problems has been analyzed in detail. Under the assumption that $\mathcal{P} \neq \mathcal{NP}$, this analysis often reveals the existence of a sharp borderline between $\mathcal{P}$ and the class of NP-complete problems that is expressible in terms of natural problem parameters. A truly remarkable feature of the theory is the large proportion of time in which a given problem in $\mathcal{NP}$ can be shown to be either in $\mathcal{P}$ or NP-complete. Moreover, the two types of problems really have proved to be quite different in character. As mentioned, extremely large instances of problems in $\mathcal{P}$ are efficiently solvable, whereas only relatively small instances of NP-complete problems admit of solution by tedious enumerative procedures. Establishing NP-completeness of a problem provides important information on the quality of the algorithm that one can hope to find, which makes it easier to accept the computational burden of enumerative methods or to face the inevitability of a heuristic approach.

In this paper we shall not attempt to present an exhaustive survey of all NP-completeness results [34, 35, 24]. Instead, we shall examine some typical NP-complete problems, demonstrate some typical proof techniques and discuss some typical open problems [1, 53, 50]. We hope that as a result the reader will be stimulated to consider the computational complexity of his or her favourite combinatorial problem and to draw the algorithmic implications.

2. Concepts of complexity theory

A formal theory of NP-completeness would require the introduction of *Turing machines* [1] as theoretical computing devices. A *deterministic* Turing machine is a classical model for an ordinary computer, which is polynomially related to more realistic models such as the *random access machine* [1]. It can be designed to *recognize languages*; the input consists of a *string*, which is *accepted* by the machine if and only if it belongs to the language. A *nondeterministic* Turing machine is an artificial model, which can be thought of as a deterministic one that can create copies of itself corresponding to different state transitions whenever convenient. In this case, a string is accepted if and only if it is accepted by one of the deterministic copies. $\mathcal{P}$ and $\mathcal{NP}$ are now defined as the classes of languages recognizable in polynomial time by deterministic and nondeterministic Turing machines, respectively.

For the purposes of exposition, we will expound the theory in terms of *recognition problems*, which require a yes/no answer. A string then corresponds to a problem *instance* and a language to a problem *type* or, more exactly, to the set of all its *feasible* instances. The feasibility of an instance is usually equivalent to the existence of an associated *structure*, whose size is bounded by a polynomial in the size of the instance; for example, the instance may be a graph and the structure a Hamiltonian circuit [35]. A recognition problem is in $\mathcal{P}$ if, for any instance, one can determine its feasibility or infeasibility in polynomial time. It is in $\mathcal{NP}$ if, for any instance, one can determine in polynomial time whether a given structure affirms its feasibility.

Problem P' is said to be *reducible* to problem P (notation: $P' \propto P$) if for any instance of P' an instance of P can be constructed in polynomial time such that solving the instance of P will solve the instance of P' as well. Informally, the reducibility of P' to P implies that P' can be considered as a special case of P, so that P is at least as hard as P'.

P is called *NP-hard* if $P' \propto P$ for every $P' \in \mathcal{NP}$. In that case, P is at least as hard as any problem in $\mathcal{NP}$. P is called *NP-complete* if P is NP-hard and $P \in \mathcal{NP}$. Thus, the NP-complete problems are the most difficult problems in $\mathcal{NP}$.

A good algorithm for an NP-complete problem P could be used to solve all problems in $\mathcal{NP}$ in polynomial time, since for any instance of such a problem the construction of the corresponding instance of P and its solution can be both effected in polynomial time. Note the following two important observations.

(i) It is very unlikely that $\mathcal{P} = \mathcal{NP}$, since $\mathcal{NP}$ contains many notorious combinatorial problems, for which in spite of a considerable research effort no good algorithms have been found so far.

(ii) It is very unlikely that $P \in \mathcal{P}$ for any NP-complete P, since this would imply that $\mathcal{P} = \mathcal{NP}$ by the earlier argument.

The first NP-completeness result is due to Cook [8]. He designed a "master reduction" to prove that every problem in $\mathcal{NP}$ is reducible to the SATISFIABILITY problem. This is the problem of determining whether a boolean expression in conjunctive normal form assumes the value *true* for some assignment of truth values to the variables; for instance, the expression

$$(x_1) \wedge (\bar{x}_1 \vee x_2 \vee \bar{x}_3) \wedge (x_3) \tag{1}$$

is satisfied if $x_1 = x_2 = x_3 = true$. Given this result, one can establish NP-completeness of some $P \in \mathcal{NP}$ by specifying a reduction $P' \propto P$ with P' already known to be NP-complete: for every $P'' \in \mathcal{NP}$, $P'' \propto P'$ and $P' \propto P$ then imply that $P'' \propto P$ as well. In the following section we shall present several such proofs.

As far as *optimization problems* are concerned, we shall reformulate a minimization (maximization) problem by asking for the existence of a feasible solution with value at most (at least) equal to a given *threshold*. It should be noted that membership of $\mathcal{NP}$ for this recognition version does not immediately imply membership of $\mathcal{NP}$ for the original optimization problem as well. In particular, proposing a systematic search over a polynomial number of threshold values, guided by positive and negative answers to the existence question, is not a valid argument. This is because a nondeterministic Turing machine is only required to give *positive* answers in polynomial time. Indeed, no *complement* of any NP-complete problem is known to be in $\mathcal{NP}$!

As an obvious consequence of the above discussion, NP-completeness can only be proved with respect to a recognition problem. However, the corresponding optimization problem might be called NP-hard in the sense that the existence of a good algorithm for its solution would imply that $\mathcal{P} = \mathcal{NP}$.

So far, we have been purposefully vague about the specific encoding of problem instances. Suffice it to say that most reasonable encodings are polynomially related. One important exception with respect to the representation of positive integers will be dealt with in Section 3.5.

The classes $\mathcal{P}$ and $\mathcal{NP}$ are certainly not the only classes of interest to complexity theorists. There is, for instance, the class $\mathcal{P}$SPACE, which contains all languages recognizable in polynomial space. This class is the same for both deterministic and nondeterministic Turing machines. There is a notion of $\mathcal{P}$SPACE-completeness analogous to NP-completeness. The standard $\mathcal{P}$SPACE-complete problem is "quantified" SATISFIABILITY or QSATISFIABILITY [56, 1]. An instance of this problem results from the quantification of a boolean expression by both existential and universal quantifiers, *e.g.*

$$\forall x_1 \exists x_2 \forall x_3 [(x_1 \vee x_2) \wedge (\bar{x}_1 \vee \bar{x}_2 \wedge \bar{x}_3)].$$

The QSATISFIABILITY problem can be viewed as defining a game between two players: an "existential" player who tries to select values to make the expression *true* and a "universal" player who tries to defeat him. This insight has suggested a rich lode of simply-structured combinatorial games for which the problem of determining the outcome of optimal play is $\mathscr{P}$SPACE-complete [54]. One example of such a game is "generalized hex" [15].

Clearly $\mathscr{NP} \subset \mathscr{P}$SPACE. It has not been proved that $\mathscr{NP} \neq \mathscr{P}$SPACE. However, it seems reasonable to conjecture that this is the case and that $\mathscr{P}$SPACE-complete problems are more difficult than NP-complete ones.

We should also mention that there are problems which have been shown to be inherently more difficult than any problem in $\mathscr{P}$SPACE. For example, consider the "reachability" problem for vector addition systems: given a finite set of vectors with integer components, an initial vector u and a final vector v, is it possible to add vectors from the given set to u, with repetition allowed, so as to reach v, while always staying within the positive orthant? This problem has been shown to be decidable [51] but to require exponential space [46]. Some other combinatorial problems have been shown to require exponential space as well [56].

3. NP-Completeness results

In this section we shall establish some basic-NP-completeness results according to the scheme given in Fig. 1, and we shall mention similar results for related

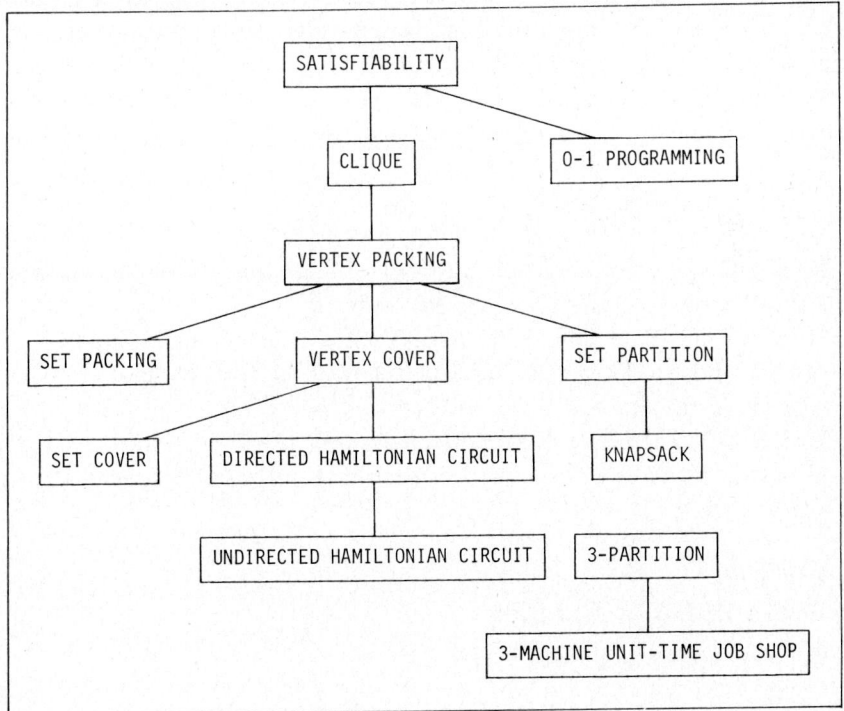

Fig. 1. Scheme of reductions.

problems. Our proofs will be sketchy; for instance, it will be left to the reader to verify the membership of $\mathcal{NP}$ for the problems considered and the polynomial-boundedness of the reductions presented.

3.1. *Satisfiability*

SATISFIABILITY: Given a *conjunctive normal form expression*, i.e. a conjunction of *clauses* $C_1, \ldots, C_s$, each of which is a disjunction of *literals* $x_1, \bar{x}_1, \ldots, x_t, \bar{x}_t$ where $x_1, \ldots, x_t$ are boolean variables and $\bar{x}_1, \ldots, \bar{x}_t$ denote their complements, is there a truth assignment to the variables such that the expression assumes the value *true*?

NP-completeness

It has already been mentioned that SATISFIABILITY was the first problem shown to be NP-complete. The proof of this key result is quite technical and beyond the scope of this paper; we refer to [8, 1]. We shall take (1) as an example of an instance of SATISFIABILITY to illustrate subsequent reductions.

Related results

Even the 3-SATISFIABILITY problem, i.e. SATISFIABILITY with at most three literals per clause, is NP-complete [8]. The 2-SATISFIABILITY problem, however, belongs to $\mathcal{P}$. Often, the borderline between easy and hard problems is crossed when a problem parameter increases from two to three. This phenomenon will be encountered on various occasions below, and is held by some to explain the division of mankind in two and not three sexes.

3.2. *Clique, vertex packing & vertex cover*

CLIQUE: Given an undirected graph $G = (V, E)$ and an integer k, does G have a set of at least k pairwise adjacent vertices?

VERTEX PACKING (INDEPENDENT SET): Given an undirected graph $G' = (V', E')$ and an integer k', does G' have a set of at least k' pairwise non-adjacent vertices?

VERTEX COVER: Given an undirected graph $G = (V, E)$ and an integer k, does G have a set of at most k vertices such that every edge is incident with at least one of them?

NP-completeness

SATISFIABILITY $\propto$ CLIQUE:

$V = \{(x, i) \mid x \text{ is a literal in clause } C_i\};$

$E = \{\{(x, i), (y, j)\} \mid x \neq \bar{y}, i \neq j\};$

$k = s.$

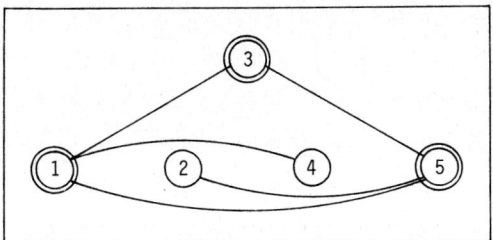

Fig. 2. Instance of CLIQUE for the example.

Cf. Fig. 2. We have created a vertex for each occurrence of a literal in a clause and an edge for each pair of literals that can be assigned the value *true* independently of each other. A clique of size k corresponds to s literals (one in each clause) that satisfy the expression and *vice versa* [8]. The NP-completeness of CLIQUE now follows from (i) its membership of $\mathcal{NP}$, (ii) the polynomial-boundedness of the reduction, and (iii) the NP-completeness of SATISFIABILITY.

CLIQUE $\propto$ VERTEX PACKING:

$V' = V$;
$E' = \{\{i, j\} \mid i \neq j, \{i, j\} \notin E\}$;
$k' = k$.

Cf. Fig. 3. A set of vertices is independent in G' if and only if it is a clique in the complementary graph G. This relation between the two problems belongs to folklore.

VERTEX PACKING $\propto$ VERTEX COVER:

$V = V'$;
$E = E'$;
$k = |V'| - k'$.

Cf. Fig. 4. It is well known that a set of vertices covers all edges if and only if its complement is independent.

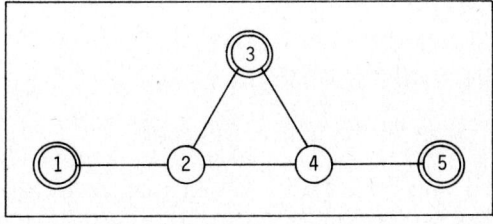

Fig. 3. Instance of VERTEX PACKING for the example.

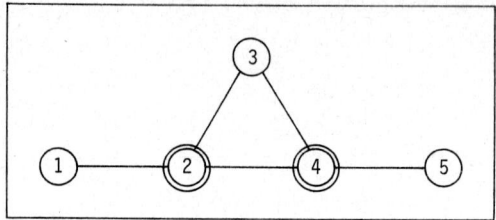

Fig. 4. Instance of VERTEX COVER for the example.

Related results

Given the above results, it is not surprising (though less easy to prove) that the problems of determining whether the vertex set of a graph can be covered by at most k cliques or, after complementation, by at most k independent sets are NP-complete [34]. These problems are known as CLIQUE COVER and GRAPH COLORABILITY respectively. In fact, it is already an NP-complete problem to determine if a planar graph with vertex degree at most 4 is 3-colorable [27], whereas 2-colorability is equivalent to bipartiteness and can be checked in polynomial time.

3.3. *Set packing, set cover & set partition*

SET PACKING: Given a finite set S, a finite family $\mathcal{S}$ of subsets of S and an integer l, does $\mathcal{S}$ include a subfamily $\mathcal{S}'$ of at least l pairwise disjoint sets?

SET COVER: Given a finite set S, a finite family $\mathcal{S}$ of subsets of S and an integer l, does $\mathcal{S}$ include a subfamily $\mathcal{S}'$ of at most l sets such that $\bigcup_{S' \in \mathcal{S}'} S' = S$?

SET PARTITION (EXACT COVER): Given a finite set S and a finite family $\mathcal{S}$ of subsets of S, does $\mathcal{S}$ include a subfamily $\mathcal{S}'$ of pairwise disjoint sets such that $\bigcup_{S' \in \mathcal{S}'} S' = S$?

NP-completeness

VERTEX PACKING $\propto$ SET PACKING:

$S = E'$;
$\mathcal{S} = \{\{\{i, j\} \mid \{i, j\} \in E'\} \mid i \in V'\}$;
$l = k'$.

VERTEX COVER $\propto$ SET COVER:

delete the primes in the above reduction.

VERTEX PACKING and VERTEX COVER are easily recognized as special cases of SET PACKING and SET COVER respectively, and these reductions require no further comment.

VERTEX PACKING ∝ SET PARTITION:

$$S = E' \cup \{1, \ldots, k'\};$$
$$\mathcal{S} = \{S_{ih} \mid i \in V', h = 1, \ldots, k'\} \cup \{S_{\{i,j\}} \mid \{i, j\} \in E'\},$$

where

$$S_{ih} = \{\{i', j\} \mid \{i', j\} \in E, i' = i\} \cup \{h\},$$
$$S_{\{i,j\}} = \{\{i, j\}\}.$$

Cf. Fig. 5. Suppose that G' has an independent set $U' \subset V'$ of size k', say, $U' = \{v_1, \ldots, v_{k'}\}$. Then the sets $S_{v_1 1}, \ldots, S_{v_{k'} k'}$ are pairwise disjoint, and the elements of S not contained in any of them belong to E'. It follows that a partition of S is given by

$$\{S_{v_1 1}, \ldots, S_{v_{k'} k'}\} \cup \{S_{\{i,j\}} \mid \{i, j\} \in E', i \notin U', j \notin U'\}.$$

Conversely, suppose that there exists a partition $\mathcal{S}'$ of S. Then $\mathcal{S}'$ contains k' pairwise disjoint sets $S_{v_1 1}, \ldots, S_{v_{k'} k'}$, and the vertices $v_1, \ldots, v_{k'}$ clearly constitute an independent set of size k' in G'.

This reduction simplifies the NP-completeness proof given in [34].

Related results

Even the EXACT 3-COVER problem, where all subsets in $\mathcal{S}$ are constrained to be of size 3, is NP-complete, since it is an obvious generalization of the 3-DIMENSIONAL MATCHING problem, proved NP-complete in [34]. An EXACT 2-COVER corresponds to a perfect matching in a graph, which can be found in polynomial time. The existence of good matching algorithms proves that EDGE PACKING and EDGE COVER problems are members of $\mathcal{P}$.

S	S_{11}	S_{21}	S_{31}	S_{41}	S_{51}	S_{12}	S_{22}	S_{32}	S_{42}	S_{52}	S_{13}	S_{23}	S_{33}	S_{43}	S_{53}	$S_{\{1,2\}}$	$S_{\{2,3\}}$	$S_{\{2,4\}}$	$S_{\{3,4\}}$	$S_{\{4,5\}}$
{1,2}	⊙					•	•				•	•				•				
{2,3}		•	•				•	⊙				•	•				•			
{2,4}		•		•			•		•			•		•				⊙		
{3,4}			•	•				•	⊙				•	•					•	
{4,5}				•	•				•	•				•	⊙					•
1	⊙	•	•	•	•															
2						•	•	⊙	•	•										
3											•	•	•	•	⊙					

Fig. 5. Instance of SET PARTITION for the example.

3.4. Directed & undirected hamiltonian circuit

DIRECTED HAMILTONIAN CIRCUIT: Given a directed graph $H = (W, A)$, does H have a directed cycle passing through each vertex exactly once?

UNDIRECTED HAMILTONIAN CIRCUIT: Given an undirected graph $G = (V, E)$, does G have a cycle passing through each vertex exactly once?

NP-completeness

VERTEX COVER $\propto$ DIRECTED HAMILTONIAN CIRCUIT:

$$W = \{(i, j), \{i, j\}, (j, i) \mid \{i, j\} \in E\} \cup \{1, \ldots, k\};$$
$$A = \{((i, j), \{i, j\}), (\{i, j\}, (i, j)), ((j, i), \{i, j\}), (\{i, j\}, (j, i)) \mid \{i, j\} \in E\}$$
$$\cup \{((h, i), (i, j)) \mid \{h, i\}, \{i, j\} \in E, h \neq j\}$$
$$\cup \{((i, j), h), (h, (i, j)), ((j, i), h), (h, (j, i)) \mid \{i, j\} \in E, h = 1, \ldots, k\}.$$

Cf. Fig. 6. For each edge $\{i, j\}$ in G we have created a configuration in H consisting of three vertices $(i, j), \{i, j\}, (j, i)$ and four arcs, as shown in the figure. The configurations are linked by arcs from (h, i) to (i, j) for $h \neq j$. Further, we have added k vertices $1, \ldots, k$ and all arcs between them and the vertices of type (i, j).

Suppose that G has a vertex cover $U \subset V$ of size k, say, $U = \{v_1 \ldots, v_k\}$. The edge set E can be written as

$$E = \{\{v_h, w_{h1}\}, \ldots, \{v_h, w_{hl_h}\} \mid h = 1, \ldots, k\}$$

and it is easily checked that a hamiltonian circuit in H is given by

$$(1, (v_1, w_{11}), \{v_1, w_{11}\}, (w_{11}, v_1), \ldots, (v_1, w_{1l_1}), \{v_1, w_{1l_1}\}, (w_{1l_1}, v_1), \ldots,$$
$$k, (v_k, w_{k1}), \{v_k, w_{k1}\}, (w_{k1}, v_k), \ldots, (v_k, w_{kl_k}), \{v_k, w_{kl_k}\}, (w_{kl_k}, v_k), 1).$$

Conversely, suppose that H has a hamiltonian circuit. By deletion of all arcs incident with vertices $1, \ldots, k$, the circuit is decomposed into k paths. A path starting at (i, j) for $\{i, j\} \in E$ has to go on to visit $\{i, j\}$ and (j, i); then it ends or goes on to visit $(i, j'), \{i, j'\}, (j', i)$ for some $\{i, j'\} \in E$, etc. Thus, this path corresponds to a specific vertex $i \in V$, covering edges $\{i, j\}, \{i, j'\}$, etc. Since the circuit passes through each $\{i, j\}$ exactly once, each edge $\{i, j\} \in E$ is covered by one of k specific vertices, which therefore constitute a vertex cover of size k in G.

The above reduction is a modification of the original construction due to E.L. Lawler [34], based on ideas of M. Fürer [55] and P. van Emde Boas.

DIRECTED HAMILTONIAN CIRCUIT $\propto$ UNDIRECTED HAMILTONIAN CIRCUIT:

$$V = \{(i, in), (i, mid), (i, out) \mid i \in W\};$$
$$E = \{\{(i, in), (i, mid)\}, \{(i, mid), (i, out)\} \mid i \in W\}$$
$$\cup \{\{(i, out), (j, in)\} \mid (i, j) \in A\}.$$

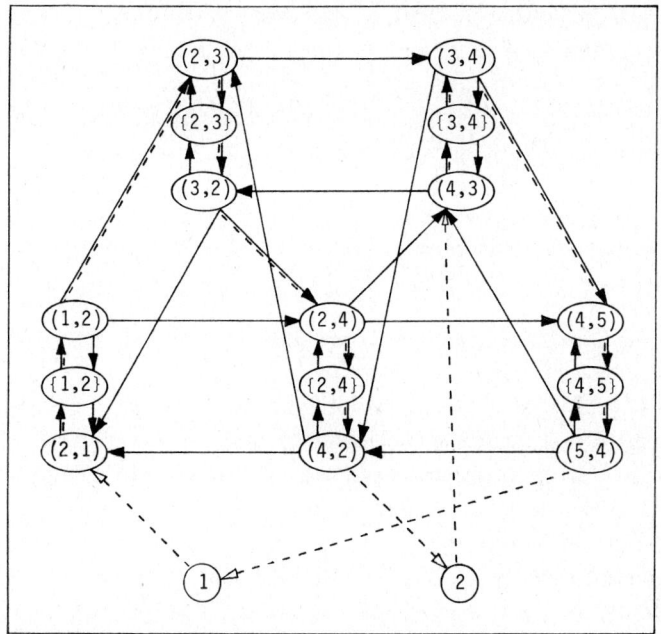

Fig. 6. Instance of DIRECTED HAMILTONIAN CIRCUIT for the example. Not all arcs incident with vertices $1, \ldots, k$ have been drawn.

The one-one correspondence between undirected hamiltonian circuits in G and directed hamiltonian circuits in H is evident. This reduction is due to R.E. Tarjan [34].

Related results

The above results have been strengthened in various ways. For instance, the UNDIRECTED HAMILTONIAN CIRCUIT problem remains NP-complete if G is planar, triply-connected and regular of degree 3 [28] or if G is bipartite [40]. The latter result is a simple extension of the last reduction given above and we recommend it as an exercise.

NP-hardness of the (general) TRAVELING SALESMAN problem is another obvious consequence. Intricate NP-hardness proofs for the EUCLIDEAN TRAVELING SALESMAN problem can be found in [19, 48]. It is well known that TRAVELING SALESMAN is a special case of the problem of finding a maximum weight independent set in the intersection of three matroids. Thus, the 3-MATROID INTERSECTION problem is NP-hard, whereas 2-MATROID INTERSECTION problems, such as finding an optimal linear assignment or spanning arborescence, can be solved in polynomial time [41].

The TRAVELING SALESMAN problem serves as a prototype for a whole class of routing problems where, given a mixed graph consisting of a set V of vertices, a set E of (undirected) edges and a set A of (directed) arcs, a salesman has to find a minimum-weight tour passing through subsets $V' \subset V$, $E' \subset E$ and $A' \subset A$. If $V' = \emptyset$, $E' = E$ and $A' = A$, we have the CHINESE POSTMAN problem, which can be solved in polynomial time in the undirected or directed case ($A = \emptyset$ or $E = \emptyset$) [12, 13], but is NP-hard in the mixed case [47]. For the case that only $V' = \emptyset$, NP-hardness has been established for the UNDIRECTED and DIRECTED RURAL POSTMAN problems ($A = \emptyset$ and $E = \emptyset$ respectively) [43] and for the STACKER-CRANE problem ($E' = \emptyset$, $A' = A$) [17].

3.5. 0–1 Programming, knapsack & 3-partition

0–1 PROGRAMMING: Given an integer matrix A and an integer vector b, does there exist a 0–1 vector x such that $Ax \geq b$?

KNAPSACK: Given positive integers $a_1, \ldots, a_t, b$, does there exist a subset $T \subset \{1, \ldots, t\}$ such that $\sum_{j \in T} a_j = b$?

3-PARTITION: Given positive integers $a_1, \ldots, a_{3t}, b$, do there exist t pairwise disjoint 3–element subsets $S_i \subset \{1, \ldots, 3t\}$ such that $\sum_{j \in S_i} a_j = b$ ($i = 1, \ldots, t$)?

NP-completeness

SATISFIABILITY $\propto$ 0–1 PROGRAMMING:

$$a_{ij} = \begin{cases} 1 & \text{if } x_j \text{ is a literal in clause } C_i, \\ -1 & \text{if } \bar{x}_j \text{ is a literal in clause } C_i, \\ 0 & \text{otherwise} \end{cases} \quad (i = 1, \ldots, s, \; j = 1, \ldots, t);$$

$$b_i = 1 - |\{j \mid \bar{x}_j \text{ is a literal in clause } C_i\}| \quad (i = 1, \ldots, s).$$

Cf. Fig. 7 and [34].

$$\begin{array}{r} x_1 \geq 1 \\ -x_1 + x_2 - x_3 \geq -1 \\ x_3 \geq 1 \end{array}$$

Fig. 7. Instance of 0–1 PROGRAMMING for the example.

SET PARTITION $\propto$ KNAPSACK:
Given $S = \{e_1, \ldots, e_s\}$ and $\mathcal{S} = \{S_1, \ldots, S_t\}$, we define

$$\varepsilon_{ij} = \begin{cases} 1 & \text{if } e_i \in S_j \\ 0 & \text{if } e_i \notin S_j \end{cases} \quad (i = 1, \ldots, s, \; j = 1, \ldots, t),$$

$u = t + 1$,

and specify the reduction by

$$a_j = \sum_{i=1}^{s} \varepsilon_{ij} u^{i-1} \quad (j = 1, \ldots, t);$$

$$b = (u^s - 1)/t.$$

Cf. Fig. 8. The one-one correspondence between solutions to KNAPSACK and SET PARTITION is easily verified [34].

Given this result, the reader should have little difficulty in establishing NP-completeness for the PARTITION problem, *i.e.* KNAPSACK with $\sum_{j=1}^{t} a_j = 2b$.

3-PARTITION has been proved NP-complete through a complicated sequence of reductions, which can be found in [20].

Binary vs. unary encoding

KNAPSACK was the first example of an NP-complete problem involving numerical data. The size of a problem instance is $O(t \log b)$ in the standard *binary* encoding and $O(tb)$ if a *unary* encoding is allowed. Readers will have noticed that the reduction SET PARTITION $\propto$ KNAPSACK is polynomial-bounded only with respect to a binary encoding. Indeed, KNAPSACK can be solved by dynamic programming in $O(tb)$ time [3], which might be called a *pseudopolynomial* algorithm in the sense that it is polynomial-bounded only with respect to a unary encoding. Thus, the *binary NP-completeness* of KNAPSACK and its *unary membership of* $\mathcal{P}$ are perfectly compatible results, although it tends to make us think of KNAPSACK as less hard than other NP-complete problems.

3-PARTITION was the first example of a problem involving numerical data that remains NP-complete even if we measure the problem size by using the actual numbers involved instead of their logarithms. This *strong* or *unary NP-completeness* of 3-PARTITION indicates that already the existence of a pseudopolynomial algorithm for its solution would imply that $\mathcal{P} = \mathcal{NP}$ [25].

ε_{ij} $\hookrightarrow$	①	2	3	4	5	6	7	⑧	9	10	11	12	13	14	⑮	16	17	⑱	19	20	a_j	b
1	①	1	0	0	0	1	1	0	0	0	1	1	0	0	0	1	0	0	0	0	$\varepsilon_{1j} \cdot u^0 +$	$u^0 +$
2	0	1	1	0	0	0	1	①	0	0	0	1	1	0	0	0	1	0	0	0	$\varepsilon_{2j} \cdot u^1 +$	$u^1 +$
3	0	1	0	1	0	0	1	0	1	0	0	1	0	1	0	0	0	①	0	0	$\varepsilon_{3j} \cdot u^2 +$	$u^2 +$
4	0	0	1	1	0	0	0	①	1	0	0	0	1	1	0	0	0	0	1	0	$\varepsilon_{4j} \cdot u^3 +$	$u^3 +$
5	0	0	0	1	1	0	0	0	1	1	0	0	0	1	①	0	0	0	0	1	$\varepsilon_{5j} \cdot u^4 +$	$u^4 +$
6	①	1	1	1	1	0	0	0	0	0	0	0	0	0	0	0	0	0	0	0	$\varepsilon_{6j} \cdot u^5 +$	$u^5 +$
7	0	0	0	0	0	1	1	①	1	1	0	0	0	0	0	0	0	0	0	0	$\varepsilon_{7j} \cdot u^6 +$	$u^6 +$
8	0	0	0	0	0	0	0	0	0	1	1	1	1	①	0	0	0	0	0	0	$\varepsilon_{8j} \cdot u^7$	u^7

Fig. 8. Instance of KNAPSACK for the example, where $s = 8$, $t = 20$, $u = 21$.

Quite often, a binary NP-completeness proof involving KNAPSACK or PARTITION can be converted to a unary NP-completeness proof involving 3-PARTITION in a straightforward manner. Occasionally, however, the polynomial-boundedness of a reduction depends essentially on allowing a unary encoding for 3-PARTITION. An example of such a reduction is given in the next section.

3.6. *3-Machine unit-time job shop*

3-MACHINE UNIT-TIME JOB SHOP: Given 3 machines M_1, M_2, M_3 each of which can process at most one job at a time, n jobs $J_1, \ldots, J_n$ where $J_j (j = 1, \ldots, n)$ consists of a chain of unit-time operations, the h-th of which has to be processed on machine μ_{jh} with $\mu_{jh} \neq \mu_{j,h-1}$ for $h > 1$, and an integer k, does there exist a schedule with length at most k?

NP-completeness

3-PARTITION $\propto$ 3-MACHINE UNIT-TIME JOB SHOP:

$$n = 3t + 2;$$
$$\mu_j = (M_1, M_3, [M_1, M_2]^{a_j}, M_3) \quad (j = 1, \ldots, 3t);$$
$$\mu_{n-1} = ([M_2, M_3, M_2, M_1, M_2, M_1, [M_2, M_3]^b, M_1, M_2, M_1]^t);$$
$$\mu_n = ([M_3, M_2, M_3, M_2, M_1, M_2, [M_3, M_1]^b, M_2, M_1, M_2]^t);$$
$$k = (2b + 9)t;$$

where $[s]^h = s, [s]^{h-1}$ for $h > 1$ and $[s]^1 = s$.

Note that both J_{n-1} and J_n consist of a chain of operations of length equal to the threshold k. We may assume the h-th operations of these chains to be completed at time h, since otherwise the schedule length would exceed k. This leaves a pattern of idle machines for the other jobs that can be described as

$$([[M_1]^3, [M_3]^3, [M_1, M_2]^b, [M_3]^3]^t)$$

(*cf.* Fig. 9). We will show that this pattern can be filled properly if and only if 3-PARTITION has a solution.

Suppose that 3-PARTITION has a solution $(S_1, \ldots, S_t)$. In this case, processing J_j with $j \in S_i$ entirely within the interval $[(2b+9)(i-1), (2b+9)i]$ $(j = 1, \ldots, 3t, i = 1, \ldots, t)$ yields a schedule with length k.

Conversely, suppose that there exists a schedule with length k. We will prove that in such a schedule exactly three jobs are started in $[0, 2b+9]$ and that they are completed in this interval as well; clearly, these jobs indicate a 3-element subset S_1 with $\sum_{j \in S_1} a_j = b$. One easily proves by induction that S_i is similarly defined by the jobs started and completed in $[(2b+9)(i-1), (2b+9)i]$ $(1 < i \leq t)$.

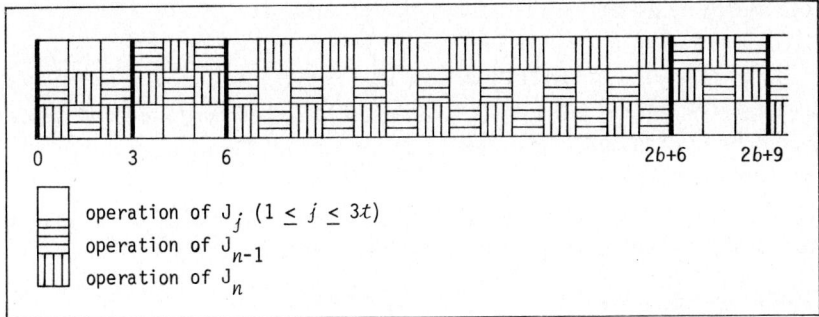

Fig. 9. First part of 3-MACHINE UNIT-TIME JOB SHOP schedule corresponding to an instance of 3-PARTITION with $b = 7$.

If J_j starts in $[0, 2b+9]$, its subchain of operations completed in that interval is one of four types:

type 1: (M_1);
type 2: $(M_1, M_3, [M_1, M_2]^h)$ $\quad (0 \leq h \leq a_j)$;
type 3: $(M_1, M_3, [M_1, M_2]^h, M_1)$ $\quad (0 \leq h < a_j)$;
type 4: $(M_1, M_3, [M_1, M_2]^{a_j}, M_3)$.

Let x_i denote the number of subchains of type i and y_i the number of operations on M_2 in subchains of type i. We have to prove that $x_1 = x_2 = x_3 = 0$, $x_4 = 3$. Observing that a schedule of length k contains no idle unit-time periods, we have

(1) $\quad x_1 + x_2 + y_2 + 2x_3 + y_3 + x_4 + y_4 = b + 3$;
(2) $\quad y_2 + y_3 + y_4 = b$;
(3) $\quad x_2 + x_3 + 2x_4 = 6$.

Subtracting (1) from the sum of (2) and (3), we obtain $-x_1 - x_3 + x_4 = 3$ and therefore $x_4 \geq 3$. Also, (3) implies that $x_4 \leq 3$. It follows that $x_4 = 3$, and $x_1 = x_2 = x_3 = 0$.

Related results

The complexity of the 2-MACHINE UNIT-TIME JOB SHOP problem is unknown; to introduce a competitive element we shall be happy to award a chocolate windmill to the first person establishing membership of $\mathcal{P}$ or NP-completeness for this problem. If the processing times of the operations are allowed to be equal to 1 or 2, the 2-machine problem can be proved NP-complete by a reduction similar to (but simpler than) the above one; this improves upon related results given in [26, 45]. If each job has at most two operations, the 2-machine problem belongs to $\mathcal{P}$ even for arbitrary processing times [32].

These results form but a small fraction of the extensive complexity analysis carried out for scheduling problems. We refer to [57, 20, 6, 26, 45, 44] for further details and to [29] for a concise survey of the field.

4. Concluding remarks

We hope that the preceding section has conveyed some of the flavor and elegance of NP-completeness results. In only a few years an impressive amount of results has been obtained. Nevertheless, there are still plenty of *open problems*, for which neither a polynomial algorithm nor an NP-completeness proof is available. We shall mention four famous ones, on whose complexity status little or no progress has been made so far.

(a) Graph isomorphism

This is the problem of determining whether there exists a one-one mapping between the vertex sets of two graphs which preserves the adjacency relation. The essential nature of the problem does not change if we restrict our attention to graphs of certain types such as bipartite or regular ones; these problems are polynomially equivalent to the general case [4]. The status of the problem is totally unknown and we do not dare to guess the final outcome.

(b) Matroid parity

This problem is interesting because it generalizes both the matroid intersection problem and the nonbipartite matching problem [41]. Despite serious investigation, its status is far from clear. A special case of the matroid parity problem is as follows. Given a connected graph G with an even number of edges, arbitrarily paired (*i.e.*, each edge e has a uniquely defined *mate* $\bar{e}$), does G have a spanning tree T with the property that if an edge is contained in T, then its mate is in T as well? An NP-completeness proof for this special case would, of course, resolve the question for the general problem. On the other hand, a polynomial-time algorithm for this special case would probably suggest a similar procedure for the general problem.

(c) 3-Machine unit-time parallel shop

This problem involves the scheduling of unit-time jobs on three identical parallel machines subject to precedence constraints between the jobs, so as to meet a common deadline of the jobs. For a variable number of machines, the problem is NP-complete [57, 44]; the special case of tree-type precedence constraints can be solved in polynomial time [30]. The 2-machine problem belongs to $\mathcal{P}$ [7], even if for each job a time-interval is specified in which it has to be processed [23]. The 3-machine problem has remained open in spite of vigorous attacks. In this case we would be willing to extrapolate on the magic quality of three-ness and conjecture NP-completeness.

(d) Linear programming

This is perhaps the most vexing open problem. The simplex method performs very well in practice and usually requires time linear in the number of constraints.

On certain weird polytopes, however, it takes exponential time [38]. Fortunately, in this case there is circumstantial evidence against NP-completeness. Thanks to duality theory, determining the existence or nonexistence of a feasible solution are equally hard problems, and NP-completeness of LINEAR PROGRAMMING would therefore imply NP-completeness for the complements of all other NP-complete problems as well. However, as mentioned, it is not even known whether the complement of any NP-complete problem belongs to $\mathcal{NP}$. In addition to the above rather technical argument, it seems highly unlikely that all NP-complete problems would allow a polynomial-bounded linear programming formulation.

Interpretation of NP-completeness results as more or less definite proofs of computational intractability has stimulated the design and analysis of fast *approximation algorithms*.

With respect to the *worst-case analysis* of such algorithms, a wide variety of outcomes is possible. We give the following examples.

(1) For the optimization version of the KNAPSACK problem, a solution within an arbitrary percentage ε from the optimum can be found in time polynomial in t and $1/\varepsilon$ [31, 42].

(2) For the EUCLIDEAN TRAVELING SALESMAN problem, a solution within 50% from the optimum can be found in polynomial time [5, 9].

(3) For the GRAPH COLORABILITY problem, a solution within 100% from the optimum cannot be found in polynomial time unless $\mathcal{P} = \mathcal{NP}$ [21].

(4) For the general TRAVELING SALESMAN problem, a solution within any fixed percentage from the optimum cannot be found in polynomial time unless $\mathcal{P} = \mathcal{NP}$ [52].

We refer to [22] for a survey of this area. Impressive advances have been made and more can be expected in the near future.

The above approach to *performance guarantees* may be accused of being overly pessimistic — cf. the simplex method with its exponential worst-case behavior! The *probabilistic analysis* of average or "almost everywhere" behavior, however, requires the specification of a probability distribution over the set of all problem instances. For some problems, a natural distribution function is available and some intriguing results have been derived [36], although technically this approach seems to be very demanding.

The worst-case analysis of approximation algorithms shows that there are significant differences in complexity within the class of NP-complete problems. These problems might be classifiable according to the best possible polynomial-time performance guarantee that one can get [2,49]. Another refinement of the complexity measure may be based on the way in which numerical problem data are encoded, *i.e.* on the distinction between binary and unary encoding mentioned in Section 3.5. Several other ways of measuring problem size could be devised and each of them could be subjected to a complexity analysis, producing new information on the best type of algorithm that is likely to exist.

The concluding remarks above were intended to confirm to the reader that the field of computational complexity is still very much alive. In the first place, however, the theory of NP-completeness has yielded highly useful tools for the analysis of combinatorial problems that deserve to find acceptance in a wide circle of researchers and practitioners.

Acknowledgments

We have received valuable comments from several participants to "Discrete Optimization 1977" in Vancouver and are particularly grateful for various constructive suggestions by E.L. Lawler.

References

[1] A.V. Aho, J.E. Hopcroft and J.D. Ullman, The Design and Analysis of Computer Algorithms (Addison-Wesley, Reading, MA, 1974).
[2] G. Ausiello, A. D'Atri, On the structure of combinatorial problems and structure preserving reductions, in: A. Salomaa and M. Steinby, eds., Automata, Languages and Programming, Lecture Notes in Computer Science 52 (Springer, Berlin, 1977) 45–60.
[3] R.E. Bellman and S.E. Dreyfus, Applied Dynamic Programming (Princeton University Press, Princeton, NJ, 1962).
[4] K.S. Booth, Problems polynomially equivalent to graph isomorphism, in: J.F. Traub, ed., Algorithms and Complexity: New Directions and Recent Results (Academic Press, New York, 1976) 435.
[5] N. Christofides, Worst-case analysis of a new heuristic for the travelling salesman problem. Math. Programming, to appear.
[6] E.G. Coffman, Jr., ed., Computer and Job-Shop Scheduling Theory (Wiley, New York, 1976).
[7] E.G. Coffman, Jr. and R.L. Graham, Optimal scheduling for two-processor systems, Acta Informat. 1 (1972) 200–213.
[8] S.A. Cook, The complexity of theorem-proving procedures, Proc. 3rd Annual ACM Symp. Theory of Computing (1971) 151–158.
[9] G. Cornuejols and G.L. Nemhauser, Tight bounds for Christofides' traveling salesman heuristic, Math. Programming 14 (1978) 116–121.
[10] E.A. Dinic, Algorithms for solution of a problem of maximum flow in a network with power estimation, Soviet Math. Dokl. 11 (1970) 1277–1280.
[11] J. Edmonds, Paths, trees and flowers, Canad. J. Math. 17 (1965) 449–467.
[12] J. Edmonds, The Chinese postman's problem, Operations Res. 13 Suppl. 1 (1965) B73.
[13] J. Edmonds and E.L. Johnson, Matching, Euler tours and the Chinese postman, Math. Programming 5 (1973) 88–124.
[14] S. Even, The max flow algorithm of Dinic and Karzanov: an exposition, Department of Computer Science, Technion, Haifa (1976).
[15] S. Even and R.E. Tarjan, A combinatorial problem which is complete in polynomial space, J. Assoc. Comput. Mach. 23 (1976) 710–719.
[16] M.L. Fisher, Private communication (1976).
[17] G.N. Frederickson, M.S. Hecht and C.E. Kim, Approximation algorithms for some routing problems, Proc. 17th Annual IEEE Symp. Foundations of Computer Science (1976) 216–227.
[18] H.N. Gabow, An efficient implementation of Edmonds' algorithm for maximum matching on graphs, J. Assoc. Comput. Mach. 23 (1976) 221–234.
[19] M.R. Garey, R.L. Graham and D.S. Johnson, Some NP-complete geometric problems, Proc. 8th Annual ACM Symp. Theory of Computing (1976) 10–22.

[20] M.R. Garey and D.S. Johnson, Complexity results for multiprocessor scheduling under resource constraints, SIAM J. Comput. 4 (1975) 397–411.
[21] M.R. Garey and D.S. Johnson, The complexity of near-optimal graph coloring, J. Assoc. Comput. Mach. 23 (1976) 43–49.
[22] M.R. Garey and D.S. Johnson, Approximation algorithms for combinatorial problems: an annotated bibliography, in: J.F. Traub, ed., Algorithms and Complexity: New Directions and Recent Results (Academic Press, New York, 1976) 41–52.
[23] M.R. Garey and D.S. Johnson, Two-processor scheduling with start-times and deadlines, SIAM J. Comput. 6 (1977) 416–426.
[24] M.R. Garey and D.S. Johnson, Computers and Intractability: a Guide to the Theory of NP-Completeness (Freeman, San Francisco, to appear).
[25] M.R. Garey and D.S. Johnson, "Strong" NP-completeness results: motivation, examples and implications, J. Assoc. Comput. Mach.,
[26] M.R. Garey, D.S. Johnson and R. Sethi, The complexity of flowshop and jobshop scheduling, Math. Operations Res. 1 (1976) 117–129.
[27] M.R. Garey, D.S. Johnson and L. Stockmeyer, Some simplified NP-complete graph problems, Theoret. Comput. Sci. 1 (1976) 237–267.
[28] M.R. Garey, D.S. Johnson and R.E. Tarjan, The planar Hamiltonian circuit problem is NP-complete, Siam J. Comput. 5 (1976) 704–714.
[29] R.L. Graham, E.L. Lawler, J.K. Lenstra and A.H.G. Rinnooy Kan, Optimization and approximation in deterministic sequencing and scheduling: a survey, Ann. Discrete Math. 5, next volume.
[30] T.C. Hu, Parallel sequencing and assembly line problems, Operations Res. 9 (1961) 841–848.
[31] O.H. Ibarra and C.E. Kim, Fast approximation algorithms for the knapsack and sum of subset problems, J. Assoc. Comput. Mach. 22 (1975) 463–468.
[32] J.R. Jackson, An extension of Johnson's results on job lot scheduling, Naval Res. Logist. Quart. 3 (1956) 201–203.
[33] D.S. Johnson, Near-optimal bin packing algorithms. Report MAC TR-109, Massachusetts Institute of Technology, Cambridge, MA (1973).
[34] R.M. Karp, Reducibility among combinatorial problems, in: R.E. Miller and J.W. Thatcher, eds., Complexity of Computer Computations (Plenum Press, New York, 1972) 85–103.
[35] R.M. Karp, On the computational complexity of combinatorial problems, Networks 5 (1975) 45–68.
[36] R.M. Karp, The probabilistic analysis of some combinatorial search algorithms, in: J.F. Traub, ed., Algorithms and Complexity: New Directions and Recent Results (Academic Press, New York, 1976) 1–19.
[37] A.V. Karzanov, Determining the maximal flow in a network by the method of preflows, Soviet Math. Dokl. 15 (1974) 434–437.
[38] V. Klee and G.J. Minty, How good is the simplex algorithm?, in: O. Shisha, ed., Inequalities III (Academic Press, New York, 1972) 159–175.
[39] D.E. Knuth, A terminological proposal, SIGACT News 6. 1 (1974) 12–18.
[40] M.S. Krishnamoorthy, An NP-hard problem in bipartite graphs, SIGACT News 7. 1 (1975) 26.
[41] E.L. Lawler, Combinatorial Optimization: Networks and Matroids (Holt, Rinehart and Winston, New York, 1976).
[42] E.L. Lawler, Fast approximation algorithms for knapsack problems, Math. Operations Res., to appear.
[43] J.K. Lenstra and A.H.G. Rinnooy Kan, On general routing problems, Networks 6 (1976) 273–280.
[44] J.K. Lenstra and A.H.G. Rinnooy Kan, Complexity of scheduling under precedence constraints, Operations Res. 26 (1978) 22–35.
[45] J.K. Lenstra, A.H.G. Rinnooy Kan and P. Brucker, Complexity of machine scheduling problems, Ann. Discrete Math. 1 (1977) 343–362.
[46] R. Lipton, The reachability problem requires exponential space. Theoret. Comput. Sci. to appear.
[47] C.H. Papadimitriou, On the complexity of edge traversing, J. Assoc. Comput. Math. 23 (1976) 544–554.

[48] C.H. Papadimitriou, The Euclidean traveling salesman problem is NP-complete, Theoret. Comput. Sci. 4 (1977) 237–244.
[49] A. Paz and S. Moran, Non-deterministic polynomial optimization problems and their approximation: abridged version, in: A. Salomaa and M. Steinby, eds., Automata, Languages and Programming, Lecture Notes in Computer Science 52 (Springer, Berlin, 1977) 370–379.
[50] E.M. Reingold, J. Nievergelt and N. Deo, Combinatorial Algorithms: Theory and Practice (Prentice-Hall, Englewood Cliffs, NJ, 1977).
[51] G.S. Sacerdote and R.L. Tenney, The decidability of the reachability problem for vector addition systems, Proc. 9th Annual ACM Symp. Theory of Computing (1977) 61–76.
[52] S. Sahni and T. Gonzalez, P-complete approximation problems, J. Assoc. Comput. Mach. 23 (1976) 555–565.
[53] J.E. Savage, The Complexity of Computing (Wiley, New York, 1976).
[54] T.J. Schaefer, Complexity of decision problems based on finite two-person perfect-information games, Proc. 8th Annual ACM Symp. Theory of Computing (1976) 41–59.
[55] P. Schuster, Probleme, die zum Erfüllungsproblem der Aussagenlogik polynomial äquivalent sind, in: E. Specker and V. Strassen, eds., Komplexität von Entscheidungsproblemen: ein Seminar, Lecture Notes in Computer Science 43 (Springer, Berlin, 1976) 36–48.
[56] L.J. Stockmeyer and A.R. Meyer, Word problems requiring exponential time: preliminary report, Proc. 5th Annual ACM Symp. Theory of Computing (1973) 1–9.
[57] J.D. Ullman, NP-complete scheduling problems, J. Comput. System Sci. 10 (1975) 384–393.

GRAPH THEORY AND INTEGER PROGRAMMING

L. LOVÁSZ

József Attila University, Szeged, Hungary

It has become clear that integer programming and combinatorics are tied together much more closely than thought before. A very large part of combinatorics deals, or can be formulated as to deal, with optimization problems in discrete structures. Generally, the constraints and the objective function are linear forms of certain variables, which are restricted to integers or — mostly — to 0 and 1. Thus the combinatorial problem is translated to a linear integer programming problem. Of course, the value of such a translation depends on whether or not it provides new insight or new methods for the solution. We hope that we shall be able to show several examples where the translation to integer programming is very useful, if not essential, for the solution. We shall restrict ourselves to problems concerning graphs.

Historically, the first theorem in graph theory with integer programming flavour was the Marriage Theorem [29]. Linear programming didn't exist at that time and it took another half century until the famous book of Ford and Fulkerson appeared [17], in which one of the first applications of linear programming to graph theory was given. Another early hint of this relationship is in the paper of Gallai [21]. Now the linear programming formulation of combinatorial problems is a common approach.

In this paper we first survey some of the most important results in integer programming which have been successfully applied to graph theory and then discuss those fields of graph theory where an integer programming approach has been most effective. Due to the limited space we shall not go into some of the more advanced fields like the theory of blocking and anti-blocking polyhedra, and shall touch those fields of graph theory where the merging of combinatorial, algorithmic and discrete programming ideas has produced new disciplines, like flow theory or matching theory. We have to ignore the algorithmic aspects completely (although these may be the most important). Since our main concern is the interaction between graph theory and discrete optimization, we shall give only those proofs where this is most characteristic. On the other hand, we list many graph theoretical results which have a linear programming flavour but no explicit treatment this way so far.

1. Some preliminaries from discrete programming

Discrete (linear) programming deals with the problem of finding (say) the maximum of $c^T x$, where c is a given vector and x is a *lattice point* (a point with integral coordinates) in a convex polyhedron described by

$$Ax \le b. \tag{1}$$

Dropping the condition that x is an integer we obtain a new problem, the *linear relaxation*, which is relatively easily solvable; at any rate, a minimax formula for the optimum using the Duality Theorem can be derived.

Since the linear optimum of $c^T x$ is always attained at a vertex (provided (1) describes a bounded polyhedron, which we assume in this discussion for sake of simplicity), if the vertices of (1) are lattice points then the linear relaxation has the same optimum as the original discrete problem. So it is extremely important to know when are the vertices of (1) lattice points. The following theorem is implicite in results of Gomory [23], Chvátal [10], Hoffman [24], Fulkerson [18], Edmonds and Giles [15].

Theorem 1.1. *The vertices of* (1) *are lattice points if and only if* $\max\{c^T x \mid Ax \le b\}$ *is an integer for every integral vector* c.

Using the Duality Theorem, this condition can be rephrased as: $\min\{b^T y \mid A^T y = c, y \ge 0\}$ is an integer for every integral vector c. Usually b is an integer vector and then what can be verified is that the vertex of $\{A^T y = c, y \ge 0\}$ which minimizes $b^T y$ is a lattice point.

Sometimes it suffices to consider a smaller set of vectors c to test in this problem [31, 32].

Theorem 1.2. *Assume that* A *is a* 01-*matrix. Then* $\{Ax \le 1, x \ge 0\}$ *has integral vertices iff* $\max\{c^T x \mid Ax \le 1, x \ge 0\}$ *is an integer for every* 01-*vector* c.

This result has many versions, most of them discovered in order to find applications in graph theory. See Lovász [34, 35].

The most important and most widely used observation which yields polyhedra with integral vertices is due to Hoffman and Kruskal [25].

Theorem 1.3. *Assume that* A *is a totally unimodular matrix* (i.e. *every square submatrix has determinant* ± 1 *or* 0), *and* b *an integral vector. Then* $Ax \le b$ *has integral vertices.*

(The proof of this assertion is very easy: any vertex is the intersection of some hyperplanes of the form $a_i^T x = b_i$ where a_i^T is a row of A and b_i is the corresponding entry of b. Determine this point by solving this system of equations by Cramer's rule. The denominator is a non-zero subdeterminant of A and hence ± 1; the numerators are integers. So the solution has integral entries.)

Assume now that the polyhedron P described by (1) does *not* have integral vertices. Then the polyhedron $\hat{P}$ which is the convex hull of lattice points in P is properly contained in P. If we can write up the system of linear inequalities characterizing $\hat{P}$ we are through with most difficult part of our task again. A general scheme to obtain these inequalities is the following. Assume that $a^T x = b$ is a hyperplane which touches P at a vertex and, moreover, a^T is integral, but b is not. Then $a^T x \le [b]$ is an inequality which eliminates a piece of P but certainly does not cut off anything from $\hat{P}$. Inequalities obtained by this construction are called *cuts*. It follows from the famous algorithm of Gomory [23], and was proved independently and more directly by Chvátal [10] that

Theorem 1.4. *If P is a (bounded) polytope then adding appropriate cuts repeatedly we obtain the convex hull $\hat{P}$ of lattice points in P in a finite number of steps.*

The Gomory algorithm also specifies how to choose these cuts, but in general it is very complicated. In some cases, e.g. for matchings, the appropriate cuts can be described in a very nice way [13].

Some notation from graph theory

$V(G)$: set of vertices,
$E(G)$: set of edges,
$\bar{G}$: complement of G,
$\nu(G)$: maximum number of disjoint edges (matching number),
$\alpha(G)$: maximum number of independent points (stability number),
$\tau(G)$: minimum number of points covering all edges,
$\rho(G)$: minimum number of edges covering all points,
$\chi(G)$: chromatic number,
$\chi'(G)$: chromatic index,
$\omega(G)$: maximum number of pairwise adjacent points (clique number).

2. Flow theory

We begin with a classical application of discrete optimization ideas to graph theory (or vice versa).

Let G be a directed graph with two specified vertices a, b, and a "capacity" $c(x, y) \ge 0$ assigned to every edge (x, y). A *flow* is a function f that assignes an "intensity" $f(x, y) \ge 0$ to each edge such that Kirchoff's law is satisfied, i.e.

$$\sum_x f(x, p) = \sum_x f(p, x) \tag{1}$$

for every point $p \ne a, b$. The amount

$$v(f) = \sum_x f(a, x) - \sum_x f(x, a) \tag{2}$$

is called the *value* of the flow. The flow is *feasible* if

$$0 \le f(x, y) \le c(x, y) \tag{3}$$

for all (x, y). The task is to find the maximum possible value of a feasible flow.

It is immediately seen that what we have is a linear programming problem with constraints (1) and (3) and objective function (2). There is even no integrality constraint involved. Let us consider the dual:

$$\text{minimize} \sum_{x,y} g(x, y) c(x, y)$$

subject to

$$g(x, y) \ge h(x) - h(y),$$
$$g(x, y) \ge 0,$$
$$h(a) = 1, \quad h(b) = 0.$$

Now it is very easy to see that the matrix of this l.p. problem is totally unimodular, so looking for an optimal solution we may assume that the variables $h(x)$, $g(x, y)$ are integers. Also trivially $0 \le h(x) \le 1$ and $g(x, y) = \max(0, h(x) - h(y))$. Thus an optimal solution is defined by specifying a set $S \subset V(G)$ such that $a \in S$, $b \notin S$, and setting

$$h(x) = \begin{cases} 1 & \text{if } x \in S, \\ 0 & \text{if } \notin S, \end{cases}$$

$$g(x, y) = \begin{cases} 1 & \text{if } x \in S, y \notin S, \\ 0 & \text{otherwise} \end{cases}$$

and so the minimum value of flow is

$$\min_{a \in S \subseteq V(G) - b} \sum_{\substack{x \in S \\ y \notin S}} c(x, y).$$

This result is the famous *max-flow-min-cut theorem*. Since the original problem also has a totally unimodular matrix, whenever the capacities are integral then one of the optimum flows is integral.

The max-flow-min-cut theorem, with the supplement given, implies a variety of important-graph-theoretical results, such as Menger's Theorem (see e.g. [7]), but has applications to seemingly distant combinatorial problems such as the construction of certain resolvable designs [4].

We cannot go into the numerous interesting and extremely important extensions of flow theory, but refer to Hu [27], Rothschild and Whinston [51], Hoffman [24] and the monograph by Hu [26].

3. Further minimax theorems

The following minimax result was conjectured by Robertson and Younger, and proved in 1973 by Lucchesi and Younger. See also Lovász [34, 35].

We say that a set S determines a *directed cut* D if $S \neq \emptyset$, $S \neq V(G)$ and no edge connects $V(G) - S$ to S. Then D is defined as the set of edges connecting S to $V(G) - S$.

Theorem 3.1. *Let G be a digraph. Then the maximum number of edge-disjoint directed cuts in G equals to the minimum number of edges covering all directed cuts.*

If we wrote up the two linear programs whose integral optima were the two numbers in question, we could not apply any of the basic discrete programming methods directly. So we have to do some combinatorial preparation.

We say that the directed cuts D_1 determined by S_1 and D_2 determined S_2 are *crossing* if $S_1 \cap S_2 \neq \emptyset$, $S_1 \cup S_2 \neq V(G)$, $S_1 \not\subseteq S_2$ and $S_2 \not\subseteq S_1$. In this case $S_1 \cap S_2$ and $S_1 \cup S_2$ also determine directed cuts, which will be denoted by $D_1 \wedge D_2$ and $D_1 \vee D_2$.

Consider now the directed-cut-packing program. We have a variable x_D assigned to each directed cut D and

$$\text{minimize} \sum_D x_D \tag{1}$$

subject to

$$x_D \geq 0 \quad (\forall D),$$

$$\sum_{D \ni e} x_D \leq c(e) \quad (\forall e \in E(G)), \tag{2}$$

where the $c(e)$ are arbitrary integers.

Let D_1, D_2 be the two crossing cuts with $0 < x_{D_1} \leq x_{D_2}$. Define

$$x_D' = \begin{cases} 0 & \text{if } D = D_1, \\ x_{D_2} - x_{D_1} & \text{if } D = D_2, \\ x_{D_1 \wedge D_2} + x_{D_1} & \text{if } D = D_1 \wedge D_2, \\ x_{D_1 \vee D_2} + x_{D_1} & \text{if } D = D_1 \vee D_2, \\ x_D & \text{otherwise.} \end{cases}$$

Then

$$x_D' \geq 0 \quad (\forall D),$$

$$\sum_{D \ni e} x_D' = \sum_{D \ni e} x_D \quad (\forall e),$$

and

$$\sum_D x_D' = \sum_D x_D.$$

So if (x_D) is an optimum solution of (1)–(2) then so is (x_D'). Repeating this procedure it is easy to see that we never get to a cycle and in a finite number of steps we end up with an optimum solution (x_D) of (1)–(2) such that the set

$$\mathcal{D} = \{D : x_D > 0\}$$

consists of pairwise non-crossing directed cuts. So (1)–(2) is replaced by the program

$$\text{maximize} \sum_{D \in \mathcal{D}} x_D \tag{3}$$

subject to

$$x_D \geq 0 \quad (\forall D \in \mathcal{D}),$$

$$\sum_{D \ni e} x_D \leq c(e) \quad (\forall e \in E(G)), \tag{4}$$

where $\mathcal{D}$ is a family of non-crossing directed cuts. But it is well known that the matrix of (4) is totally unimodular and so, by the Hoffman–Kruskal Theorem, the maximum of $\sum x_D$ is attained at integral values of x_D. In particular if $c(e) = 1$ then this maximum is the maximum number of edge-disjoint directed cuts.

Consider now the dual of (1)–(2); we have a variable y_e for each edge e and want to

$$\text{minimize} \sum_e c(e) y_e \tag{5}$$

subject to

$$y_e \geq 0 \quad (\forall e),$$

$$\sum_{e \in D} y_e \geq 1 \quad (\forall D).$$

We know by the above that the minimum of $\sum_e c(e) y_e$ is an integer for all integral $c(e)$. Hence by Theorem (1.2), the polytope (6) has integral vertices and hence the minimum is attained for an integral choice of the y_i's. Again if $c(e) = 1$ then this minimum is the minimum number of edges covering all directed cuts.

Generalizations of the Lucchesi–Younger Theorem were given by Edmonds and Giles [15]. The "Chinese Postman Theorem" of Edmonds and Johnson [16] can be viewed as an analogue for undirected graphs (cf. [36]).

The packing of other kinds of subgraphs has also been a favourite problem in graph theory. Very little is known about packing circuits or directed circuits; it is an old problem of Gallai [22] whether the ration between the (discrete) optima of the directed circuit packing problem and its dual remains bounded.

Some results are known about the problem of packing circuits through a given point [34, 35]. This is strongly related to the problem of packing paths with both endpoints in a specified subset. This problem was considered by Nash–Williams [41], Gallai [22], Lovász [30, 34, 35], Mader (1977), Karzanov–Lomonosov (1977), and Seymour (1977).

The problems of packing spanning trees in a graph [56], covering a graph by spanning trees [42], packing spanning arborescences rooted at a given point [14], or rooted cuts [20] are solved.

It is my impression that the common nature of these results is not yet completely understood, especially in terms of discrete programming.

4. Matching and point packing

The following are four important optimization problems in graph theory.

(a) *matching*: given a graph G, determine the maximum number $\nu(G)$ of disjoint edges.

(b) *point-packing*: determine the maximum number $\alpha(G)$ of independent points.

(c) *point-covering*: determine the minimum number $\tau(G)$ of points covering all edges.

(d) *edge-covering*: determine the minimum number $\rho(G)$ of edges covering all points. (Here we assume that there are no isolated points.)

In all cases, we may consider a weighting of the elements (points or edges) and may want to determine the optimum total weight of the sets in consideration.

The problems above are not independent. The complement of a point-cover is an independent set of points and vice versa, therefore

$$\alpha(G)+\tau(G)=|V(G)|.$$

The following assertion is less trivial:

$$\gamma(G)+\rho(G)=|V(G)|.$$

These formulas are called *Gallai's identities*. They justify that we shall restrict ourselves to problems (a) and (b).

Even these problems are not independent: the matching problem is the point-packing problem for the line-graph. But the matching problem is much easier in the sense that much more can be said about it.

If we want to use linear programming we have to introduce the following notation. Let $V(G)=\{v_1,\ldots,v_n\}$, $E(G)=\{e_1,\ldots,e_m\}$. Let $A=(a_{ij})_{i=1 j=1}^{n\ m}$ be the point-edge incidence matrix of G, i.e.

$$a_{ij} = \begin{cases} 1 & \text{if } v_i \in e_j, \\ 0 & \text{otherwise}. \end{cases}$$

Let $x_1,\ldots,x_n, y_1,\ldots,y_m$ be variables. We shall identify a subset $S\subset V(G)$ with its characteristic vector (x_i), where

$$x_i = \begin{cases} 1 & \text{if } v_i \in S, \\ 0 & \text{otherwise}, \end{cases}$$

and similarly for subsets of edges.

Let M(G) and PP(C) denote the convex hulls of all matchings and point-packings, respectively. These are clearly polytopes. If we could write up the systems of linear inequalities which describe these polytopes, in other words if we could find their facets, then we could find the optima of problems (a) and (b), at least in a sense. In fact, maximizing — say — the form $\sum_{j=1}^{m} y_j$ over the polytope M(G) we get the maximum number of independent edges, since there always exists an optimum feasible solution which is a vertex of M(G), i.e. matching.

So let us try to find — first — valid inequalities for these polytopes and then see how close we are to the solution. Clearly every matching $y = (y_j)_{j=1}^{m}$ satisfies

$$Ay \leq 1, \tag{1}$$
$$y \geq 0,$$

or, written out in detail,

$$\sum_{e_j \ni v_i} y_j \leq 1 \quad (i = 1, \ldots, n), \tag{1'}$$
$$y_j \geq 0 \quad (j = 1, \ldots, m).$$

Similarly, every independent set $x = (x_i)_{i=1}^{n}$ of points satisfies

$$xA \leq 1, \tag{2}$$
$$x \geq 0.$$

In general, the polytopes (1), (2) properly contain the polytopes M(G), PP(G). Let e.g. G be a triangle, then the polytope described by (1) contains the vertex $(\frac{1}{2}, \frac{1}{2}, \frac{1}{2})^T$ which is not a convex combination of matchings (Fig. 1).

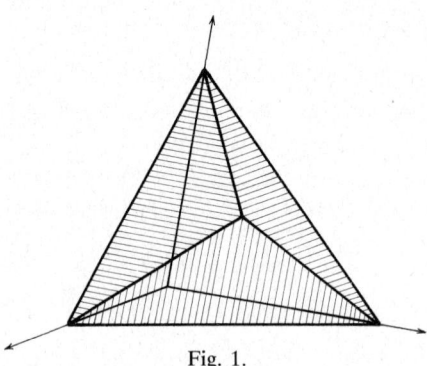

Fig. 1.

However, each *lattice point* satisfying (1) is a matching; therefore it is natural to call the points in (1) and (2) *fractional matchings* and *fractional point packings*, resp. Denote the polytopes (1) and (2) accordingly by FM(G) and FPP(G).

The case of bipartite graphs is very simple and well-known:

Theorem 4.1. *The following are equivalent*:
(i) *G is bipartite*;
(ii) *its point-edge incidence matrix A is totally unimodular*;
(iii) *FM(G) has integral vertices, i.e. FM(G) = M(G)*;
(iv) *FPP(G) has integral vertices, i.e. FPP(G) = PP(G)*.

Theorem 4.1 has important consequences for bipartite graphs. $\nu(G)$ is the integral optimum of $1 \cdot x$, subject to (1). Since FM(G) has integral vertices, this is the same as the real optimum. The dual problem gives $\tau(G)$. So the Duality Theorem implies

König's Theorem 4.2. *In a bipartite graph* $\nu(G) = \tau(G)$.

If we want to proceed to non-bipartite graphs, we face two questions:
(a) describe the vertices of the polytopes FM(G), FPP(G).
(b) describe the facets of M(G), PP(G).
The first problem is easier, and completely solved [1, 2, 43].

Theorem 4.3. *Let $C_1, \ldots, C_p$ be disjoint odd circuits and M a matching, vertex-disjoint from the C_i's in the graph G. Set*

$$x_1 = \begin{cases} \frac{1}{2} & \text{if } e_i \in C_1 \cup \cdots \cup C_p, \\ 1 & \text{if } e_i \in M, \\ 0 & \text{otherwise.} \end{cases}$$

Then (x_i) is a vertex of FM(G) and all vertices of FM(G) arise this way.

Theorem 4.4. *Let G be a graph and A, B, C a partition of V(G) such that (a) A is independent, (b) B contains all neighbors of A and (c) no component of the subgraph induced by C is bipartite. Then*

$$x_i = \begin{cases} 1 & \text{if } v_i \in A, \\ 0 & \text{if } v_i \in B, \\ \frac{1}{2} & \text{if } v_i \in C, \end{cases}$$

defines a vertex of FPP(G), and every vertex of FPP(G) arises this way.

These results have a variety of graph theoretical applications. We only mention a few. Tutte (1953) calls a set of edges a *q-factor*, if it forms vertex-disjoint circuits and edges, and covers all points.

Theorem 4.5. *A graph has a q-factor iff every independent set A of points has at least $|A|$ neighbors.*

The following version of this result was observed by J. Edmonds; it follows by the same kind of argument. A *2-matching* is a collection of edges such that any point belongs to at most two of them. A *2-cover* is a collection of points such that every edge contains at least two of them.

Theorem 4.6. *The maximum number of edges in a 2-matching equals to the minimum number of points in a 2-cover.*

Now we turn to the question of describing the facets of $M(G)$. This was answered by Edmonds [13]:

Theorem 4.7. $M(G)$ *is described by the following set of inequalities*:

(I) $\quad x_j \geq 0 \qquad\qquad\qquad (j = 1, \ldots, m)$,

(II) $\quad \sum_{e_j \ni v_i} x_j \leq 1 \qquad\quad (i = 1, \ldots, n)$,

(III) $\quad \sum_{e_j \subseteq S} x_j \leq \dfrac{|S|-1}{2} \qquad (S \subseteq V(G), |S| \text{ odd})$.

Moreover, for every linear objective function with integral coefficients the dual problem has an integral optimum solution.

So to obtain $M(G)$ from $FM(G)$ only the cuts (III) are needed; not even all of them, because Edmonds and Pulleyblank [49] proved:

Theorem 4.8. $\sum_{e_i \subseteq S} x_i \leq (|S|-1)/2$ *is a facet of* $M(G)$ *iff* S *spans a 2-connected graph* G_S *such that* $G_S - v$ *has a perfect matching for every* $v \in S$.

The last two theorems can be derived from an analysis of the Edmonds Matching Algorithm. Let us sketch here a short direct proof of the first part of Theorem (3.9).

Let

$$\sum_{i=1}^{m} a_i x_i \leq b \qquad\qquad (3)$$

be a facet of $M(G)$; we want to show that it is of one of the forms (I), (II) or (III).

Case 1. Assume that there is an i_0 such that $a_{i_0} < 0$. Then every matching (x_i) which gives equality in (3) must satisfy $x_{i_0} = 0$. Consequently, (3) must be identical with the facet $x_{i_0} \geq 0$.

Case 2. Assume that there is a point v_j such that every matching with equality in (3) contains an edge adjacent to v_j. Then every such matching satisfies $\sum_{e_i \ni v_j} x_i = 1$ and, consequently, (3) must be identical with the facet $\sum_{e_i \ni v_j} x_i \leq 1$.

Case 3. Assume that $a_i \geq 0$ for all i and that for each point j, there is a matching avoiding v_j and giving equality in (3). Let G' be the graph formed by those edges e_i for which $a_i > 0$. It is easily seen that G' is connected. Next one shows that no matching which gives equality in (3) misses two points of G.

This can be seen as follows. Assume that M_1 is a matching missing both points $x, y \in V(G')$. We use induction on the distance of x and y in G'. Trivially, x and y cannot be adjacent in G', so we may choose a point z on the shortest path connecting x to y. Let M_2 be a matching which gives equality in (3) and misses z. By the induction hypothesis, M_1 covers z and M_2 covers both of x and y. Consider the connected components of $M_1 \cup M_2$ containing x and y, respectively. These are paths which may be identical, but at least one of them, P say, misses z. Set

$$M' = M_1 - (M_1 \cap P) \cup (M_2 \cap P)$$

and

$$M'' = M_2 - (M_2 \cap P) \cup (M_1 \cap P)$$

Then M' and M'' are matchings; furthermore, they are extreme matchings because

$$2b \geq \sum_{e_i \in M'} a_i x_i + \sum_{e_i \in M''} a_i x_i = \sum_{e_i \in M_1} a_i x_i + \sum_{e_i \in M_2} a_i x_i = 2b.$$

But now M'' is an extremal matching which misses z and one of x and y, contrary to the induction hypothesis.

Now it follows that if $V(G') = S$ then $|S|$ is odd and every matching with equality in (3) satisfies

$$\sum_{e_i \subseteq S} x_i = \frac{|S| - 1}{2}.$$

Consequently, facet (3) is of the form (III).

Edmonds' theorem has various applications in graph theory. First, let us remark that Tutte's famous condition for the existence of a 1-factor can be derived from it. Let us show another one:

The *chromatic index* of a graph is defined as the minimum number of colors necessary to color the edges so that adjacent edges have different colors. This number is clearly at least as large as the maximum degree and quite often larger. The following result is due to Edmonds [13].

Theorem 4.9. *Let G be a k-regular k-edge-connected graph with an even number of points. Then there exists a number t such that if we replace every edge by t parallel edges the resulting graph G has chromatic index kt.*

Proof. *Consider the point $(1/k, \ldots, 1/k)^T$.* Straightforward computation shows that it satisfies all 3 types of inequalities in Edmonds' theorem, and hence it belongs to $M(G)$. Therefore it is a convex combination of vertices of $M(G)$:

$$\left(\frac{1}{k}, \ldots, \frac{1}{k}\right)^T = \sum_{n=1}^{N} \alpha_i F_1, \quad \sum_{i=1}^{N} \alpha_i = 1, \quad \alpha_i \geq 0.$$

where the F_i are matchings. Clearly one may assume that every α_i is rational. Let P be a common denominator of the numbers $k\alpha_i$ and $k\alpha_i = P_i/P$. Then

$$(P, \ldots, P)^T = \sum P_i F_i, \quad \sum_{i=1}^{N} P_i = kp.$$

So if we multiply edge e_i by P then the resulting graph can be decomposed into the union of $\sum P_i = kp$ matchings.

Remark. The least value of P here is not known. A conjecture of Fulkerson states $P \leq 2$ for $k = 3$. Seymour [52] proved that if $k = 3$ then P can be chosen a power of 2.

A second application in which the full strength of Edmonds' theorem can be used is the Chinese Postman Problem (Edmonds and Johnson [16]). A postman must traverse every street of a town and return to the post office. He wants to minimize the total length of his walk. In other words: Given a connected graph G with a "length" c_i assigned to every edge. Find a closed walk which contains every edge at least once and has minimum length. Note that if the graph is Eulerian then any Eulerian trail is such a walk. This motivates the following reformulation: Find a collection $\mathcal{T}$ of edges of G with minimum total length such that the number of edges of $\mathcal{T}$ incident with a point x of G has the same parity as the degree of x.

Let us divide the set of points of odd degree in pairs (x_1, y_i) $(i = 1, \ldots, t)$ and take a path P_i connecting x_i and y_i. Then $\mathcal{T} = \sum_{i=1}^{t} P_i$ is a collection with the desired property. It is easy to see that, conversely, the optimal $\mathcal{T}$ arises this way. The total length of edges in $\mathcal{T}$ is equal to the sum of "lengths" of the P_i. So the problem is transformed to the following: let $d(x, y)$ denote the minimal length of (x, y)-paths in G. Partition the points of G with odd degree into pairs $(x_1, y_1), \ldots, (x_t, y_t)$ such that $\sum_{i=1}^{t} d(x_i, y_i)$ is minimal.

But this is the problem of finding a maximum-weight matching of the complete graph on the set of points of G with odd degree, and can be solved by the methods of matching theory.

Let an *odd cut* be defined as the set of edges connecting some $S \subseteq V(G)$ to $V(G) - S$, provided it has odd cardinality.

Theorem 4.10. *The minimum number of edges whose doubling results in an Eulerian graph equals half of the maximum number of odd cuts, containing each edge at most twice.*

The point-packing polytope has a much more difficult structure. As remarked before, $PP(G) = FPP(G)$ if the graph G is bipartite. If G is not bipartite then $PP(G)$ must have facets of more complicated form than $FPP(G)$. The following class of facets is immediately seen:

$$\sum_{i \in C} x_i \leq 1, \tag{1}$$

where C is an arbitrary clique [18]. The class of graphs for which these are all is wider than bipartite graphs, and is very interesting from the graph-theoretical point of view.

A graph G is called *perfect* if every induced subgraph G' of G satisfies $\chi(G') = \omega(G')$. This notion has been introduced by Berge [6]. There are many examples of perfect graphs. The most important is probably the following:

Let $V(G)$ be a partially ordered set and connect two points iff they are comparable.

Let G be a graph and let H_x ($x \in V(G)$) be a vertex disjoint graphs. Consider $\bigcup H_x$ and add all edges between H_x and H_y if $(x, y) \in E(G)$. We shall say that the resulting graph is obtained by substituting the graphs H_x for the points of G. The following theorem uses this construction [31, 32]; its proof is quite easy and elementary.

Theorem 4.11. *Substituting perfect graphs for the points of perfect graphs we obtain a perfect graph.*

The next result provides further examples [31, 32]:

Theorem 4.12. *The complement of a perfect graph is perfect.*

The result was conjectured by Berge [6]. Its strong connection to linear programming was discovered by Fulkerson [18, 19], who proved it in a slightly weaker form.

We prove (3.13) together with the following theorem [11, 18, 19],

Theorem 4.13. *G is perfect iff the polytope*

$$x_i \geq 0 \quad (i = 1, \ldots, n), \tag{5}$$

$$\sum_{i \in C} x_i \leq 1 \quad (\forall \text{cliques } C \text{ in } G)$$

has integral vertices.

We shall show that G is perfect $\Rightarrow$ (5) has integral vertices $\Rightarrow \bar{G}$ is perfect. The cycle closes by interchanging the role of G and $\bar{G}$.

I. Assume first that G is perfect. Let x be a vertex of (5). Clearly x is rational and so for some integer $k > 0$, $kx = (p_1, \ldots, p_n)$ is a lattice point. Substitute a complete p_i-graph for v_i ($i = 1, \ldots, n$). Let C' be any clique in the resulting graph G' and let C be its "projection" to G. Then

$$|C'| \leq \sum_{i \in C} |P_i| = k \sum_{i \in C} x_i \leq k.$$

Since G' is perfect by 3.12, $\chi(G') = \omega(G') \leq k$. Let $\{A'_1, \ldots, A'_k\}$ be a k-coloration of G', let A_i be the "projection" of A'_i to G and a_i the corresponding vector. Then

$$\frac{1}{k} \sum_{i=1}^{n} a_i = x.$$

Since x is a vertex, this implies that $x = a_1 = \cdots = a_k$, i.e. x is a lattice point.

II. Assume now that (5) has integral vertices. Note that the induced subgraphs of G have similar property; since the corresponding polytopes are faces of (5).

The face $\mathbf{1} \cdot x \leq \alpha(G)$ of (5) is contained in several facets, at least one of which is of the form $\sum_{i \in C} x_i \leq 1$. This means combinatorially that the clique C intersects every maximum independent set, i.e. $\alpha(G - C) < \alpha(G)$. Repeating the argument we get $\alpha(G)$ cliques which cover all points of G. This proves that $\chi(\bar{G}) = \alpha(G) = \omega(\bar{G})$. Since this follows for the induced subgraphs of G similarly, we conclude that G is perfect.

It is natural to study those graphs which are not perfect but all of their induced subgraphs are. Let us call these graphs *critically imperfect*. The famous Strong Perfect Graph Conjecture of Berge asserts that every critically imperfect graph is a chordless odd circuit of length ≥ 5 or the complement of such a circuit. It is known [32] that

Theorem 4.14. *A critically imperfect graph G has $\alpha(G)\omega(G) + 1$ points.*

Padberg [45, 46] proved, among many interesting structural results on these graphs, that

Theorem 4.15. *If G is a critically imperfect graph then the polytope (5) has precisely one non-integral vertex (the point $1/\omega(G) \cdot \mathbf{1}$) and this is contained in precisely n facets.*

For further important results on perfectness see Chvátal [11], Meyniel [39], Parthasaraty and Ravindra [48].

As the Berge conjecture suggests, further classes of facets of PP(G) can be derived from chordless odd circuits C in G or $\bar{G}$. Assume first that G is itself an odd circuit. Then

$$\sum_{i=1}^{n} x_i \leq \frac{n-1}{2} \tag{6}$$

is a facet. Similarly, if G is the complement of an odd circuit then

$$\sum_{i=1}^{n} x_i \leq 2 \qquad (7)$$

is a facet. Now if $C \neq G$ then (6) yields a facet of $PP(G)$; in fact, there is a general "lifting" procedure which derives a facet of $PP(G)$ from a facet of $PP(G')$ where G' is an induced subgraph of G. Geometrically, this can be described as follows. It suffices to consider the case $G' = G - v_n$. Let F be a facet of $PP(G')$.

Since $PP(G') = PP(G) \cap \{x_n = 0\}$, F is an $(n-2)$-face of $PP(G)$ and hence, F is the intersection of two facets of $PP(G)$. One of these is clearly $x_n = 0$, the other is the *lift* of F. For example, lifting the facet $x_1 + \cdots + x_5 \leq 2$ of the pentagon, we obtain the facet $x_1 + \cdots + x_5 + 2x_6 \leq 2$ of the 5-wheel (Fig. 2).

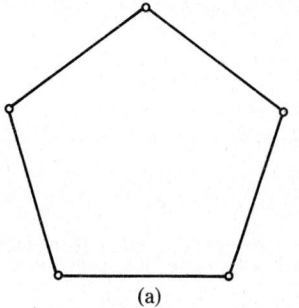 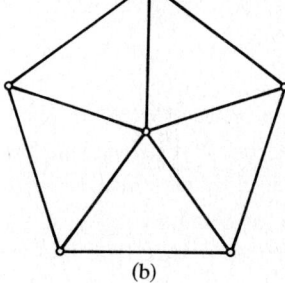

Fig. 2.

So in a sense it suffices to study those facets which are *essential* in the sense that they do not arise by lifting. This procedure is due to Padberg [47] and Nemhauser and Trotter [43].

An interesting class of facets has been described by Chvátal [11]. Call an edge e of a graph G α-critical if $\alpha(G-e) > \alpha(G)$.

Theorem 4.16. *Let G be a graph whose α-critical edges form a spanning connected graph. Then*

$$\sum_{i=1}^{n} x_i \leq \alpha(G)$$

is a facet of $PP(G)$.

Proof. It is trivial that this is a valid inequality. To show that it is a facet it suffices to prove that if

$$a_i x_i \leq b \qquad (8)$$

is a facet of $PP(G)$ such that every maximum matching satisfies it with equality then $a_1 = \cdots = a_n$. Assume indirectly $a_1 \neq a_2$ (say). We may assume that v_1 and v_2

are adjacent. Let T be an independent $(\alpha+1)$-set in $G\text{-}(v_1, v_2)$. Obviously $v_1, v_2 \in T$. Let $T - v_i$ have characteristic vector $x^{(i)}$ ($i = 1, 2$). Then

$$1 \cdot x^{(1)} - 1 \cdot x^{(2)} = a_2 - a_1 \neq 0,$$

so it is impossible that both $x^{(1)}$ and $x^{(2)}$ give equality in (S).

These facets may and may not be essential. The structure of graphs to which theorem (3.17) applies can be very complicated, e.g. all connected α-critical graphs are among them. This fact already indicates that a general description of the facets of PP(G) in the spirit of Edmonds' theorem is hopeless.

However, there may be important classes of graphs for which the facets of PP(G) can be described. A very recent result of Minty [40] seems to point out such a class. Minty has a polynomial-bounded algorithm to determine $\alpha(G)$ provided G contains no 3-star (claw) as an induced subgraph. Since all line-graphs are claw-free this result considerably extends Edmond's algorithm for finding maximum matchings. One would expect that, analogously, Edmonds' description of the matching polytope (which is equivalent to a description of the facets of PP(G) if G is a line-graph) could be extended to a description of the facets of claw-free graphs. It is interesting that there is another result which points out the importance of claw-free graphs when independence number is investigated: this is the theorem of Parthasarathy and Ravindra [48] stating that the strong perfect graph conjecture is true for claw-free graphs.

We close this topic with the following interesting result [43] which underlines the importance of considering FPP together with PP:

Theorem 4.17. *Let x_0 be a vertex of* FPP *which maximizes* $1 \cdot x$. *Then there is a vertex x_1 of* PP *which maximizes* $1 \cdot x$ *(i.e. a maximum independent set) such that x_1 coincides with x_0 in all coordinates which are integral in x_0.*

Concluding remarks

1. One further field where linear programming has been applied in graph theory is the problem of Hamilton circuits. This is related to the well-known Travelling Salesman problem. We only refer here to the papers of Chvátal [10], Bellmore and Nemhauser [5], Christofides [9].

2. The chromatic theory of graphs is a field of possible further development. One can define the chromatic number as the discrete optimum of a linear program, and investigate the corresponding linear optimum. The two are equal for perfect graphs, but very little is known about their relationship in general. See e.g. M. Rosenfeld [50], Stahl [53], Lovász [36] and Chen and Hilton [8].

3. It is often an interesting question when are two vertices of a combinatorially defined polytope neighboring. This is solved for M(G) and PP(G) [11], and some

other classes (Hausman and Korte 1976). This points out how little is known about *faces* of these polyhedra even if the *facets* are known.

4. If there is no hope to fully describe the facets of a polytope (e.g. of PP(G)) it is still a possibility to find some kind of classification. This hope is supported by the fact that α-critical graphs, which provide a large variety of facets by (3.17), do have an interesting classification [33]. It might be the case that some classes of facets cut down FPP(G) to a polytope "not much larger" than PP(G).

References

[1] M.L. Balinski, Establishing the matching polytope, J. Combinatorial Theory 13 (1972) 1–13.
[2] M.L. Balinski and K. Spielberg, Methods of integer programming: algebraic, combinatorial and enumerative, in: Progress in Operations Research (Wiley, New York, 1969) 195–292.
[3] E. Balas and M. Padberg, Set partitioning: a survey, SIAM Rev. 18 (1976) 710–760.
[4] Zs. Baranyai, On the factorization of the complete uniform hypergraph, in: Infinite and Finite Sets, Coll. Math. Soc. J. Bolyai 10 (North-Holland, Amsterdam, 1974) 91–108.
[5] M. Bellmore and G.L. Nemhauser, The travelling salesman problem: a survey, Operations Res. 16 (1968) 538–558.
[6] C. Berge, Färbung von Graphen, deren sämtliche bzw. deren ungerade Kreise starr sind, Wiss. Z. Martin–Luther–Univ. Halle–Wittenberg (1961) 114.
[7] C. Berge, Graphs and Hypergraphs (North-Holland, Amsterdam, 1973).
[8] C.C. Chen and A.W. Hilton, A 4-color conjecture for planar graphs, in: Combinatorics, Coll. Math. Soc. J. Bolyai 18 (North-Holland, Amsterdam, 1978).
[9] N. Christofides, Graph Theory: An Algorithmic Approach (Academic Press, New York, 1975).
[10] V. Chvátal, Edmonds polytopes and a hierarchy of combinatorial problems, Discrete Math. 4 (1973) 305–337.
[11] V. Chvátal, On certain polytopes associated with graphs, J. Combinatorial Theory (B) 18 (1975) 138–154.
[12] V. Chvátal, Edmonds polytopes and weakly Hamiltonian graphs, Math. Programming 5 (1973) 29–40.
[13] J. Edmonds, Maximum matching and a polyhedron with 0, 1 vertices. J. Res. Nat. Bur. Standards 69B (1965) 125–180.
[14] J. Edmonds, Submodular functions, matroids, and certain polyhedra, in: Combin. Structures and their Applications (Gordon and Breach, New York, 1970).
[15] J. Edmonds and R. Giles, A min-max relation for submodular functions on graphs, Ann. Discrete Math. I (1977) 185–204.
[16] J. Edmonds and E.L. Johnson, Matching, Euler tours and the Chinese postman, Math. Programming 5(1973) 88–124.
[17] L.R. Ford, Jr. and D.R. Fulkerson, Flows in Networks (Princeton University Press, Princeton, NY, 1962).
[18] D.R. Fulkerson, Blocking and anti-blocking pairs of polyhedra, Math. Programming 1 (1971) 168–194.
[19] D.R. Fulkerson, Anti-blocking polyhedra, J. Combinatorial Theory 12 (1972) 50–71.
[20] D.R. Fulkerson, Packing rooted directed cuts in weighted directed graphs, Math. Programming, 6 (1974) 1–13.
[21] T. Gallai, Maximum-minimum Sätze über Graphen, Acta Math. Acad. Sci. Hung. 9 (1958) 395–434.
[22] T. Gallai, Maximum-minimumsätze und verallgemeinerte Faktoren von Graphen, Acta Math. Acad. Sci. Hungar. 12 (1961) 131–173.
[23] R. Gomory, An algorithm for integer solutions of linear programs, in: Recent Advances in Mathematical Programming (McGraw-Hill, New York, 1963) 269–302.

[24] A.J. Hoffman, A generalization of max flow-min cut, Math. Programming 6 (1974) 352–359.
[25] A.J. Hoffman and J.B. Kruskal, Integral boundary points of convex polyhedra, in: Linear Inequalities and Related Systems, Annals of Math. Studies 38 (1956).
[26] T.C. Hu, Integer Programming and Network Flows (Addison-Wesley, Reading, MA, 1970).
[27] T.C. Hu, Multicommodity network flows, J.ORSA 11 (1963) 344–360.
[28] Karzanov and Lomonosov, preprint.
[29] D. König, Vonalrendszerek és determinánsok, Math. és Term. Értesitö 33 (1915) 221–229.
[30] L. Lovász, Generalized factors of graphs, Comb. Theory and its Appl. Coll. Math. Soc. J. Bolyai 4 (1970) 773–781.
[31] L. Lovász, Normal hypergraphs and the perfect graph conjecture, Discrete Math. 2 (1972) 253–268.
[32] L. Lovász, A characterization of perfect graphs, J. Combinatorial Theory B 13 (1972) 253–268.
[33] L. Lovász, Certain duality principles in integer programming, Ann. Discrete Math. 1 (1977) 363–374.
[34] L. Lovász, On two minimax theorems in graph theory, J. Combinatorial Theory B (1976) 96–103.
[35] L. Lovász, Edge-connectivity in Eulerian graphs, Acta Math. Acad. Sci. Hungar. 28 (1976) 129–138.
[36] L. Lovász, 2-matchings and 2-covers of hypergraphs, Acta Math. Acad. Sci. Hungar. 26 (1975) 433–444.
[37] L. Lovász, Some finite basis theorems in graph theory, in: Combinatorics, Coll. Math. Soc. J. Bolyai 18 (North-Holland, Amsterdam, 1978) 717–729.
[38] C. Lucchesi and D.H. Younger, A minimax theorem for directed graphs, J. London Math. Soc., 17 (1978) 369–374.
[39] H. Meyniel, On the perfect graph conjecture, Discrete Math. 16 (1976) 339–342.
[40] G. Minty, Maximum independent sets of vertices in claw-tree graphs (submitted to J. Combinatorial Theory).
[41] C.St.J.A. Nash-Williams, Decomposition of finite graphs into open chains, Canad. J. Math. 13 (1961) 157–166.
[42] C.St.J.A. Nash-Williams, Decomposition of finite graph into forests, J. London Math. Soc. 39 (1964) 12.
[43] G.L. Nemhauser and L.E. Trotter, jr: Properties of vertex packing and independence system polyhedra, Math. Programming 6 (1974) 48–61.
[44] G.L. Nemhauser and L.E. Trotter, Vertex packings: structural properties and algorithms, Math. Programming 8 (1975) 232–248.
[45] M. Padberg, Perfect zero-one matrices, Math. Programming 6 (1974) 180–196.
[46] M. Padberg, Characterizations of totally unimodular, balanced and perfect matrices, in: Combinatorial Programming: Methods and Applications, (Reidel, Boston, 1975) 275–286.
[47] M. Padberg, On the facial structure of set packing polyhedra, Math. Programming 5 (1973) 199–216.
[48] Parthasaratly and Ravindra, The strong perfect graph conjecture for $K_{1,3}$-free graphs, J. Combinatorial Theory (B) 21 (1976) 212–223.
[49] W. Pulleyblank and J. Edmonds, Facets of 1-matching polyhedra, in: Hypergraph Seminar, Lecture Notes in Math. 411 (Springer, Berlin, 1974) 214–242.
[50] M. Rosenfeld, On a problem of C.E. Shannon in graph theory, Proc. Amer. Math. Soc. 18 (1967).
[51] B. Rothschild and A. Whinston, Feasibility of two commodity network flows, Operations Res. 14 (1966) 1121–1129.
[52] P. Seymour, On multi-colorings of cubic graphs and conjectures of Fulkerson and Tutte (to appear).
[53] S. Stahl, n-tuple colorings and associated graphs, J. Combinatorial Theory B 20 (1976) 185–203.
[54] W.T. Tutte, The factorisation of linear graphs, J. London Math. Soc. 22 (1947) 107–111.
[55] W.T. Tutte, The 1-factors of oriented graphs, Proc. Amer. Math. Soc. 4 (1953) 922–931.
[56] W.T. Tutte, On the problem of decomposing a linear graph into n connected factors, J. London Math. Soc. 142 (1961) 221–230.

BLOCKING AND ANTIBLOCKING POLYHEDRA

Jørgen TIND

IFOR, Aarhus Universitet, c/o Matematisk Institut, bygn. 530, Ny Munkegade, DK-8000 Aarhus C, Denmark

This paper describes the fundamental aspects of the theory of blocking and antiblocking polyhedra as developed by D.R. Fulkerson. The theory is an excellent framework for a joint consideration of a class of continuous problems and a class of combinatorial problems, and it unifies several notions that are similar for both classes of problems. The present paper presents shortly some of the many examples of this sort together with several references to related works.

Introduction

This paper describes some of the results that have been obtained within the powerful blocking and antiblocking theory, introduced by Fulkerson [12, 13, 14]. This theory has appeared to be an excellent framework for the discussion of problems of a discrete nature as well as problems of continuous origin. It relates the duality theories known from convex analysis and linear programming to equivalent theories for combinatorial problems in a very elegant way.

The present paper states the basic results by Fulkerson and mentions later works, based on these results.

Although the theory has been motivated by the study of certain problems in combinatorics, the theory itself is modelled over continuous problems. The actual presentation is primarily focused on problems of this last sort, but several references have been given to results in combinatorics, and there the unsatisfied reader might hopefully find help for the omissions made here. One might also consult other basic descriptions of blocking and antiblocking theory, especially the original one by Fulkerson [14]. But, additionally, some later papers contain a description of the most basic properties of the theory together with some references, e.g. [18, 40]. See also [20], where further references are included.

There has been made no attempt to actually unify the two theories. They contain many mutual similarities and developments, but there are also dissimilarities, important enough to justify the separate presentation, which has been followed here.

The first two sections present the basic material without any proofs, but contain some basic illustrating examples. In Section 3 the relation to similar concepts in convex analysis is discussed, and Section 4 is devoted to the relation towards integer programming and combinatorics. Section 5 mentions very briefly some of the connections between the two theories.

1.

1.1. Blocking polyhedra

Let A be an m by n non-negative matrix and consider the polyhedron

$$\mathcal{B} = \{x \in \mathbf{R}^n_+ \mid Ax \geq 1\},$$

where $1 = (1, \ldots, 1)$ with m components, and $\mathbf{R}^n_+$ denotes the non-negative orthant. It is seen that $\mathcal{B}$ is a convex, closed polyhedron, not containing 0, and unbounded, except for the degenerate case where A has no rows or contains a zero row.

The rows of A may not all represent facets of $\mathcal{B}$, which means that some of the constraints in $Ax \geq 1$ may be superfluous. A row vector a^i of A is called *inessential*, if a^i dominates, i.e. is greater than or equal to, a convex combination of other rows of A. As a consequence of Farkas' Lemma, we have that a row vector is inessential, if and only if it does not represent a facet. We call a non-negative matrix *proper*, if and only if there are no inessential rows. If A is a proper 0–1 matrix, then no row is contained in another row. This means that A can be considered as the incidence matrix of a *clutter*, i.e. a set of non-comparable elements.

The *blocker* $\hat{\mathcal{B}}$ of $\mathcal{B}$ is defined as

$$\hat{\mathcal{B}} = \{x^* \in \mathbf{R}^n_+ \mid x^* \cdot x \geq 1 \text{ for all } x \in \mathcal{B}\},$$

where $x^* \cdot x$ denotes the inner product.

We include the degenerate cases, where a proper A is either a one rowed zero matrix, which implies that $\mathcal{B}$ is empty and $\hat{\mathcal{B}} = \mathbf{R}^n_+$, or where A has no rows, which implies that $\mathcal{B} = \mathbf{R}^n_+$ and $\hat{\mathcal{B}}$ is empty.

The following theorem describes the basic involutory polar properties between $\mathcal{B}$ and $\hat{\mathcal{B}}$.

Theorem 1.1 (Fulkerson [12]). *Let $\mathcal{B}$ have extreme points $b^1, \ldots, b^r$, and let B be the r by n matrix with rows $b^1, \ldots, b^r$. Let also A be a proper matrix with rows $a^1, \ldots, a^m$. Then*
  (i) $\hat{\mathcal{B}} = \{x \in \mathbf{R}^n_+ \mid Bx \geq 1\}$,
  (ii) *B is proper*,
  (iii) *the extreme points of $\hat{\mathcal{B}}$ are $a^1, \ldots, a^m$, and*
  (iv) $\hat{\hat{\mathcal{B}}} = \mathcal{B}$.

Since $\mathcal{B}$ and $\hat{\mathcal{B}}$ play a symmetric role in the theorem, we call $\mathcal{B}$ and $\hat{\mathcal{B}}$ a *blocking pair of polyhedra*, and A and B a *blocking pair of matrices*.

An example [12] that illustrates the above theorem, is given in Fig. 1.1.

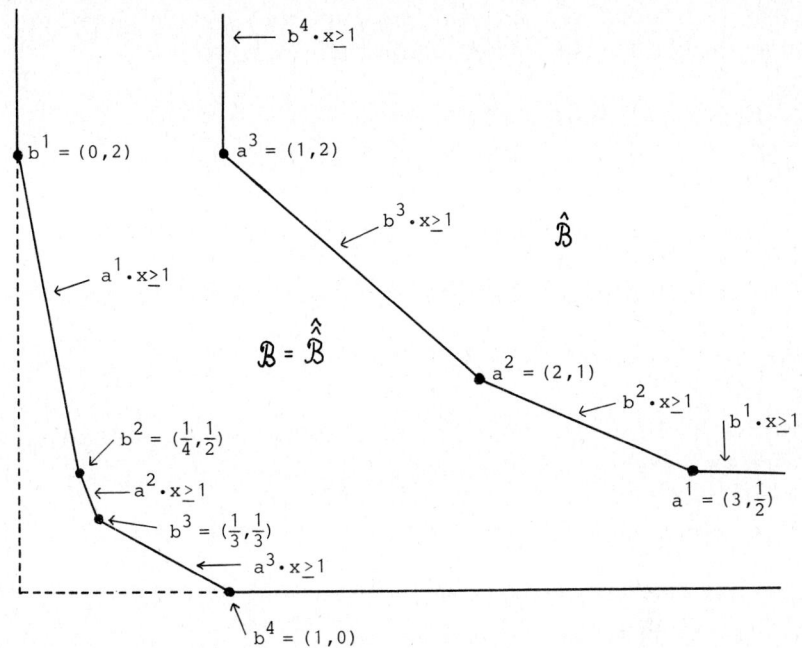

Fig. 1.1.

Now consider the following packing problem:

$$\max_{y \in \mathbf{R}^m} 1 \cdot y,$$

$$yA \leq w, \tag{1.1}$$

$$y \geq 0,$$

where $w \in \mathbf{R}^n_+$, $0 = (0, \ldots, 0)$ and $1 = (1, \ldots, 1)$, each with m components.

Again, let A be a non-negative matrix with rows $a^1, \ldots, a^m$, and let B be a non-negative matrix with rows $b^1, \ldots, b^r$.

We now say that the *max-min equality* holds for A and B (in this order), if and only if for every $w \in \mathbf{R}^n_+$ the packing problem (1.1) has an optimal solution y such that

$$1 \cdot y = \min_{1 \leq j \leq r} b^j \cdot w. \tag{1.2}$$

Similarly we say that the *min-min inequality* holds for the (unordered) pair A, B, if and only if for every $l \in \mathbf{R}_+^n$ and $w \in \mathbf{R}_+^n$, we have

$$\left(\min_i a^i \cdot l\right)\left(\min_j b^j \cdot w\right) \leq l \cdot w.$$

The following theorems show the intimate relationship between those two notions and the blocking relation.

Theorem 1.2 (Fulkerson [12, 14]). *The max-min equality holds for the ordered pair of matrices A, B, if and only if A and B are a blocking pair of matrices. Hence it follows, that if the max-min equality holds for A, B, it also holds for B, A.*

Theorem 1.3 (Fulkerson [12, 14]). *The pair of matrices, A, B, is a blocking pair of matrices, if and only if*
  (i) $a^i \cdot b^j \geq 1$ *for all rows a^i of A and b^j of B, and*
  (ii) *the min-min inequality holds for A, B.*

1.2. An example (paths in a two-terminal network)

Actually, the present example originally gave Fulkerson a part of the motivation for the introduction and the invention of the theory of blocking polyhedra.

Let the rows of A denote the incidence vectors of all paths of edges in a graph between two terminals. Additionally, let the rows of B be the incidence vectors of all cuts of edges separating the terminals.

These conditions are illustrated in Fig. 1.2, where s and t denote the two terminals. The edges have been numbered. B only contains the essential rows corresponding to cuts constituting a clutter.

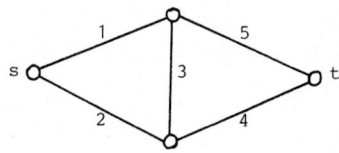

$$A = \begin{bmatrix} 1 & 0 & 0 & 0 & 1 \\ 0 & 1 & 0 & 1 & 0 \\ 0 & 1 & 1 & 0 & 1 \\ 1 & 0 & 1 & 1 & 0 \end{bmatrix} \qquad B = \begin{bmatrix} 1 & 1 & 0 & 0 & 0 \\ 1 & 0 & 1 & 1 & 0 \\ 0 & 0 & 0 & 1 & 1 \\ 0 & 1 & 1 & 0 & 1 \end{bmatrix}$$

Fig. 1.2.

Now, let $w = (w_1, \ldots, w_n) \in \mathbf{R}_+^n$ be given capacities of the edges. Then the maximal flow between the terminals is found to be the optimal objective function value of (1.1), where y is indicating the flow values of each path. On the other hand, the minimal cut capacity is found as indicated at the right hand side of the max-min equality (1.2). By the max flow-min cut theorem [11] the two numbers coincide. Hence the max-min equality is holding, and A and B are a blocking pair of matrices.

But Theorem 1.2 says also that the max-min equality holds the other way as well. In the present context of two terminal networks it has the following meaning. Let w_i denote the length of the ith edge. Then the max-min equality for B, A in this order says: Consider assigning a non-negative weight to each cut so that the total weight of all cuts containing any edge does not exceed the length of the edge. Then the maximum possible total weight of all cuts is equal to the length of the shortest path between s and t.

The min-min inequality is the analogue of the length-width inequality for source-sink networks [23].

1.3. Contractions and deletions

Let A be a non-negative matrix. By a *contraction* of coordinate i is meant the following: drop the ith column of A, and then drop all inessential rows in the resulting matrix. *Deleting* coordinate i means dropping the ith column from A and also dropping all rows from A containing a positive entry in this column. These operations are analogous to the similar operations of "contracting an element" or "deleting an element" in a graph. The two operations are illustrated in Fig. 1.3.

The following theorem shows that they are in the blocking context dual to each other.

Theorem 1.4 (Fulkerson [12]). *Let A and B be a blocking pair of matrices. Define A' and B' as the matrices obtained by contracting coordinate i of A and deleting coordinate i of B. Then A' and B' constitute a blocking pair of matrices.*

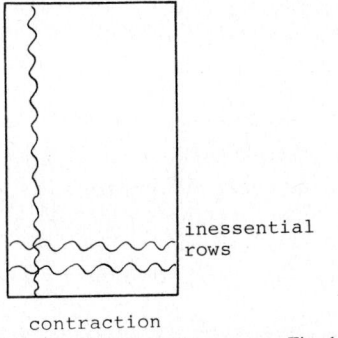
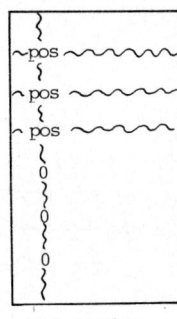

Fig. 1.3.

1.4. *Further examples of blocking pairs*

In addition to the example of paths in a two terminal network there are several other examples. Most of them have been mentioned and explained in detail by Fulkerson. Therefore, they shall only here be listed by name. Again they are cases for which a blocking matrix has been explicitly determined. As most of the examples are of combinatorial origin, A is typically a 0–1 matrix, describing the combinatorial structure in question.

Elementary vectors in orthogonal subspaces [12, 5]. *Matroid intersections* [14, 27, 7, 21, 9, 39]. *Permutations, Matroid bases, Lines in an array, Spanning arborescences* [12]. *Rooted directed cuts* [16]. *Supply-Demand networks* [18, 40]. *Circulation networks* [38].

2.

2.1 *Antiblocking polyhedra*

The theory of antiblocking polyhedra follows a development similar to the one for blocking polyhedra. The basic difference lies in the definition, where the inequality sign has been reversed. Here we will follow the development in the previous section and at the same time, and also later (in Section 5), explain some of the main differences and similarities between the two theories.

Consider again an m by n matrix A with non-negative elements and consider the polyhedron

$$\mathcal{A} = \{x \in \mathbf{R}^n_+ \mid Ax \leq 1\},$$

where $1 = (1, \ldots, 1)$ with m elements. $\mathcal{A}$ is a convex, closed polyhedron containing 0. Additionally, $\mathcal{A}$ is bounded if and only if A contains no column consisting entirely of zeros.

The Farkas lemma implies that a row vector a^i of A is *inessential* in defining $\mathcal{A}$ if and only if a^i is dominated by a convex combination of rows of A.

The *antiblocker* $\bar{\mathcal{A}}$ of $\mathcal{A}$ is defined as

$$\bar{\mathcal{A}} = \{x^* \in \mathbf{R}^n_+ \mid x^* \cdot x \leq 1 \quad \text{for all } x \in \mathcal{A}\}.$$

The following theorem contains the basic involutory polar properties of $\mathcal{A}$ and $\bar{\mathcal{A}}$.

Theorem 2.1 (Fulkerson [13]). *Let A be a non-negative matrix having no zero columns. (This implies that $\mathcal{A}$ is the convex hull at its extreme points). Let $b^1, \ldots, b^r$ denote the extreme points. Let the matrix B have rows $b^1, \ldots, b^r$. Then*
  (i) $\bar{\mathcal{A}} = \{x \in \mathbf{R}^n_+ \mid Bx \leq 1\}$,
  (ii) B *has no zero columns,*
  (iii) $\bar{\bar{\mathcal{A}}} = \mathcal{A}$.

$\mathcal{A}(A)$ and $\bar{\mathcal{A}}(B)$ are called an antiblocking pair of polyhedra (matrices).

It is remarked that there is no 1–1 correspondence between facets and extreme points as in Theorem 1.1. Hence the notion of proper matrices has been dropped here. Otherwise, the situation for extreme points has been more fully described in the following theorem.

Theorem 2.2 (Fulkerson [13]). *Let A be a non-negative matrix defining the polyhedron $\mathcal{A} = \{x \in \mathbf{R}^n_+ \mid Ax \leq 1\}$ and let $b, b^1, \ldots, b^s$ be points of $\mathcal{A}$ such that b is an extreme point of $\mathcal{A}$ and dominated by a convex combination of $b^1, \ldots, b^s$. Then b can be obtained from some b^i by projection, i.e. by setting certain components of b^i equal to zero.*

Fig. 2.1 [40] illustrates the previous two theorems. Note that the first three rows of the B matrix are inessential and can be obtained as projections of other rows, according to Theorem 2.2.

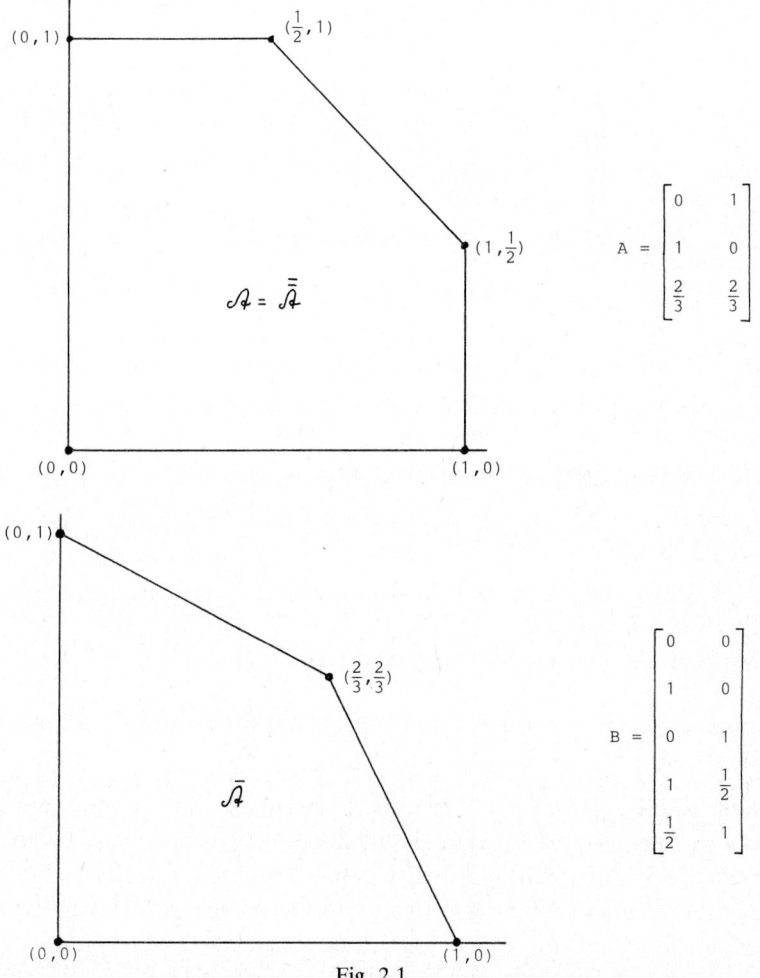

Fig. 2.1.

In the study of blocking polyhedra was included the consideration of a certain packing problem. Instead, we shall here study the following covering problem:

$$\min_{y \in \mathbf{R}^m} 1 \cdot y,$$
$$yA \geq w, \quad (2.1)$$
$$y \geq 0,$$

where $w \in \mathbf{R}_+^n$ and $1 = (1, \ldots, 1)$ with m components.

Let again A be a non-negative matrix with rows $a^1, \ldots, a^m$, and let B be a non-negative matrix with rows $b^1, \ldots, b^r$. Both matrices contain no zero columns. Similarly to the blocking situation we now say that the *min-max equality* holds for A and B (in this order), if and only if for every $w \in \mathbf{R}_+^n$ the covering problem (2.1) has an optimal solution y such that

$$1 \cdot y = \max_{1 \leq j \leq r} b^j \cdot w.$$

We also say that the *max-max inequality* holds for the (unordered) pair A, B, if and only if for every pair of non-negative vectors $1 \in \mathbf{R}_+^n$, $w \in \mathbf{R}_+^n$, we have

$$\left(\max_i a^i \cdot 1 \right) \left(\max_j b^j \cdot w \right) \geq 1 \cdot w.$$

We have the following

Theorem 2.3 (Fulkerson [13]). *The min-max equality holds for the ordered pair A, B, if and only if A and B are an antiblocking pair of matrices. Hence, if the min-max equality holds for A, B it also holds for B, A.*

Similar to the blocking situation we also have

Theorem 2.4 (Fulkerson [13]). *The pair A, B is an antiblocking pair of matrices, if and only if*
  (i) $a^i \cdot b^j \leq 1$ *for all rows a^i of A and b^j of B,*
  (ii) *the max-max inequality holds for A, B.*

2.2. *An example (The rational form of the Dilworth theorem on partially ordered sets)*

Consider a finite partially ordered set of elements. Let the rows of A denote the incidence vectors of the maximal dependent sets (chains), and let the rows of B be incidence vectors of all maximal independent sets (antichains). Fig. 2.2 [13], illustrates the situation for a case of five elements, where the dependences are indicated by the chains of vertices in the graph attached.

Here A and B constitute a pair of antiblocking matrices.

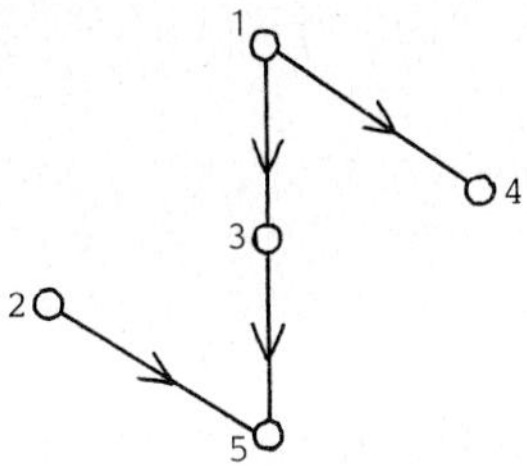

$$A = \begin{bmatrix} 1 & 0 & 1 & 0 & 1 \\ 1 & 0 & 0 & 1 & 0 \\ 0 & 1 & 0 & 0 & 1 \end{bmatrix}$$

$$B = \begin{bmatrix} 1 & 1 & 0 & 0 & 0 \\ 0 & 1 & 1 & 1 & 0 \\ 0 & 0 & 0 & 1 & 1 \end{bmatrix}$$

Fig. 2.2.

The min-max equality for A, B (in this order) is the rational form of the Dilworth Theorem, which can be expressed as follows. Let w denote a vector of certain given numbers on the elements, where w_i is the number on the ith element. The objective is to minimize a sum of weights y, where each element of y denotes the weight of the corresponding dependent set (chain of elements), such that w_i is less than or equal to the weighted sum of dependent sets covering each element i. The result is then equal to the maximal value for the sum of numbers of elements in an independent set.

For example, consider Fig. 2.3 where some numbers w_i have been put on the elements of the example in Fig. 2.2. $y = (y_1, y_2, y_3) = (9, 8, 11)$ is here an optimal solution for the covering problem (2.2) with the result $y_1 + y_2 + y_3 = 28$. The antichain corresponding to the third row of matrix B contains the maximal sum of numbers. This sum is $w_4 + w_5 = 8 + 20 = 28$, which is equal to the result of the covering problem.

Dually, the min-max equality for B, A says the following. Let the elements of y denote weights of each independent set (antichain). The objective is now to select y in order to minimize the sum of weights, such that w_i is less than or equal to the weighted sum of independent sets covering each element i. The result is then equal to the maximal sum of numbers of elements in a dependent set.

Again, considering the example in Fig. 2.3 we obtain for the covering problem with the B matrix instead of A that $y = (y_1, y_2, y_3) = (9, 7, 20)$ is an optimal solution with the result $y_1 + y_2 + y_3 = 36$. The chain corresponding to the first row of A has the same value for the maximal sum of weights, which is $w_1 + w_3 + w_5 = 36$.

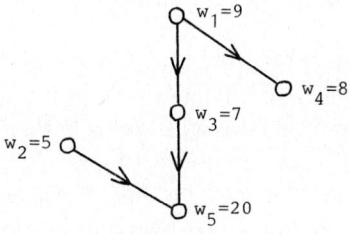

Fig. 2.3.

It is noted that the optimal solutions in both cases are here integral. See Section 4, in particular Theorem 4.1.

The max-max inequality says that for two weight vectors, w and l, we have that the product of the largest total weight for a dependent set computed by using l, and the largest total weight of an independent set, computed using w, is at least equal to $l \cdot w$.

It is finally noted that in order to establish the companion to Theorem 1.4, we only need the dropping operation, and the corresponding theorem is as follows:

Theorem 2.5 (Fulkerson [14]). *Let A and B be an antiblocking pair of matrices. Define A' and B' as the matrices obtained by dropping the ith column of A and B respectively. Then A' and B' are an antiblocking pair of matrices.*

2.3. *Further examples on antiblocking pairs*

Most of the previous subjects, mentioned in Section 1, have also been considered in the antiblocking framework. Additionally, one can mention the study of *matching problems* in a graph [13, 8] and the *clique problem* [13]. The study of the last problem lead to the proof of the *perfect graph theorem* [24, 15].

3. The relation to convex analysis

As it turns out, much of the material in the preceding sections is also valid in more general situations. However, we shall here primarily restrict the discussion to the antiblocking framework. We are going to study sets instead of polyhedra, and we shall consider a generalization of the assumption on non-negativity. This is done by a modification of the classic *Minkowski polarity*. Hence, we shall first state the basic properties of this polarity (see e.g. [32]).

Consider a non-empty set $C \subseteq \mathbf{R}^n$. Let C^* denote the *polar set* of C, i.e.

$$C^* = \{x^* \in \mathbf{R}^n \mid x \cdot x^* \leq 1, \forall x \in C\}.$$

If C is closed, convex and containing 0, then we have the involutory property, i.e.

$$C^{**} = C,$$

which means that C is the polar of C^*.

The idea is now to modify the definition of polar sets by intersection of the polar sets and a given set, D. As the polar set is basically a set which contains all normals for the supporting hyperplanes of the original set, we may loose some normals by this intersection. In general, this destroys the involutory property. Therefore, we shall impose some additional conditions in order to save the involutory property.

Let $\mathcal{A} \subseteq \mathbf{R}^n$ and $D \subseteq \mathbf{R}^n$ be non-empty sets. Define *the antiblocking set $\bar{\mathcal{A}}$ of $\mathcal{A}$ with respect to D* as follows

$$\bar{\mathcal{A}} = \mathcal{A}^* \cap D.$$

The basic question is now to examine the conditions under which

$$\bar{\bar{\mathcal{A}}} = \mathcal{A}, \tag{3.1}$$

i.e., when $\mathcal{A}$ and $\bar{\mathcal{A}}$ constitute *a pair of antiblocking sets*.

It is remarked that if $\mathbf{R}^n = D$, we are back in the Minkowski polarity. Hence, we are actually dealing with a generalization of this polarity.

The following theorem gives a necessary and sufficient condition for equation (3.1) to be valid.

Theorem 3.1 (Tind [34, 35]). *Let D be a closed, convex set containing 0. Then $\bar{\bar{\mathcal{A}}} = \mathcal{A}$, if and only if there exists a closed, convex set $C \subseteq \mathbf{R}^n$, such that $\mathcal{A} = C \cap D$ and such that $D^* \subseteq C$.*

If $C = \{x \in \mathbf{R}^n \mid Ax \leq 1\}$, where A is a non-negative matrix, and $D = \mathbf{R}_+^n$, then $D^* = \{x \in \mathbf{R}^n \mid x \leq 0\}$ and $D^* \subseteq C$. Now, $\mathcal{A} = C \cap D$ together with $\bar{\mathcal{A}}$ constitute a pair of antiblocking polyhedra (with respect to $\mathbf{R}_+^n$). Hence Theorem 3.1 is a generalization of statement (iii) in Theorem 2.1.

Araóz [1] and Johnson [22] have studied generalizations in which D is a closed, convex and polyhedral cone. $\bar{\mathcal{A}}$ is defined with respect to D as before, but $\bar{\bar{\mathcal{A}}}$ is defined with respect to the negative polar cone of D, i.e. $\bar{\bar{\mathcal{A}}} = \bar{\mathcal{A}} \cap (-D^*)$. This leads for polyhedra to generalizations of the min-max equality and to a generalization of antiblocking pairs of matrices, and to similar generalizations for blockers.

Griffin [19] has recently made a study of general polarities including the following development. Let $W \in \mathbf{R}^{n \times m}$, $u \in \mathbf{R}^n$, $v \in \mathbf{R}^m$ and $\alpha \in \mathbf{R}$ be constants. Consider the following inequality with variables $x \in \mathbf{R}^n$ and $x^* \in \mathbf{R}^m$,

$$xWx^* + x \cdot u + v \cdot x^* \leq \alpha.$$

This is a generalization of the basic inequality $x \cdot x^* \leq 1$ used in the definitions of an antiblocker. In the case of polyhedra, polarities are investigated under this inequality leading to results that generalize previous results for antiblockers, including Theorem 3.1; and similarly for blockers. See also [10].

Since sets in general cannot be described as the intersection of a finite number of halfspaces, given by a finitely dimensional matrix, no attempt has been made to generalize the notion of antiblocking matrices and the relating results for sets. However, it should be noted that a generalization of the max-max inequality is going to be almost identical to similar inequalities for support functions of a set and its polar set. See [32], pp. 129–130.

Parallel to the development of antiblocking sets based on a certain modification of the Minkowski polarity, we can also develop a notion of antiblocking functions based on a certain modification of the idea of polar functions [32].

Consider a concave, non-negative, closed and non-decreasing function $f(x): \mathbf{R}_+^n \to \mathbf{R}$.

Define the following function $\bar{f}(x^*): \mathbf{R}_+^n \to \mathbf{R}$ as

$$\bar{f}(x^*) = \sup\{y^* \in \mathbf{R} \mid -x \cdot x^* + y^* f(x) \le 1 \quad \text{for all } x \in \mathbf{R}_+^n\}.$$

Call $\bar{f}(x^*)$ the *antiblocking function* of f. $\bar{f}(x^*)$ becomes also concave, non-negative, closed and non-decreasing, and we have the involutory property [34],

$$\bar{\bar{f}}(x) = f(x), \qquad (3.2)$$

i.e., f and $\bar{f}$ constitute a pair of antiblocking functions.

The equation (3.2) is implied by the fact that the subgraphs of f and $\bar{f}$ are a pair of antiblocking sets with respect to $D = \mathbf{R}_+^{n+1}$ (with an inessential modification due to a change of sign).

Equation (3.2) can be given an economic interpretation like the one Williams [41] has given for conjugate functions. Especially in this economic context it makes sense to study the polar relationship in the non-negative orthant, as it has been done here. For details, see [34].

As often mentioned, similar developments exist for the blocking case. This involves the study of the socalled *reverse polar set* C^0 of a non-empty set $C \subseteq R^n$, where C^0 is defined by

$$C^0 = \{x^0 \in \mathbf{R}^n \mid x \cdot x^0 \ge 1 \quad \text{for all } x \in C\}.$$

See [1, 2, 33].

4. Relation to integer programming

In spite of the fact that the development of the blocking and antiblocking theory grew out of the study of various problems in extremal combinatorics, not much has been said about combinatorics in this paper so far. But a few of the most essential relationships will now be described.

The main question is here: When does the packing problem (1.1) (or the covering problem (2.1)) have an integer solution? We assume that w is integer, and that A is a zero-one matrix, e.g. an incidence matrix over a certain combinatorial structure. We use the expression that the *max-min (min-max) equality holds strongly* for A, B, provided that the packing (covering) problem has an integer solution vector y, whenever w is a non-negative integer vector. Now we have the following theorem for antiblocking pairs:

Theorem 4.1 (Fulkerson [13]). *Let A be a 0–1 matrix and let B be an antiblocking matrix of A. The min-max equality holds strongly for A, B, if and only if each* essential row of B is a 0–1 vector. Hence if the min-max equality holds strongly for A, B, it also holds strongly for B^*, A, where B^* consists of all essential rows of B.

The same theorem cannot directly be transferred to blocking polyhedra, as shown and illustrated by Fulkerson [14].

However, for the two main examples in Section 1.2 and Section 2.2, the max-min and the min-max equality holds strongly for each corresponding problem, respectively.

Consider first the example in Section 1.2. For $w = (1, \ldots, 1)$ the strong form of the max-min equality says for A, B (in this direction): The maximum number of pairwise edge-disjoint paths joining the terminals is equal to the minimum number of edges in a cut separating the two terminals [11]. Dually, for B, A (in this direction) the max-min equality says: The maximum number of pairwise edge disjoint cuts separating the terminals is equal to the minimum number of edges in a path joining the terminals.

Consider next the example of Section 2.2, and let again $w = (1, \ldots, 1)$. First we get for A, B the wellknown Dilworth theorem for partially ordered sets: The minimum number of chains required to cover all elements of the set is equal to the maximum number of elements in an antichain. Dually, the other way we get: The minimum number of antichains required to cover all elements is equal to the maximum number of elements in a chain.

Theorem 4.1 plays a central role for a number of combinatorial results. In particular, one can mention the perfect graph theorem, mentioned in Section 2.3.

Let us here state the contents of this theorem. In a graph G, a *clique* is a collection of mutually adjacent vertices, and an *anticlique* — or an independent set of vertices — is a collection of mutually non-adjacent vertices. We are looking for the number of cliques necessary to cover all vertices. Call this number $\pi(G)$. Let $w(G)$ denote the size of the largest anticlique. Obviously, $\pi(G) \geq w(G)$. If the two numbers coincide for G and all of its (vertex-) induced subgraphs, G is called *perfect*. (It is remarked that this is not the original definition of a perfect graph as defined by Berge [3]. Here we use the same framework as in Lovász [25] and Huang [20].)

The complement of a graph $G = (V, E)$, where V denotes the vertices and E the edges, is the graph $\bar{G} = (V, \bar{E})$ where

$$\bar{E} = \{e \in (u, v) \mid u, v \in V, u \neq v, (u, v) \notin E\}.$$

(u, v) denotes an edge between u and v. We are now ready to state the *perfect graph theorem*:

Theorem 4.2. *G is perfect if and only if $\bar{G}$ is perfect.*

Berge [3] lists several classes of perfect graphs. They include the wellknown class of bipartite graphs.

The proof of the perfect graph theorem was made by Lovász [24], based on the work of Fulkerson, in particular Theorem 4.1, to which the perfect graph theorem is intimately related [15]. In fact the essential rows (in the antiblocking sense) of a matrix A that satisfies the contents of Theorem 4.1 are the incidence vectors of the maximal cliques of a perfect graph.

Returning to the question, when integer solutions are obtained for the packing problem (1.1) and the covering problem (2.1), further results have been obtained by Berge [4] and Fulkerson et al. [17] for these and similar problems. For further developments of this sort one may also look into [24, 26, 29, 30, 31, 37, 28, 6].

5. On the relation between blocking and antiblocking polyhedra

As can be seen from the preceding material, there are many similarities and parallel developments between the theory of blocking polyhedra and the theory of antiblocking polyhedra. But there are also significant differences. To mention a couple, we can again note the differences between Theorem 1.4 and Theorem 2.5, and we can note the non-existence of a blocking theorem equivalent to Theorem 4.1.

But there has been made additional research in order to clarify the relations between the two theories. Fulkerson has described a connection between blocking and antiblocking for 0–1 matrices. This relation determines the antiblocking matrix of the complement matrix $E-A$ of A, if the blocking matrix of A is known. E is the matrix all of whose elements are equal to one. Todd [36] has made a generalization of this theorem. See also McDiarmid [27].

Additionally, Huang and Trotter [21] have provided a method for determining blocking and antiblocking results for a given matrix A, when the rows of A are the extreme points of a polyhedron.

Finally, consider the polar and the reverse polar of a set, as defined in Section 3. Johnson [22] has by means of the concept of a penumbra (for a definition, see Rockafellar [32], Section 3) established necessary and sufficient conditions for two sets to be the polar and the reverse polar, respectively of a joint set. Similar results, also shown in [22], are transferred to the blocking and antiblocking case, and are thus relating the two theories to each other.

References

[1] J.A. Araóz, Polyhedral neopolarities, Dissertation, Department of Applied Analysis and Computer Science, University of Waterloo, Waterloo, Canada (1973).
[2] E. Balas, Disjunctive programming: properties of the convex hull of feasible points, Graduate School of Industrial Administration, Management Science Research Report N° 348, Carnegie–Mellon University, Pittsburg, PA 15213, U.S.A. (1974).
[3] C. Berge, Graphs and Hypergraphs (North-Holland, Amsterdam, 1973).

[4] C. Berge, Balanced Matrices, Math. Programming 2 (1972) 19–31.
[5] R.G. Bland, Elementary vectors and two polyhedral relaxations, European Institute of Advanced Studies, Working Paper N° 76–60, Brussels, Belgium (1976).
[6] V. Chvátal, On certain polytopes associated with graphs, J. Combinatorial Theory Ser. B 18 (1975) 138–154.
[7] W.H. Cunningham, An unbounded matroid intersection polyhedron, Linear Algebra and Appl. 16 (1977) 209–215.
[8] J. Edmonds, Paths, trees and flowers, Canad. J. Math. 17 (1965) 449–467.
[9] J. Edmonds and R. Giles, A min-max relation for submodular functions on graphs, in: Hammer et al., eds, Studies In Integer Programming, Ann. Discrete Math. 1 (1977) 185–204.
[10] J. Edmonds, V. Griffin and J. Araóz, Polarities given by a system of bilinear inequalities, Working Paper N° 25, Departemento de Matematicas y Ciencia de la Computacion, Universidad Simon Bolivar, Valle de Sartenejas, Edo. Miranda, Venezuela (1977).
[11] L. Ford and D.R. Fulkerson, *Flows in Networks* (Princeton University Press, Princeton, NJ, 1962).
[12] D.R. Fulkerson, Blocking polyhedra, in: B. Harris, ed, Graph Theory and Its Applications (Academic Press, New York, 1970) 93–112.
[13] D.R. Fulkerson Anti-blocking polyhedra, J. Combinatorial Theory 12 (1972) 50–71.
[14] D.R. Fulkerson, Blocking and anti-blocking pairs of polyhedra, Math. Programming 1 (1971) 168–194.
[15] D.R. Fulkerson, On the perfect graph theorem, in: T.C. Hu and S.M. Robinson, eds., Mathematical Programming (Academic Press, New York, 1973) 69–77.
[16] D.R. Fulkerson, Packing rooted directed cuts in a weighted directed graph, Math. Programming 6 (1974) 1–14.
[17] D.R. Fulkerson, A.J. Hoffman and R. Oppenheim, On balanced matrices, in: M.L. Balinski, ed., Pivoting and Extensions, Math. Programming Study 1 (1974) 120–133.
[18] D.R. Fulkerson and D.B. Weinberger, Blocking pairs of polyhedra arising from network flows, J. Combinatorial Theory 18 (1975) 265–283.
[19] V. Griffin, Polyhedral polarity, Ph.D. thesis, Research Report CORR 77/33, Department of Combinatorics and Optimization, University of Waterloo, Waterloo, Ontario, Canada (1977).
[20] H.-C. Huang, Investigations on combinatorial optimization, Dissertation, Department of Operations Research, Technical report N° 308, Cornell University, Ithaca, NY (1976).
[21] H.-C. Huang and L.E. Trotter, Jr. A technique for determining blocking and anti-blocking polyhedral descriptions, Report N° 7763–OR, Institut für Ökonometrie und Operations Research, Universität Bonn, Bonn, W. Germany (1977).
[22] E.L. Johnson, Support functions, blocking pairs and anti-blocking pairs, IBM T.J. Watson Research Center, Yorktown Heights, New York, (1977).
[23] A. Lehman, On the width length inequality (Mimeo, 1965).
[24] L. Lovász, Normal hypergraphs and the perfect graph conjecture, Discrete Math. 2 (1972) 253–267.
[25] L. Lovász, A characterization of perfect graphs, J. Combinatorial Theory Ser. B 13 (1972) 95–98.
[26] L. Lovász, Certain duality principles in integer programming, in: Hammer et al. eds., Studies in Integer Programming, Ann. Discrete Math. 1 (1977) 363–374.
[27] C. McDiarmid, Blocking, antiblocking, and pairs of matroids and polymatroids, Department of Operations Research, Technical Report N° 282, Cornell University, Ithaca, NY (1976).
[28] G.L. Nemhauser and L.E. Trotter, Jr., Vertex packings: Structural properties and algorithms, Math. Programming 8 (1975) 232–248.
[29] M.W. Padberg, Perfect zero-one matrices, Math. Programming 6 (1974) 180–196.
[30] M.W. Padberg, Characterizations of totally unimodular, balanced and perfect matrices, in: B. Roy, ed., Combinatorial Programming: Methods and Applications (Reidel, Boston, 1975) 275–284.
[31] M.W. Padberg, Almost integral polyhedra related to certain combinatorial optimization problems, Linear Algebra and Appl. 15 (1976) 69–88.
[32] R.T. Rockafellar, Convex Analysis (Princeton University Press, Princeton, NJ, 1970).
[33] J. Tind, Blocking and antiblocking sets. Math. Programming 6 (1974) 157–166.

[34] J. Tind, On antiblocking sets and polyhedra, in: Hammer et al. eds., Studies in Integer Programming, Ann. Discrete Math. 1 (1977) 507–516.
[35] J. Tind, Certain kinds of polar sets and their relation to mathematical programming, Institut for Operationsanalyse, Aarhus Universitet, Aarhus, Denmark (1976).
[36] M. Todd, Dual families of linear programs, Department of Operations Research, technical report no. 197, Cornell University, Ithaca, NY (1973).
[37] L. Trotter, Solution characteristics and algorithms for the vertex packing problem, Dissertation, technical report no. 168, Department of Operations Research, Cornell University, Ithaca, NY (1973).
[38] L. Trotter and D.B. Weinberger, Symmetric blocking and antiblocking relations for generalized circulations, Department of Operations Research, technical report no. 300, Cornell University, Ithaca, NY (1976).
[39] D.B. Weinberger, Transversal matroid intersections and related packings, Math. Programming 11 (1976) 164–176.
[40] D.B. Weinberger, Network flows, minimum coverings, and the four-color conjecture, Operations Res. 24 (1976) 272–290.
[41] A.C. Williams, Nonlinear activity analysis, Management Sci. 17 (1970) 127–139.

Report of the Session on
COMPLEXITY OF COMBINATORIAL PROBLEMS

R.L. GRAHAM (Chairman)
D. Hausmann (Editorial Associate)

Computational complexity theory as developed by theoretical computer scientists started to have a major impact on the field of discrete optimization in the early 1970's. In two remarkable papers, S.A. Cook and R.M. Karp demonstrated that a large number of combinatorial problems, notorious for their computational intractability, were all equivalent up to a polynomial transformation. This implies that a good (i.e., polynomial-time) algorithm for any of them could be used to solve all others in polynomial time as well, which would establish the equality of the problem classes $\mathcal{P}$ and $\mathcal{NP}$ [3]. In general, this equality is considered to be very unlikely, and NP-completeness of a combinatorial problem has become commonly accepted as being indicative of its inherent difficulty.

Computer scientists again must take a lot of credit for establishing membership of $\mathcal{P}$ for many traditional problems. However, the notion of a good algorithm is originally due to Edmonds [1], whose work in matroid optimization theory has been essential in getting this concept accepted and appreciated. We refer to another session report [2] for more extensive comments on the development of good algorithms. Within this area, the challenge to have upper and lower bounds on the time complexity of a problem meet each other remains as large as ever, even for seemingly simple problems. For example, finding the median of n numbers is now known to be possible in $O(n)$ time, but a lot of ingenuity seems to be required to decrease the upper bound or increase the lower bound on the multiplicative constant. It may even be possible to apply mathematical programming on a metalevel to determine optimal algorithms!

The class of problems that have been proved NP-complete is still expanding rapidly and extremely refined results are available (see, e.g. [3]). Typically, what is lacking for those problems is a good characterization for the optimality of feasible solutions, such as the one that duality theory yields for linear programming. The duality theorem provides strong evidence against the NP-completeness of linear programming, since (unlike $\mathcal{P}$) $\mathcal{NP}$ is suspected not to be closed under complementation. Linear programming, graph isomorphism and other problems that still defy classification may well turn out to be neither in $\mathcal{P}$ nor NP-complete, which has been shown to be possible, provided, of course, that $\mathcal{P} \neq \mathcal{NP}$.

An often-heard and justified complaint against NP-completeness theory is that it only yields a very coarse measure of problem complexity. Despite the inadequacy of the usual testing procedures for enumerative algorithms, it is obvious to every practitioner that some NP-complete problems are harder than others. Refinements of the complexity measure would be very welcome. Within the class of NP-complete problems, one might obtain such refinement by investigating various ways of encoding the problem data or by examining the trade-off between running time and worst-case performance of approximation algorithms. A lot of work has been done in the latter area. Questions that need much further study is how to analyze properly the average-case behavior of heuristics (which presupposes a probability distribution over the set of problem instances) and how to relate it to the worst-case behavior.

Computer science has more to offer in terms of theoretical models for the analysis of problem complexity. $\mathcal{NP}$ is included in the class of problems solvable in polynomial space. This class is the same for deterministic and nondeterministic Turing machines. Polynomial-space completeness has been established, for example, for a modification of the well-known game of hex. Beyond that, some problems have been shown to require at least exponential space, such as the "vector reachability" problem: given a finite set of integer vectors, an initial vector u and a final vector v, is it possible to add vectors from the set to u (with repetition allowed) so as to reach v, while remaining within the positive orthant? Another question of obvious interest is to extend all these concepts to parallel machine models.

Among practitioners one often encounters a lack of enthusiasm for these new tools, which probably only an equality proof of $\mathcal{P}$ and $\mathcal{NP}$ could remove completely. It is unfortunately true that many of the results obtained in this area are asymptotic and yield little concrete information for problems of "real world" size; they still have to stand the test of practical applicability. Yet we would argue that the evident success of complexity theory in determining factors that influence the inherent difficulty of a problem and in bordering the best algorithmic performance that we may expect to get, will be of ultimate benefit to theoreticans and practitioners alike.

Acknowledgements

J. Edmonds, R.L. Graham, B. Korte, E.L. Lawler and G.L. Nemhauser were among the discussants in Banff. A summary of the points they raised was prepared by J.K. Lenstra and A.H.G. Rinnooy Kan and edited by D. Hausmann.

References

[1] J. Edmonds, Paths, trees, and flowers. Canad. J. Math. 17 (1965) 449–467.
[2] J.K. Lenstra, Report of the session on "Algorithms for special classes of combinatorial optimization problems", Ann. Discrete Math. 4 (1979) this volume.
[3] J.K. Lenstra and A.H.G. Rinnooy Kan, Computational complexity of discrete optimization problems, Ann. Discrete Math. 4 (1979) this volume.

Report of the Session on
STRUCTURAL ASPECTS OF DISCRETE PROBLEMS

P. HANSEN (Chairman)
B. Simeone (Editorial Associate)

1. Structure and structured problems

Some variance is observed regarding the precise meaning of the terms "structure" and "structured problems". It is not unreasonable to expect that the more structure a problem has, the easier it is to solve. However, there are several ways of defining structure in discrete optimization problems. *Well-solved problems*, i.e. problems for which there exists a *polynomially bounded algorithm* are clearly structured problems. The class of structured problems could be limited to those which are not *NP-complete*. However, all NP-complete problems are not, in all respects, equivalent in difficulty. For some problems, all constraints are of a particular type, e.g. inequality constraints on sums of 0–1 variables. This is the case for the *packing, covering* and *partitioning problems*, for the *travelling salesman problem* and for many others. Special techniques, not applicable to general integer programming problems, are available to solve these problems. It can therefore be said that they have some structure. For similar reasons, the *vertex packing problem*, the *chromatic number problem*, and many other problems defined on graphs may be considered as structured problems.

Integer programming problems for which only a part of the constraints, have a simple form can be said to possess some structure. Finally, all integer programming problems do possess some underlying structure, i.e. the *convex hull of the feasible integer points*, which it may be worthwhile to explore both for the insight it provides and for computational reasons.

2. Recognizing structured problems

Two broad questions have been discussed:
(a) how can a structured problem be recognized as such?
(b) is the effort necessary to do so worthwhile?
Regarding question (a) let us first note the interest for the analyst in constructing his model so as to capture the relevant underlying structure. If, however, no structure is immediately apparent it may be possible to find an equivalent

structured model. This may arise during the initial modelling phase, or during later analysis of the original model.

Well-solved problems to which other problems may sometimes be reduced are the *shortest path problem*, the *network flow problem*, the *matching problem*, and the *two matroid intersection problem*. A recent example of an insightful direct reduction is the expression of the *shortest elementary even path problem* as a matching problem by Edmonds. It has also been shown by Edmonds and Johnson that the *Chinese postman problem* is reducible to the matching problem. Minimum cost network flow problems are among the best solved combinatorial optimization problems. Perhaps the reduction problem which is the most important in practice is to discover when a linear programming problem $\min cX$ subject to the constraints $AX = b$ is equivalent to a minimum cost network flow problem. This is of course the case if the Heller–Tompkins condition holds for the matrix of coefficients A. More generally, if the matrix A can be put into the form RUD, R nonsingular, U node-arc incidence matrix, and D with positive diagonal elements and 0's elsewhere the linear program can be converted into a network flow problem with constraints $UY = R^{-1}b$ where $Y = DX$. Iri, as well as Cunningham and Bixby have proposed polynomial algorithms to test if A can be put in the above-described form, which yields the equivalent minimum cost network flow problem, if it exits. A related unsolved, important and apparently difficult problem is to *find the largest submatrix of a linear program which is equivalent to a network flow problem*.

Some combinatorial optimization problems defined on graphs are better solved for planar graphs than in the general case. An example is the *maximum cut problem*, which is well-solved for planar graphs but NP-complete for non-planar graphs, as shown by Hadlock. Another example is the *maximum flow problem* for which Cherkasky has recently obtained an algorithm, which combines those of Dinits and of Karzanov, and works in $O(M^{1/2}N^2)$ i.e. in $O(N^{5/2})$ for planar graphs, while Karzanov's algorithm takes $O(N^3)$ operations both for planar and for non-planar graphs. Testing a graph for planarity can be done in $O(N)$ operations with an algorithm of Hopcroft and Tarjan.

If a graph has the "König property", i.e. the maximum cardinality of a matching is equal to the minimum cardinality of a vertex cover, a minimum vertex cover can be obtained by a matching algorithm. Gavril has shown that testing for the König property can be done in $O(M)$ when a maximum matching is known.

Recognizing whether a graph G is the *line-graph* of another graph G^* (with the result that a hamiltonian circuit in G is simply an eulerian circuit in G^*) could be of interest. Indeed, users of PERT prefer to work on a graph whose line-graph was the one that actually represented their problem. This is achieved in a not-too-formal way by adding a number of dummy arcs. Hall has proposed an algorithm for testing if a graph is a line-graph. It is also easy to test, by enumeration of all four-point subgraphs if a graph is *claw-free*. If such is the case, a recent polynomial algorithm of Minty allows finding a *maximum vertex packing*.

Regarding question (b) many discussants feel that the structure of a problem should be obvious to the analyst — but there is evidence that there are many cases when such a structure has been initially unobserved, or when a reformulation as an equivalent structured problem has only been done after some time. For instance, certain production planning problems were initially solved by a dynamic programming method due to Wagner and Whitin. Later on, Zangwill extended the models in terms of a network representation of the problems. In many cases, as the constraints were simple material balances from one period to the next, the resulting linear programs were network flow problems. The usefulness of automatic methods for detecting special structures in production codes could be ascertained only by a delicate cost-benefit analysis. Indeed, such codes include a matrix generator, an optimization routine, and a report writer. If the structure-detecting routine suggests an alternative algorithm, there may still be a substantial amount of work involved in restructuring the data in suitable form for this algorithm, and restructuring the output from the algorithm so that the report writer can read it. If the application is run frequently and the amount of time saved is substantial, in spite of the extra phases of data-processing, then the operation could be useful.

Much work is successfully done in using the structural properties of linear programming matrices for efficiently representating and updating current bases (e.g. $L-U$ decomposition). This is relevant for discrete optimization when considering linear programming relaxations of mixed-integer programs and could result in much reduced computation times for multi-project–multi-period decision models such as those currently occurring in investment planning and capital budgeting.

3. Equivalent forms of structured problems

The Cook–Karp theorem establishes an equivalence between the NP-complete problems. As some NP-complete problems are empirically known to be easier to solve than others, the question of using that equivalence for reformulating a problem is worth raising. However, a "blow up" effect is often observed; for instance a small *vertex cover problem* results in a large *hamiltonian circuit problem*. Reductions seem not to be useful for computational purposes, unless the increase factor is linear. The *knapsack problem* is often considered as easier to solve than others and some other problems may be efficiently reduced to that form. Many results from *threshold logic* are available, which give rather complicated necessary and sufficient conditions, and a number of easy necessary conditions for a system of inequalities of the covering type to be reducible to a single knapsack inequality. In the case of the *set-packing problem* Chvatal and Hammer devised a polynomial algorithm for recognizing whether it can be reduced to a knapsack problem.

A problem is sometimes expressed as a structured problem but not in the simplest possible way; for instance, a *clustering problem of* McCormick, Schweitzer and White, initially expressed as a *quadratic assignment problem* was shown to be equivalent to a *travelling salesman problem* by Lenstra. While there seems to be no systematic way to avoid overly complicated formulations, the known inclusion relations between structured problems, e.g. those used in the proof of the Cook–Karp theorem, could help to focus on perhaps equivalent simpler expressions.

The equivalence between some discrete optimization problems and *continuous nonlinear programs*, studied by Ragavachari and by Rosenberg, among others, provides interesting insights, but the resulting programs are very difficult to solve. Results can still be obtained in establishing the equivalence between discrete optimization problems and continuous *linear programs* or *network flow problems*. Indeed, Hammer's expression of the *min-cut problem* as an *unconstrained quadratic 0–1 program* has been recently exploited by Picard and Ratliff to study *cut-equivalent networks* and *quasi-symmetric networks* and to solve some *machine location* and *mine exploitation problems*.

A variety of means have been devised to exploit the underlying structure of a discrete optimization problem in order to express it in a simpler form. Some simplification may be done before applying any algorithm, in a *preprocessing phase*. As shown by Bradley, Hammer and Wolsey, in linear 0–1 programs quite often large coefficients can be reduced to smaller ones without modifying the set of feasible solutions. For Petersen's capital budgeting problems, this results in most coefficients becoming equal to 0 and the few remaining ones becoming very small, i.e. equal to 1 or 2.

For some problems, such as the 0–1 *knapsack problem*, many variables may be fixed at 0 or 1 after applying some easy tests. Quite often, more than 90% of the variables are thus eliminated. Similar results could perhaps be obtained for other structured 0–1 problems, for instance *set covering problems*. Nemhauser and Trotter have shown that for the vertex packing problem the variables equal to 0 or 1 in the optimal solution of the linear programming relaxation remain at these values in at least one optimal solution.

Simplifications are usually done by eliminating infeasible points, so that after the simplification the feasible set remains the same. In addition to that, one might consider transformations which just preserve the set of optimal solutions, and which may allow new points to be considered as "feasible" as long as they cannot be optimal. In that way the transformed problem may have less constraints than the original one.

In addition to fixing variables, logical inequalities involving few variables may often be deduced from given constraints; these logical inequalities studied by Balas, Guignard, Hammer, Jeroslow, Spielberg and others represent in some sense the essential part of the original constraints and may often be exploited to simplify the problem. The order relations between 0–1 variables are also often of interest, and may be combined to show that some problems or subproblems are infeasible or some variables may be fixed, or some other variables shown to be equal to each other.

These logical manipulations have the desirable property that they can be imbedded in almost any general solution technique for 0–1 problems, leading to improved performance. When some of the facets of the convex hull of feasible solutions of a problem can be generated, as for instance for the *travelling salesman problem* or the *multiknapsack problem*, cuts may be added, at least as long as they continue to reduce the duality gap. Good results using such an approach have been obtained by Padberg and Grotschel for the *travelling salesman problem*.

4. Some interesting structured problems

The amount of work devoted to the study of various structured problems appears to be very unevenly distributed, and not at all proportional to the number of potential applications. For instance among NP-complete problems, the *travelling salesman problem* has perhaps received a disproportionate amount of attention relative to its known applications. Other structured problems have been neglected or studied superficially. A good example of a very useful neglected problem is the *acyclic subgraph problem*: find the maximum weight subgraph of a graph containing no directed cycles. That problem has applications to multiple criteria decision making, ranking objects in paired comparison experiments, triangularizing input-output matrices, ordering archaeological objects in time, suppressing feedback in electrical networks, and many other problems.

Report of the Session on
POLYHEDRAL ASPECTS OF DISCRETE OPTIMIZATION

A. HOFFMAN (Chairman)

F. Granot (Editorial Associate)

The Linear Discrete Optimization problem is the problem of maximizing a linear objective function cx over a finite set S. For any such finite set there exists a unique bounded polyhedron P which contains S, and whose vertices are a subset of S. This polyhedron is called the convex hull of S. Every such polyhedron can be alternatively defined as the solution set of a finite system L of linear inequalities (see e.g. [14]). Since the optimal solution to an optimization problem where one maximizes a linear function over a polyhedral set occurs at a vertex of that polyhedron, one could solve the discrete optimization problem by maximizing cx over the solution set of L using linear programming techniques. These observations motivated extensive research efforts directed to characterizing and investigating the facial structure of the polyhedron associated with a given discrete optimization problem.

If we could come up with a systematic way to generate the system of linear inequalities L, or equivalently if we could produce all facets of P, then we could replace a discrete optimization problem by its equivalent linear programming problem.

The difficulties in such an approach are two-fold:

(a) Characterizing and calculating the facets of a convex hull of a discrete set S is usually very hard.

(b) Even if we would have a very efficient method for calculating facets of P, the enormous number of facets will usually discourage any attempt to generate all of them.

1. General comments on 0–1 problems

There are some special classes of 0–1 problems for which facets can be calculated efficiently. One such example is the knapsack problem, in which we maximize a linear objective function over a set of 0–1 vectors satisfying a unique linear inequality. The knapsack polytope is thus the convex hull of 0–1 points satisfying a linear inequality. There exists an algorithm (see e.g. [1]) for calculating a valid inequality for the knapsack problem in at most n comparisons. This valid inequality either defines a facet of the polytope, or else differs in each of its coefficients by at most one unit from an inequality defining a facet.

Furthermore, the question of whether the inequality defines a facet or not, can be decided by at most n comparisons. Computational experience, as well as theoretical considerations, indicate that the vast majority of inequalities in this class indeed define facets of the knapsack polytope. For other relevant work see e.g. [10, 15].

The above results can be used in solving arbitrary 0–1 programs. Indeed, inequalities in the above class derived from either one constraint or from some linear combination of the constraints can be used as cutting planes which, as stated by Balas, are usually considerably stronger than the traditional cuts.

For the set packing problem, as reported by Padberg, the concept of the intersection graph associated with a zero one matrix, has been found to be most useful. It provides information about types of linear inequalities that define facets of these problems. One class of graphs that give rise to facet-defining linear inequalities are cliques, i.e., maximal complete subgraphs of a graph. Other subgraphs which are facet producing include the regular graphs, Webs and more. For a complete analysis of this subject see e.g. [11].

The enormous number of facets required for a complete linear description of the solution set of a 0–1 polytope can perhaps be best illustrated by considering a characterization of the convex hull of 0–1 points satisfying a set of linear inequalities with nonnegative coefficients (see e.g. [7]). This characterization provides some indication as to the typical number of facets in these problems. Let the polytope P corresponding to a multiknapsack problem be defined by

$$Ax \leq b$$

with

$$A = (a_{ij}), \quad a_{ij} \geq 0, \quad i \in M, \quad j \in N$$

and

$$x_j \in \{0, 1\}, \quad j \in N.$$

A set $S \subseteq N$ is called a cover if

$$\sum_{j \in S} a_{ij} > b_i \quad \text{for at least one} \quad i \in M,$$

and is called a minimal cover if no proper subset of S is a cover. Every minimal cover gives rise to a family of facets which can be obtained by a "lifting" procedure. The number of such minimal covers for a given polytope is in general exponential. This implies that the number of facets obtained in such a way will also be exponential. Furthermore in order to account for all the facets of a 0–1 polytope with positive coefficients one has to consider not only the minimal covers of P, but also the minimal covers of all polytopes obtainable from P by complementing variables (i.e., setting $y_i = 1 - x_i$) in all possible ways that leave the right hand side positive. The number of minimal covers of such a family of polytopes depends on A, but except for the cases of very small or very large b this number is exponential in n, or may be even worse.

In that regard Jeroslow remarked that the fact that finding all facial inequalities is hard is not surprising at all. This is due to the fact that finding all facial inequalities is actually equivalent to performing a parametric analysis on the problem, or in other words solving the problem for any cost coefficients. This is obviously much more difficult than solving a unique integer problem.

But in spite of all the above observations both Balas and Padberg argued that the attempts at generating some facets of the convex hull, or approximations of facets, should not be viewed as futile or impractical. This mainly stems from the fact that, no matter how large the number of facets is, only a few of them define the optimal points.

Indeed, as Padberg argued, all we are actually interested in is a local description of the convex hull of the combinatorial optimization problem. In particular, this means that the total number of linear inequalities needed to completely describe the convex hull of the solution set, is in all likelihood somewhat irrelevant in exhibiting optimality and exploring the neighborhood of the optimizing point. For the travelling salesman problem we know that the linear description of the problem is very complicated and that the number of such inequalities needed for a complete description of the problem is enormous. However a recent computational study [12] checked the usefulness of some special facial inequalities (the generalized comb inequalities) in the solution of symmetric travelling salesman problems. In the above study the largest problem they attempted to solve was a 318 city problem whose associated linear programming problem has 318 equality constraints and 50,403 variables with upper bounds of one. Following the procedure they described, 183 automatically generated facial inequalities were iteratively added and the augmented linear program was solved. The optimal objective function value of the final augmented linear program was within 1/4 per cent of the actual optimum.

For those cases in which not all facial inequalities are considered, the question of the quality of the inequalities might be very relevant. Nemhauser suggested the worst possible ratio between the linear programming value and the integer programming value as such a measure. Balas added that in some cases the ratio itself is perhaps not sufficiently good and recommended the ratio he used in his work on the knapsack problem, i.e., the ratio of the gap of the linear programming and I.P. solution and the range of the coefficients.

There was general agreement that the difficulty of a problem should not be measured by the number of inequalities needed to specify it. For example, Hoffman pointed out that Dijkstra's algorithm for shortest paths can be viewed as working on a dual problem — the cut packing problem, and then using complementary slackness. In this case we need something of the order of 2^n inequalities to specify the problem, which might lead to the wrong conclusion that the shortest path problem is very hard. There is a temptation to nominate as a measure of difficulty the number of original inequalities needed to specify the optimum, inspired by the results of Bell and Scarf that this number for Z^n is at most $2^n - 1$.

Balas added that in practice, when we attempt to solve a zero one programming problem we usually use some heuristics in order to find a "good" feasible solution, and then we search for a better one. Once a feasible solution is known, we are only interested in the convex hull of better solutions. That convex hull may contain just a few 0–1 points, in which case the convex hull of this set would have fewer facets. This obviously raises the question of how to characterize the convex hull of those points, i.e., the convex hull of the original integer points intersected by a special half space.

2. LOCO Problems

The discussion proceeded with Edmonds presentation of the "Linear Objective Combinatorial Optimization" (LOCO) problem. A LOCO problem is the problem of maximizing a linear objective function cx over a given finite set of vectors S fitting some combinatorial description. A particular case of the LOCO problem is the integer programming problem. However in a LOCO problem the combinatorial characterization of the set of vectors S is not necessarily in terms of integer solutions to a system of linear inequalities. For example, in the matriod optimization problem the points are incidence vectors of linearly independent subsets of a set, described in terms of algebraic relations.

If we would know a good description of a system of linear inequalities whose polyhedron P contains S, has vertices only in S and bounds the objective function, then the LOCO problem would be equivalent to the linear programming problem of optimizing the function subject to these inequalities. The linear programming duality theorem applied to this linear programming problem provides a good characterization of optimality for the LOCO problem. The number of inequalities in the system and whether it includes inessential inequalities seem to be totally irrelevant to its usefulness.

There are only two distinct well known types of LOCO problems that have been well solved by algorithms which make crucial use of a complete description of P. These two problems are the matching problem and the matroid intersection problem, both by Edmonds [3, 4] or alternatively in E. Lawler [9].

For both of these problems, the polyhedral question involves an exponential number of facets, even though the description of the system of inequalities is nice and primal dual techniques have been used to obtain a well bounded algorithm.

Efforts by Jenkyns and Lawler have been put into trying to unify these two classes of problems. If these efforts had been more successful then we would theoretically have only one problem, instead of two. These efforts failed, and on the other hand these two classes of problems include various special cases, like network flow, minimal spanning tree and more. (Lovász's work connecting integer and fractional solutions of matching and packing problems might lead to a general theory.)

There are some interesting theories of special polyhedra for which the associated LOCO problems have not been well solved. Examples for such polyhedra include the totally unimodular polyhedron and the lattice polyhedron, both due to Hoffman, the perfect graph polyhedron of Fulkerson, Lovász and Padberg, and the very general Mengerian binary clutters of Paul Seymour. Nevertheless, the theoretical duality consequences exist. Also, for certain problems on lattice polyhedra, Deborah Kornblum (extending results of Johnson) has shown that a dual greedy algorithm works. She has also shown that the primal greedy algorithms for polymatroids extend to certain lattice polyhedra.

There are also some well solved LOCO problems for which we have not yet been able to describe the associated polyhedron. One example for such a problem is the shortest even path problem, i.e., in an undirected graph take the family R of incident vectors of all simple even paths between two specified nodes. The shortest even path problem is thus a linear programming problem over a polyhedron defined by the convex hull of the set of vectors R plus the positive ortant. This LOCO problem can be solved using matching theory. No good description of the linear system defining that polyhedron has been found. Another example, suggested by recent work of Minty (and also based on matching) is to find a maximum weight independent set in a claw-free graph (i.e., a graph which does not have $K_{1,3}$ as an induced subgraph).

An example for which the LOCO problem is easily solved, but for which the associated polyhedron is nontrivial can be found in some recent work by Jean Francois Maurras. The LOCO problem there can be stated as follows: in a node weighted graph find that edge for which the sum of weights of its end nodes is the largest. A good description of the convex hull associated with this problem is given by a nontrivial theory. However, there is no reason to use this theory for solving that problem. The problem can be solved by just going through the list of edges and choosing the best one.

Still however in comparison with the impressive array of special combinatorial problems which have been shown to be NP-complete, in other words shown to be *not* special but rather equivalent to general integer linear programming, it is disappointing that not more LOCO problems have been well solved by the polyhedral approach.

In some recent work, the polyhedron developed by Edmonds and Giles [5] which unifies matroid intersection theory with the Lucchesi–Yonger theorem (on the min size of a set of edges in a directed graph which meets every directed cut) will hopefully lead to further well solved LOCO problems. Further insight comes from recent work of A. Frank. Giles has also recently well solved, using the polyhedral approach, a new LOCO problem which unifies optimum matchings and optimum branching.

The fascinating topic of the facet inequalities of a LOCO polyhedron was also discussed. Identifying facet inequalities for combinatorial polyhedra where complete linear descriptions are not known is particularly useful for showing that the simpler linear systems are not complete, and for devising possibly incomplete algorithms.

3. Involutoric concepts

Tind brought up the importance of the involutary concept for discrete optimization problems. Involutary properties have been studied for a long time and include rather general settings. For example, we can mention the Minkowski polarity for sets, polarity for cones, polarity in projective spaces and more. They all contain, with some specifications, expressions like: the polar of the polar is the original set or the dual of the dual problem is the original problem.

In combinatorial problems, those rather general ideas are used in a more specific way. This inevitably brings in some essential structures of combinatorial problems like polyhedra, linear programming, 0–1 matrices and more.

The theory of Blocking and Anti-Blocking polyhedron developed by Fulkerson (see e.g. [6]) is an example of a theory of involutoric concepts. This theory encompasses a lot of nice combinatorial problems, and descriptions of nice relations between facets for relevant relaxed polyhedra and extreme points for dual polyhedra.

Still some of the nice results in combinatorial optimization are often hidden in some kind of a general duality argument. If this is not the case then the problem is simply deferred to branch and bound or implicit enumeration or in more recent years to heuristics and computational complexity.

An important fact is that the study of certain special combinatorial problems had led to discovery of certain results, which are valid in a more general setting. Examples of that sort include the Reverse Polar set studied by Balas in his research on disjunctive programming and the work by Araoz, Edmonds and Griffin on Polarity of Polyhedra. Johnson in his work "Integer Programming with Continuous Variables" [8] discovered a very nice relation between the polar and the reverse polar of a set.

On the other hand, it also happens that the development of a general theory can lead to a proof for a special structure. A nice example for that is again the development of the theory of blocking and anti-blocking polyhedra by Fulkerson. This theory was motivated by the lack of a proof for the perfect graph conjecture of Berge. The development of the general theory indeed led Lovasz to the proof of that conjecture. There is also other work going on, trying to fit in more problems into the theory of Blocking and Anti-Blocking polyhedra. This includes, among others, work by L. Trotter and D.B. Weinberger [13] and R.G. Bland [2].

4. Closing remarks

Tind raised the question: "What should we do with problems that do not fit nicely into nice framework as, for instance, the theory of Blocking and Anti-Blocking polyhedra". The answer to which we all agree is not to give up. The description of polyhedral relaxation and their extreme points might lead to certain statements about the complexity of a given problem. Balas suggests the relationship between the 0–1 polytope P and its linear programming relaxation linear programming as one way of measuring the degree of difficulty of a 0–1 program. In the set partitioning polytope, we know that every edge of P is also an edge of linear programming. This important property makes the set partitioning problems easier to solve. Can we identify other classes of problems with such a property, or else for a given class of problems is there a way to measure the maximum edge-distance on the linear programming polytope of two adjacent vertices of P? Answers to these questions might give, according to Balas, a meaningful classification of NP-complete 0–1 programs. From an algorithmic point of view, when working on the polyhedron associated with a combinatorial problem most algorithms are moving from one vertex to an adjacent one, thus, the question raised by Korte, of an "adjacency criterion" is very relevant. The general question whether two vertices of a given combinatorial polyhedron are adjacent in NP-hard. However, for some proper sub-classes, under some symmetry conditions nice conditions for adjacency were obtained. Hopefully some easy way of characterizing adjacency, if possible, might lead to some good approximate algorithms.

References

[1] E. Balas, Facets of the knapsack polytope, Math. Programming 8 (1975) 146–164.
[2] R.G. Bland, Elementary vectors and two polyhedral relaxations, European Institute of Advanced Studies, Working Paper no. 76–60, Brussels (1976).
[3] J. Edmonds, Path trees and flowers, Canad. J. Math. 17 (1965).
[4] J. Edmonds, Submodular functions, matroids and certain polyhedra, Proc. of Calgary Inter. Conf. on Combinatorial Structures (1969).
[5] J. Edmonds and R. Giles, A min-max relation for submodular functions on graphs, Ann. Discrete Math. 1 (1977) 185–204.
[6] D.R. Fulkerson, Blocking and anti-blocking pairs of polyhedra, Math. Programming 1 (1971) 168–194.
[7] Hammer, Johnson and Peled, Facets of regular 0–1 polytopes, Math. Programming 8 (1975) 179–206.
[8] E.L. Johnson, Integer programming with continuous variables, Rept. No. 7418–OR, Institut für Ökonometric und Operations Research, Universität Bonn (July 1974).
[9] E.L. Lawler, Combinatorial Optimization (Wiley, New York, 1976).
[10] M.W. Padberg, A note on zero-one programming, Operations Res. 23 (1975) 833–837.
[11] M.W. Padberg, Covering, packing, and knapsack problems, IBM Research Report (1977).

[12] M.W. Padberg and Hong, On the travelling salesman problem: a computational study, T.J. Watson Research Center Report, IBM Research (1977).
[13] L. Trotter and D.B. Weinberger, Symmetric blocking and anti-blocking relations for generalised circulations, Department of O.R., Technical Report no. 300, Cornell University (1976).
[14] H. Weyl, Elementare theorie der convexen polyheder, Comment. Math. Helv. 7 (1935) 290–306.
[15] L.A. Wolsey, Facets for a linear inequality in 0–1 variables, Math. Programming 8 (1975) 165–178.

PART 2

**SOME FUNDAMENTAL CLASSES
OF PROBLEMS**

CONTENTS OF PART 2

Surveys

R.E. BURKARD, Travelling salesman and assignment problems: a survey   193
P.C. GILMORE, Cutting stock, linear programming, knapsacking, dynamic programming and integer programming, some interconnections   217
V. KLEE and D. LARMAN, Use of Floyd's algorithm to find shortest restricted paths   237
E.L. LAWLER, Shortest path and network flow algorithms   251
M.W. PADBERG, Covering, Packing and Knapsack problems   265

Reports

Network flow, assignment and travelling salesman problems (D. DE GHELLINCK)   289
Algorithms for special classes of combinatorial optimization problems (J.K. LENSTRA)   295

TRAVELLING SALESMAN AND ASSIGNMENT PROBLEMS: A SURVEY

Rainer E. BURKARD

Mathematisches Institut der Universität Köln, Köln, West Germany

This paper surveys the main results found in the last years on assignment and travelling salesman problems. Algorithmic aspects as well as theoretical questions are considered. In particular algebraic assignment problems, the permanent conjecture, algorithms for quadratic assignment problems, travelling salesman polytopes and algorithms and worst case analysis for TSP-heuristics are treated.

1. Introduction

Travelling salesman problems (TSP) and assignment problems (AP) are classical examples of combinatorial optimization problems, which have found a considerable interest because of their relevance for practice and their simple structure. This paper surveys some of the main results found in the last years and treats algorithmic aspects as well as theoretical questions.

The research performed in the last years can be outlined as follows: The development of algorithms for the linear assignment problem (LAP) seemed to be terminated in the mid sixties. Yet some modifications in the algorithms improved their order to $O(n^3)$. Several papers dealt with the structure of the assignment polyhedron, but some of the questions in this field are still unanswered, especially the famous permanent conjecture has not yet been proved. By algebraic methods it was achieved that different looking problems can be treated in an unified way and efficient algorithms were developed to solve the algebraic assignment problem. By specializing these general results an $O(n^3)$ method for the bottleneck assignment problem was derived.

Only little progress could be achieved at quadratic assignment problems (QAP). In particular the numerical behaviour of different algorithms was studied and suboptimal procedures were developed to tackle problems with sizes which arise in practice. Some further generalizations of the classical assignment problem will be mentioned in brief.

In the last years much attention was payed to the approach of Held and Karp for solving TSP. Several papers propose improvements and modifications of this method. Further the structure of the travelling salesman polytope was studied in detail. The polyhedral structure of TSP was used in LP-based interactive methods by which a 120 cities problem could be solved optimally (Grötschel, 1976).

Further interesting work was done by probabilistic studies and in the worst case analysis of heuristic algorithms.

A comprehensive classified bibliography on integer programming, which covers also AP and TSP, was published by Kastning in 1976. Therefore only those papers are referenced which are cited in detail or which are not yet contained in this excellent bibliography.

2. Assignments and tours

Let be given two finite sets M and N with $|N| \leq |M|$. An injective (one to one) mapping $\varphi: N \to M$ is called an *assignment*. By regarding the graph of the mapping φ we get immediately the graph theoretical representation of an assignment (cf. Fig. 1)

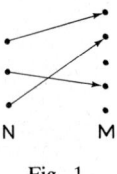

Fig. 1

Whether a partial graph of the complete bipartite graph $K_{n,m}$ contains an assignment can be tested by the *Hall condition*: Let $(V, W; E)$ be a partial graph of $K_{n,m}$ with disjunct node sets V and W and arc set $E \subseteq V \times W$. For $X \subseteq V$ we define

$$W(X) := \{w \in W \mid \exists (v, w) \in E, v \in X\}.$$

The graph $(V, W; E)$ contains an assignment, if

$$\forall X \subseteq V : |X| \leq |W(X)| \qquad \text{(Hall condition)}$$

holds.

Without loss of generality we can assume that $|M| = |N|$. Since then the elements of M and N can be identified, an assignment is a *permutation* of the set N. This is a combinatorial formulation for an assignment. We denote the set of all permutations φ of a set with n elements by S_n. S_n contains $n!$ permutations.

A linear characterization of an assignment can be made by the following system of equations and inequalities:

$$\sum_{i \in N} x_{ij} = 1 \quad (j \in N),$$

$$\sum_{j \in N} x_{ij} = 1 \quad (i \in N), \qquad (1)$$

$$x_{ij} \geq 0 \quad (i, j \in N).$$

A matrix $X = (x_{ij})$, fulfilling (1) is called *doubly-stochastic*. The set of all doubly stochastic matrices forms the *assignment polytope* P_A.

It was shown by Birkhoff (1944) that every extreme point of P_A corresponds to a permutation matrix and therefore to a permutation. The assignment polytope P_A will be studied in more detail in Section 4.

Assignment problems are special cases of three different important classes of combinatorial problems, namely of flows in networks, of matching problems and of matroid intersection problems. Therefore any algorithm for such a problem yields also an algorithm for assignment problems. Algorithms for assignment problems reflect already many features of the more general algorithms.

Travelling salesman problems are obtained from AP by the additional restriction that the permutation φ has to be cyclic. The set of all cyclic permutations of the set N is denoted by Z_n and contains $(n-1)!$ elements. The graph theoretical analogon of a cyclic permutation is a *tour*. A cycle resp. circuit $[x_1, x_2, \ldots, x_n, x_1]$ in a graph is called *Hamiltonian* or a *tour* if it passes exactly once through every node x_i of the graph (cf. Fig. 2).

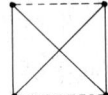

Fig. 2. A tour in K_4.

A graph is called *Hamiltonian*, if it admits a tour. The question, if an arbitrary graph is Hamiltonian or not is difficult. No necessary and sufficient condition is known for the Hamiltonicity of an arbitrary graph. Partial answers to this question can be found in Berge (1973), Nash–Williams (1975) and many others.

Let P_A be the assignment polytope. The convex hull of all of its extreme points, which correspond to *cyclic* permutations, is called *travelling salesman polytope* P_T. No linear characterization of P_T is known so far. Questions arising in connection with P_T will be discussed in Section 8.

The *symmetric assignment problem* was introduced by Murty (1967). Feasible solutions of this problem are permutations, which split in subcycles of length ≤ 2. Cruse (1975) showed that a linear characterization of symmetric assignments is given by (1) and the following two types of additional constraints

$$x_{ij} = x_{ji} \quad (i, j \in N), \qquad \text{(symmetry)} \qquad (2a)$$

$$\sum_{i \in R} \sum_{j \in R \setminus \{i\}} x_{ij} \leq 2k \quad \text{for each subset } R \subseteq N \text{ with an odd cardinality} \\ |R| = 2k+1, k > 0. \qquad (2b)$$

(2b) is related to the blossom inequalities of matching theory. Cruse showed that no constraint of type (2b) is redundant.

3. Objective functions

The classical objectives for AP's and TSP's are linear in the single assignments of node x_i to node x_j. Let $C = (c_{ij})$ be a real $(n \times n)$-matrix. Then

$$\min_{\varphi \in S_n} \sum_{i \in N} c_{i\varphi(i)}, \qquad \text{LAP} \qquad (3a)$$

$$\min_{\varphi \in Z_n} \sum_{i \in N} c_{i\varphi(i)} \qquad \text{TSP} \qquad (3b)$$

are the classical linear assignment (LAP) resp. travelling salesman problems. By replacing "+" by "max" we get linear bottleneck problems, namely the linear bottleneck AP (4a) and the bottleneck-TSP (4b)

$$\min_{\varphi \in S_n} \max_{i \in N} c_{i\varphi(i)}, \qquad \text{LBP} \qquad (4a)$$

$$\min_{\varphi \in Z_n} \max_{i \in N} c_{i\varphi(i)}. \qquad \text{Bottleneck TSP} \qquad (4b)$$

The evident analogies of problems (3) and (4) lead to the definition of the following *algebraic problem*. Let $(H, *, \leq)$ be a commutative ordered semigroup with internal composition $*$ and order relation $\leq$, that is

$$(H, *) \text{ is associative and commutative,} \qquad (5a)$$

$$(H, \leq) \text{ is totally ordered,} \qquad (5b)$$

$$\forall a, b, c \in H: a \leq b \Rightarrow a * c \leq b * c. \qquad (5c)$$

Further the following reducibility axiom (6) is required to hold in H:

$$\forall a, b \in H \text{ with } a \leq b \quad \exists c \in H: a * c = b. \qquad (6)$$

If c_{ij} $(i, j \in N)$ are elements of a semigroup $(H, *, \leq)$, which fulfills (5) and (6), then the *algebraic* AP resp. the *algebraic* TSP are defined by (7)

$$\min_{\varphi \in S_n} \underset{i \in N}{*} c_{i\varphi(i)}, \qquad \text{(algebraic AP)} \qquad (7a)$$

$$\min_{\varphi \in Z_n} \underset{i \in N}{*} c_{i\varphi(i)}. \qquad \text{(algebraic TSP)} \qquad (7b)$$

Special cases of $(H, *, \leq)$ are for example

(a) $(\mathbf{R}, +, \leq)$ leads to problems (3) with sum objectives,
(b) $(\mathbf{R}, \max, \leq)$ leads to problems (4) with bottleneck objectives,
(c) $(\mathbf{R}^n, +, \leq)$ with lexicographical order relation $\leq$ leads to lexicographical problems,
(d) $(\mathbf{R}^2, *, \leq)$ with
$$(a, b) * (c, d) := \begin{cases} (a, b) & \text{if } a > c, \\ (c, d) & \text{if } a < c, \\ (a, b+d) & \text{if } a = c, \end{cases} \quad \text{(ordinal sums),}$$
leads to time cost problems.

In Section 5 algorithms for solving algebraic AP are derived. Their specializations will lead to well known methods for LAP's and LBP's.

Often in practice the objective does not depend linearly on single assignments of x_i to x_j. An important case is that of quadratic dependence. Let d_{ijpq} be an array with n^4 real numbers.

The *quadratic assignment problem* (QAP) is defined by

$$\min_{\varphi \in S_n} \sum_{i \in N} \sum_{p \in N} d_{i\varphi(i)p\varphi(p)}. \tag{8}$$

The corresponding *quadratic bottleneck problem* runs

$$\min_{\varphi \in S_n} \max_{i \in N} \max_{p \in N} d_{i\varphi(i)p\varphi(p)}. \qquad \text{(QBP)} \tag{9}$$

(The algebraic treatment of combinatorial optimization problems started with the development of analogous algorithms for QAP and QBP by Burkard resp. with the treatment of algebraic TSP by Gabovič [1970]).

Often the coefficients d_{ijpq} are split into a product

$$d_{ijpq} = a_{ip} b_{jq} \qquad (i, j, p, q \in N). \tag{10}$$

If (10) holds the QAP (QBP) is called a *Koopmans–Beckmann problem*. The (linear) TSP can be written as a special Koopmans–Beckman problem, namely one with cost matrix A and a cyclic permutation matrix B. QAP are further studied in Section 6.

4. The assignment polytope

The hope that a thorough knowledge of the underlying polytopes of combinatorial optimization problems leads to good solution procedures stimulated the research in this field. One of the earliest results derived for the assignment polytope P_A was that P_A has only integral extreme points. This can be shown directly via the approach of Birkhoff (1944) or via the total unimodularity of its defining matrix and the well known Hoffman–Kruskal theorem. A more detailed study on the structure of P_A was done by Balinski and Russakoff (1974), who showed the following results.

Theorem (Balinski and Russakoff, 1974). *Let P_A be the assignment polytope for assignments of a set N with $|N| = n$. Then the following statements hold*

(a) $\dim P_A = (n-1)^2$.

(b) *Every of the $n!$ extreme points coincides with $\sum_{k=0}^{n-1} \binom{n}{k}(n-k-1)!$ different edges.*

(c) *From any extreme point any other extreme point can be reached over at most 2 edges, i.e. the diameter of P_A is 2.*

(d) P_A *is Hamiltonian.*

(e) *Associate with any $x \in P_A$ a bipartite subgraph $B(x)$ of $K_{n,n}$ having edges (i,j) if $x_{ij} > 0$. Two extreme points x and y of P_A are neighboured, iff $B(x) \cup B(y)$ contains exactly one circuit.*

Two results on feasible bases and the extreme points of P_A were also shown by Balinski and Russakoff. The following theorem asserts in particular that the Hirsch conjecture is true for assignment problems:

Theorem (Balinski and Russakoff, 1974).

(a) *Every extreme point of P_A corresponds to $2^{n-1} n^{n-2}$ feasible bases of the corresponding LP.*

(b) *A path of neighbouring feasible bases of length at most $2n - 1$ connects any pair of feasible bases of the assignment problem.*

A further interest in P_A stems from the famous *permanent conjecture* posed by v.d. Waerden in 1926, which is unsolved until now.

Let $C = (c_{ij})$ be a doubly stochastic matrix, i.e. $C \in P_A$. The *permanent* of C is defined by

$$\mathrm{per}\,(C) = \sum_{\varphi \in S_n} c_{1\varphi(1)} \cdot c_{2\varphi(2)} \cdots c_{n\varphi(n)}.$$

The permanent conjecture says

$$\min_{C \in P_A} \mathrm{per}\,(C) = \mathrm{per}\,(C^*) = \frac{n!}{n^n}$$

with $C^* = (c_{ij}^*)$, $c_{ij}^* = 1/n$ for all $i, j = 1, 2, \ldots, n$.

The correctness of this conjecture was shown by Markus and Newman (1959) for $n \leq 3$, by Eberlein and Mudholkar (1968) for $n = 4$ and by Eberlein (1969) for $n = 5$. Further results in this direction are stated in the following theorem.

Theorem (Markus and Newman). *Let be $Q := \{C \in P_A \mid c_{ij} > 0 \text{ for all } i, j = 1, \ldots, n\}$ and assume the minimum of $\mathrm{per}\,(C)$, $C \in P_A$ is attained for the matrix D. Then*

(a) $\exists k : D^k \in Q$,
(b) $D \in Q \Rightarrow D = C^*$,
(c) C^* *is a strict local minimum for* $\mathrm{per}\,(C)$.
(d) *If the minimization is restricted to positive semi-definite, symmetric matrices $C \in P_A$, then C^* is the only minimum.*

Permanents are surveyed by Markus and Minc (1965). Further properties of the assignment polytope were recently determined by Brualdi and Gibson (1976) by means of the permanent function.

5. Algorithms for linear assignment problems

One of the best known methods for the solution of linear assignment problems with sum objective is the Hungarian method of Kuhn (1955), which is a dual method and is based on the following theorem of König and Egerváry. m elements of a matrix $C = (c_{ij})$ are called *independent*, if no two of them are in the same row or column. A subset of rows and columns is called an 0-*covering* of C, if all 0-elements are only in a row or column of this subset.

Theorem (König, Egerváry). *Let C be a real $(n \times n)$-matrix. The maximal number of independent 0-elements is equal to the minimal number of rows and columns of an 0-covering.*

Balinski and Gomory (1964) developed a primal method for the solution of assignment problems. The advantage was that all intermediate solutions were feasible. All methods mentioned so far have a complexity of $O(n^4)$. Dinic and Kronrad (1969) were the first who gave a method for LAP with sum objective which needs only $O(n^3)$ steps. A further method with $O(n^3)$ steps was developed by Tomizawa (1971) and numerically tested by Dorhout (1973).

Recently primal methods have found again a broad interest. They belong now to the most efficient solution procedures for linear assignment problems with sum objectives. The improvements are due to a proper representation of the problems in a computer and choosing suitable combinations of subroutines (determination of the initial basic solution, selection of the variables which leave and enter the base), cf. Srinivasan and Thompson (1973), Glover et al. (1974), Gavish, Schweitzer and Shlifer (1977), Barr, Glover and Klingmann (1977). Basic solutions are stored as trees by threaded list structures which allow a fast updating in a pivot step. Gavish, Schweitzer and Shlifer (1977) propose to determine an initial basic solution by the *chain rule*, which is a modification of the row minima rule. In the chain rule trees are built up whose arcs belong to strictly positive variables. These trees are connected by choosing "zero"-arcs with the least costs. Further these authors propose to select among all basis variables which are candidates to leave the basis one with the highest cost coefficient and they give a search procedure for selecting the variable entering the basis. This search procedure reduces also the number of accesses to an auxiliary storage, when the data of a large AP are stored out of core.

De Ghellink (1977) reported quite recently favourable numerical results with large AP using the labelling method. The idea is to order the cost coefficients row-wise and to use this order for stopping the labelling and the cost change subroutines.

Linear bottleneck problems were first considered by Gross (1959). Garfinkel (1971) published a solution procedure and Pape and Schön (1970) developed a computer code for these problems. Since LAP with sum and bottleneck objective are only special cases of the algebraic LAP, we describe here a solution method for the algebraic problem which yields a special form of the Hungarian method for sum objectives and leads to the threshold method of Garfinkel for bottleneck problems.

Let $(H, *, \leq)$ be a semigroup fulfilling (5) and (6). The set dom a, defined by

$$\text{dom } a := \{b \mid a * b = a\}$$

contains besides the neutral element $e \in H$ eventually further elements, for example we get for the composition "max":

$$\text{dom } a = \{b \mid b \leq a\}.$$

An element $c \in H$ is called z-positive, if $c * z \geq z$.

Now let $C = (c_{ij})$ be a $(n \times n)$-matrix with elements $c_{ij} \in H$. A transformation $T: C \rightsquigarrow \bar{C}$ is called *admissible*, if $\exists z(T) \in H$, such that for all $\varphi \in S_n$

$$\underset{i \in N}{*} c_{i\varphi(i)} = z * \underset{i \in N}{*} \bar{c}_{i\varphi(i)}.$$

$z(T)$ is the *index* of the admissible transformation. Evidently the composition of an admissible transformation $T_1: C \rightsquigarrow \bar{C}$ with index $z(T_1)$ and of an admissible transformation $T_2: \bar{C} \rightsquigarrow \bar{\bar{C}}$ with index $z(T_2)$ leads again to an admissible transformation $T := T_2 \circ T_1: C \rightsquigarrow \bar{\bar{C}}$ with index $z(T_1) * z(T_2)$. For algebraic assignment problems the following admissible transformations play an important rôle.

Theorem (Burkard, Hahn and Zimmermann). *Let* $I \subseteq N$, $J \subseteq N$ *and* $m := |I| + |J| - n \geq 1$.

$$c := \min \{c_{ij} \mid i \in I, j \in J\}.$$

Then $T(I, J): C \rightsquigarrow \bar{C}$, *implicitly defined by*

$$c * \bar{c}_{ij} = c_{ij} \quad (i \in I, j \in J),$$
$$\bar{c}_{ij} := c_{ij} * c \quad (i \notin I, j \notin J),$$
$$\bar{c}_{ij} := c_{ij} \quad \text{otherwise},$$

is an admissible transformation with index $z(T) := c^m$.

For the solution of algebraic assignment problems the following algorithm can be used:

Algorithm (Burkard, Hahn and Zimmerman).

1. Start with $C = (c_{ij})$ and $z := e$.
2. Row reductions: For $i := 1, 2, \ldots, n$ put $I := \{i\}, J := N$; carry out $T(I, J)$ and $z := z * z(T)$.

3. Column reductions: For $j := 1, 2, \ldots, n$ put $I := N$, $J := \{j\}$; carry out $T(I, J)$ and $z := z * z(T)$.

4. Determine a maximal number of independent elements in C which belong to dom z. If their number is n, stop. An optimal solution is found.

Otherwise determine a minimal covering (I', J') for the elements in dom z and put

$$I := N \setminus I'; J := N \setminus J'.$$

5. Carry out $T(I, J)$ and $z := z * z(T)$. Go to 4.

It can be shown that after step 3 all elements $\bar{c}_{ij}$ are z-positive. Further we get

Theorem (Burkard, Hahn and Zimmermann). *The algorithm yields an optimal solution after at most $\frac{1}{2}(n^2 + 3n - 2)$ admissible transformations.*

As already mentioned this algorithm leads to the Hungarian method for the system $(\mathbf{R}, +, \leq)$ and to Garfinkel's threshold method for the system $(\mathbf{R}, \max, \leq)$, but solves quite a number of problems different from these special cases.

Since an admissible transformation has the complexity $O(n^2)$, if the complexity to compute $a * b$ is $O(1)$, we get an algorithm of order $O(n^4)$. To improve the order of the algorithm, recently weakly admissible transformations were introduced (Burkard and Zimmermann, 1977):

Let H be a semigroup which fulfills in addition to (5) and (6) the *weak cancellation rule*

$$a * c = b * c \Rightarrow a = b \quad \text{or} \quad a \in \text{dom } c, \quad b \in \text{dom } c \tag{11}$$

and is positively ordered

$$a * b \geq a \quad \forall a, b \in H. \tag{12}$$

Further let z_0 be the optimal value of $\min_{\varphi \in S_n} \ast_{i \in N} c_{i\varphi(i)}$. Then in such a semigroup a transformation $T: C \leadsto \bar{C}$ is called *weakly admissible*, if there are constants $z_1, z_2 \in H$ with $z_1 * z_0 \leq z_2 * z_0$ such that for all $\varphi \in S_n$

$$z_1 * \underset{i \in N}{\ast} c_{i\varphi(i)} = z_2 * \underset{i \in N}{\ast} \bar{c}_{i\varphi(i)}$$

holds. The pair $(z_1(T), z_2(T))$ is the index of this transformation. The composition of weakly admissible transformations yields again a weakly admissible transformation. Now the indices $(z_1(T), z_2(T))$ can be determined by means of a shortest route algorithm with complexity $O(n^2)$ such that after at most n weakly admissible transformations the optimal solution is found. This leads therefore to an $O(n^3)$ algorithm for algebraic assignment problems and its special cases. In the special case of $(\mathbf{R}, +, \leq)$ this algorithm is a modification of Tomizawa's method (1971). Derigs and Zimmermann (1977) developed a code for linear bottleneck problems using the theory of weakly admissible transformations. In numerical tests and comparisons they showed the efficiency of this new method.

There are only a few numerical investigations on symmetric assignment problems. Since these problems can be transformed into matching problems the matching algorithms can be used for their solution. Murty (1967) proposes a branch and bound method using assignment problems as relaxation. Devine and Glover (1972) report that for small values of n their branch and bound algorithm for symmetric assignment problems is more effective than the matching algorithm applied to these problems, since only a few branchings occur. A method using weakly admissible transformations for symmetric assignment problems with algebraic objective is under investigation.

6. Algorithms for quadratic assignment problems

QAP have surprisingly many applications, so they were applied recently in location theory, in hospital planning or in typewriter keyboard optimization (Burkard and Offermann, 1977). Therefore QAP found a considerable interest in the literature. Unlike LAP they are hard to solve, since they do not only belong to the class of NP-complete problems, but even the ε-approximation problem for QAP is NP-complete (Sahni and Gonzales, 1976). An algorithm is said to be ε-*approximate*, if it finds a solution $\bar{\varphi} \in S_n$ with objective value $z(\bar{\varphi})$, such that with the optimal value z^* the following relation holds:

$$\left| \frac{z^* - z(\bar{\varphi})}{z^*} \right| < \varepsilon, \quad \varepsilon > 0.$$

This result makes clear that only implicit enumeration algorithms are known for this problem. The first algorithm for Koopmans–Beckmann problems was proposed by Gilmore (1962) and extended to general QAP by Lawler (1963). Gilmore introduced the following technique which proved to be useful for computing bounds in Koopmans–Beckmann problems.

He showed that

$$\min_{\varphi \in S_n} \sum_{i \in N} a_i b_{\varphi(i)}$$

can be attained by ordering the elements a_i increasingly and the elements b_j decreasingly. A survey of different branch and bound methods until 1970 is given by Pierce and Crowston (1971). See also the survey of Lawler (1974) on QAP.

Recently good results were achieved by the perturbation method (Burkard 1973, 1976), which was extended also to quadratic bottleneck problems (Burkard, 1974). The perturbation method for solving general QAP makes use of the fact that the elements d_{ijij} can be considered as cost coefficients of a LAP, which is disturbed by the elements d_{ijpq} with $i \neq p$ and $j \neq q$. To compute good bounds the disturbing elements are minimized and as much information as possible is put into the linear part of the problem.

The perturbation method is a single assignment algorithm in which successively single assignments are made. Single assignment algorithms seem to be more efficient for QAP than pair assignment and pair exclusion algorithms (Land (1963), Gavett and Plyter (1966)). This conclusion is corroborated by a computational study of Smith (1975).

Numerical results for optimal methods showed (Burkard, 1975a) that QAP can be solved optimally only up to $n = 15$ in a reasonable time. Due to the pleasing numerical behavior of bottleneck problems, QBP can be solved optimally up to $n = 20$. These figures show that QAP, which occur in real world applications, hardly can be solved optimally. Therefore there is a special need for good heuristics.

A survey on heuristics for QAP was published by Burkard (1975b). Most of the methods were numerically tested (Burkard and Stratmann, 1976). The tests gave the following results:

In the class of heuristic construction methods the method of increasing degree of freedom (Müller–Merbach, 1970) yields favorable results, especially if it is run with different start values. In average the start phase of branch and bound methods yields better results, but only one suboptimal solution can be found in this way. Very good results for symmetric Koopmans–Beckmann problems can be obtained by the method of Gaschütz and Ahrens (1968), especially if it is combined with improving methods. In Burkard and Stratmann (1976) an improvement package was developed which yielded combined with the Gaschütz–Ahrens procedure or with the perturbation method very good results. All suboptimal solutions for the test problems in the literature could be improved in this way.

Christofides and Gerrard (1976) studied special structured QAP and obtained interesting results on solvable special cases. Let two graphs $G_f = (V_f, E_f)$ and $G_d = (V_d, E_d)$ be given. Further let a_{ip} be weights attached to the arcs (i, p) of the *flow graph* G_f and b_{jq} weights of the arcs (j, q) in the *distance graph* G_d. We consider now injective mappings $\varphi : V_f \to V_d$ with $(\varphi(i), \varphi(p)) \in E_d$ for all $(i, p) \in E_f$ and associate with them the costs

$$z(\varphi) = \underset{(i,p) \in E_f}{\text{\Large *}} a_{ip} b_{\varphi(i)\varphi(p)}. \tag{13}$$

Now an injective mapping φ is wanted which minimizes $z(\varphi)$. If both, G_f and G_d are stargraphs (cf. Fig. 3), the minimization of $z(\varphi)$ leads to the above mentioned rule of Gilmore. This method can also be used to solve by a polynomial algorithm the case G_f is a star graph and G_d is complete.

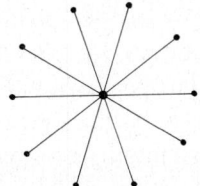

Fig. 3. A star graph $K_{1,n}$

Also in the case G_f and G_d are arbitrary trees Christofides and Gerrard derive a polynomial bounded algorithm for the minimization of (13). Since the TSP is an example for the case at which G_f is a tree and G_d is complete this case leads to a NP-complete problem. The authors derive a dynamic programming approach for this special case. The ideas sketched above can be used to compute bounds for Koopmans–Beckmann problems (Christofides, 1977). In particular the Gilmore bounds and the bounds of pair assignment algorithms can be obtained in this way.

7. Extensions of assignment problems

A straightforward generalization of LAP's are multidimensional assignment problems, which play for example a rôle at time-table construction. We restrict ourselves here to three dimensional assignment problems, which occur in two types.

Let be $N=\{1, 2, \ldots, n\}$. The solution set for the *axial problem* is defined by

$$\sum_{i \in N} \sum_{j \in N} x_{ijk} = 1 \quad (k \in N), \qquad \sum_{i \in N} \sum_{k \in N} x_{ijk} = 1 \quad (j \in N),$$

$$\sum_{j \in N} \sum_{k \in N} x_{ijk} = 1 \quad (i \in N), \qquad x_{ijk} \in \{0, 1\} \quad (i, j, k \in N).$$

The feasible solutions for the *planar problem* are given by

$$\sum_{i \in N} x_{ijk} = 1 \quad (j, k \in N), \qquad \sum_{j \in N} x_{ijk} = 1 \quad (i, k \in N),$$

$$\sum_{k \in N} x_{ijk} = 1 \quad (i, j \in N), \qquad x_{ijk} \in \{0, 1\} \quad (i, j, k \in N).$$

In both cases $\sum_{i \in N} \sum_{j \in N} \sum_{k \in N} c_{ijk} x_{ijk}$ is to minimize.

Since the axial as well as the planar problem are NP-complete, only implicit enumeration techniques exist for their solution. Hansen and Kaufman (1973) propose a combination of the Hungarian method applied to hypergraphs and of the branch and bound scheme to solve the axial problem. Other branch and bound algorithms were proposed by Pierskalla (1968) and Leue (1972) for the axial problem and by Vlach (1967) for the planar problem. Pierskalla (1967) proposed also a suboptimal procedure for the axial problem. Frieze (1974) showed that the axial problem is equivalent to a bilinear programming problem.

Ross and Soland (1975) studied the following generalized assignment problem and gave a branch and bound algorithm for its solution which works for problems up to 4000 Boolean variables.

Minimize $\sum_{i \in M} \sum_{j \in N} c_{ij} x_{ij}$,

Subject to $\sum_{j \in N} r_{ij} x_{ij} \leq b_i \quad (i \in M)$,

$\sum_{j \in M} x_{ij} = 1 \quad (j \in N)$,

$x_{ij} \in \{0, 1\} \quad (i \in M, j \in N)$.

Sahni and Gonzales (1976) showed that the corresponding ε-approximation problem is NP-complete.

Assignments in ordered sets were considered by Gale (1968) and by Gaul and Heinecke (1975). Let N be a finite ordered set and let Φ be a mapping from N to subsets of a set M. A subset $A \subseteq N$ is called *assignable*, if there is an injection φ from A to M that $\varphi(a) \in \Phi(a)$ for all $a \in A$. Gale showed that assignable sets form a matroid on N. Therefore there exists an *optimal* assignable set A^* in the following sense: if A is any assignable set then there is an injection $f: A \to A^*$ with $f(a) \geq a$ for all $a \in A$. There are many generalizations of AP to different kinds of matching problems, network flow problems and matroid intersection problems, but a detailed description of these is beyond the scope of this survey.

8. Travelling salesman polytopes

Since any tour is an assignment, the tours correspond to extreme points of the assignment polytope P_A. More exactly it can be shown [Heller (1955), Murty (1969), Padberg and Rao (1974)] that the tours are neighbours of the vertex which corresponds to the identity matrix. The extreme points of P_A which do not correspond to tours can be cut off by the so called *subtour elimination constraints* [Dantzig, Fulkerson and Johnson (1954)]:

$$\sum_{(i,j) \in S \times S} x_{ij} \leq |S| - 1 \quad \left(S \subseteq N; 2 \leq |S| \leq \left[\frac{n}{2}\right]\right). \tag{14}$$

The subtour elimination constraints are equivalent with the *loop constraints*

$$\sum_{i \in S} \sum_{j \in V \setminus S} x_{ij} \geq 1 \quad \left(S \subseteq N; 2 \leq |S| \leq \left[\frac{n}{2}\right]\right). \tag{15}$$

The polytope which is obtained from P_A by adding the constraints (14) is not P_T, the convex hull of all tours. It has, however, the property that every integral extreme point corresponds to a tour. The TSP is therefore an integer program with restrictions (1), (14) and $x_{ij} \in \{0, 1\}$ for $1 \leq i, j \leq n$. Many of the restrictions in (1) and (14) are redundant, therefore the question was raised which inequalities lead to a face of the polytope P_T. Grötschel and Padberg (1974) showed that the subtour elimination constraints with $|S|=2$ and $|S|=3$ lead to faces of P_T for all TSP with $n \geq 5$. Since the polytope defined by (1) and (14) has extreme points with fractional components, one is interested in inequalities which are fulfilled by all tours but not by such an extreme point. Several classes of such *valid inequalities* were derived by Grötschel and Padberg (1974) who showed in addition that some of them define faces of P_T.

In an abstract Heller (1953) gave the dimension of P_T by n^2-3n+1 ($n \geq 3$). A proof of this result can be found in Grötschel and Padberg (1974a). Padberg and Rao (1974) showed that the diameter of P_T is less or equal 2 for all n and equal to 2 for $n \geq 6$. Further these authors showed that a weak form of the famous Hirsch conjecture is also true for TSP.

If the cost matrix of a linear TSP is symmetric, the tours $(x_1, x_2, \ldots, x_n)$ and $(x_n, x_{n-1}, \ldots, x_1)$ give the same objective value, therefore they can be identified. This leads to the concept of a *symmetric* TSP. A symmetric TSP is the problem to find a shortest tour in an undirected graph. Only variables x_{ij} with $i<j$ play a rôle in symmetric TSP, therefore a tour corresponds to a point in $\mathbf{R}^m$ with $m = \frac{1}{2}(n^2-n)$. The polytope Q_T of the symmetric TSP is again defined as convex hull of the tours in $\mathbf{R}^m$. Norman (1955) gave its dimension with $\frac{1}{2}(n^2-3n)$, ($n \geq 3$). Grötschel and Padberg (1974b) show that for $n \geq 4$ all subtour elimination constraints define faces of Q_T. Chvátal (1973) derived valid inequalities for the symmetric TSP, the so-called *comb-inequalities*. Further results on the polytope Q_T are given by Maurras (1974). Grötschel and Padberg (1977) generalized in a recent paper the notion of a comb inequality and proved that these inequalities define faces of Q_T. Further they showed that their total number is a combinatorial function of n which increases much faster with n than 2^n. This is of considerable interest, because a classical argument against the use of linear programming in the TSP was the long-known fact that the total number of subtour-elimination constraints grows exponentially with n.

The linear characterization of the polytopes P_T and Q_T is of interest for LP-based algorithms for the TSP. If the LP stops with a fractional extreme point, a valid inequality is added to the present linear program which cuts off this extreme point. In this way large TSP have been solved, for example the classical 42 cities problem of Dantzig, Fulkerson and Johnson (1954) and the 120 cities problem of Grötschel (1976). Recently Padberg and Hong (1977) determined in this way a suboptimal solution for the 318 cities problem of Lin and Kernighan (1970). This suboptimal solution differs at most $\frac{1}{4}$ percent from the optimal solution. The Padberg Hong code generates subtour elimination constraints and comb-inequalities automatically. It is done there in a more or less heuristic manner and the open problem in this area is to find always such a constraint if it exists.

A different linear programming approach for TSP was studied by Miliotis (1976, 1977), who starts with a subset of the constraints (14) and uses Gomory cuts to obtain an integral solution. Schiebel, Unger and Terno (1976) propose to accumulate loop constraints. But by adding up loop constraints one produces a new constraint that no longer defines a facet of Q_T.

9. Algorithms for asymmetric TSP

TSP belong to the class of NP-complete problems. Sahni and Gonzales (1976) showed moreover that the corresponding ε-approximation problem is NP-complete. Therefore only implicit enumeration algorithms are known for solving TSP's. The implicit enumeration depends strongly on the form of the cost matrix C. Symmetric matrices C lead to symmetric TSP which show a completely different numerical behaviour from asymmetric TSP with a nonsymmetric cost matrix C. For asymmetric problems a relaxation to assignment problems yields good results, whereas symmetric problems can be solved efficiently by using minimal spanning trees as relaxations. Because of the different numerical behaviour we survey in this section algorithms for asymmetric TSP and in the next section algorithms for symmetric TSP.

The basic principle for solving asymmetric TSP is the following. Relax the TSP to a LAP. If the optimal solution of the LAP is not cyclic, use a cycle with minimal length for branching. Bounds for the subproblems can now be computed in different ways.

In this connection Bellmore and Malone (1971) proved the following interesting result.

Theorem (Bellmore and Malone, 1971). *Assuming that the cost coefficients $c_{ij}(i \neq j)$ of a LAP are identically and independently distributed and that the solutions of the LAP are equally likely to be optimal, then the probability that the optimal solution of the LAP is a tour is*

$$\frac{(n-1)!}{n!\,(e^{-1}+0.5)} \approx \frac{e}{n} \quad \text{for large } n.$$

Further these authors showed that by adding a constraint of the form

$$\sum_{(i,j)\in S'} x_{ij} \leq |S'|-1$$

where S' is the arc set of a subcycle the number of feasible solutions of the corresponding special LAP is reduced by $[k!\,e^{-1}+0.5]$, whereas by adding a loop constraint (15) $[(n-k)!\,e^{-1}+0.5]\cdot[k!\,e^{-1}+0.5]$ solutions of the corresponding LAP are eliminated. Therefore it is more favorable to use the loop constraints instead of constraints of the above form for $|S'|\geq 3$.

Christofides (1972) gives a special procedure for computing strong bounds for the subproblems. Assume a LAP has a noncyclic solution φ with corresponding costmatrix $C^* = (c_{ij}^*)$. Let $S_1, \ldots, S_k$ be the sets of indices of the different subcycles of φ. Then define a new $(k \times k)$-cost matrix $\bar{C}$ by

$$\bar{c}_{rs} = \min_{\substack{i \in S_r \\ j \in S_s}} c_{ij}^* \qquad \text{(contraction)}$$

and replace every element $\bar{c}_{rs}$ by

$$\bar{\bar{c}}_{rs} := \min_t (\bar{c}_{rt} + \bar{c}_{ts}). \qquad \text{(compression)}$$

Now solve the LAP with cost matrix $\bar{\bar{C}}$. If its solution is cyclic, stop, otherwise define a new problem by contraction and compression and solve it again as LAP. Christofides showed that the sum of objective values of the LAP's obtained during the solution — contraction — compression process (up to the stage when the contracted problem becomes a single vertex) is a lower bound for the TSP. A new method for computing strong bounds and using disjunctive programming techniques for enumeration was recently studied by Balas and Christofides (1976) and looks very promising.

Several authors [Bellmore and Malone (1971), Garfinkel (1973), Thompson (1975)] point out that branching schemes which construct subproblems with disjoint sets of feasible solutions are superior to other branching rules.

Hong (1976) applied the Held and Karp approach to the asymmetric case by relaxing to spanning arborescences. Until now, however, no better results were obtained in this way than by the above mentioned methods.

10. Algorithms for symmetric TSP

Symmetric TSP's are difficult to solve by a LAP-approach since the solutions of the LAP's have in general many cycles of length two. Proposals for overcoming this difficulty stem from Bellmore and Malone (1971) who use the weighted b-matching problem as relaxation. Some other improvements of the solution of symmetric TSP by LAP were reported by Steckhan and Thome (1972).

Further references for algorithmic approaches for the symmetric TSP which can be combined with branch and bound for exact solutions are Dantzig, Fulkerson and Johnson (1954), Hong (1971), Grötschel (1976) and Padberg and Hong (1977).

An interesting new approach was suggested by Held and Karp (1970) who introduced the concept of a 1-tree as a relaxation for symmetric TSP. A 1-*tree* is a graph (V, T) with $V = \{x_1, \ldots, x_n\}$ consisting of a tree on the vertices $\{x_2, \ldots, x_n\}$ together with two edges incident with vertex x_1. The Held–Karp approach is based on the following three observations:

(a) a tour is a 1-tree in which each vertex has degree 2,

(b) a minimum 1-tree can easily be computed by a minimum spanning tree and the two edges incident with vertex x_1 which have minimal weights,

(c) the transformation $c_{ij} \rightsquigarrow c_{ij} + \pi_i + \pi_j$ leaves the relative order of the solutions of the TSP invariant but changes the minimum 1-tree.

Let (V, T) be a minimal 1-tree with respect to the transformed cost coefficients $c_{ij} + \pi_i + \pi_j$. Then the optimal value of a tour is greater or equal to the weight of the 1-tree (V, T). Therefore we get

$$z + 2 \sum_{i=1}^{n} \pi_i \geq z(T) + \sum \pi_i d_i \qquad (16)$$

where z is the optimal value of the TSP, $z(T)$ is the weight of (V, T) with respect to c_{ij} and d_i are the vertex-degrees of (V, T).

We get

$$z(T) + \sum_{i=1}^{n} \pi_i (d_i - 2) \leq z.$$

Therefore

$$w(\pi) := z(T) + \sum_{i=1}^{n} \pi_i (d_i - 2)$$

is a family of lower bounds for the TSP; the best bound is $\max_\pi w(\pi)$.

Held and Karp showed that $\max w(\pi) < z$ can occur and related this phenomenon to the existence of non integral extreme points of the polyhedron Q_s. Q_s is obtained by intersecting the underlying 2-matching polyhedron with all half spaces corresponding to subtour elimination constraints.

To get a sharp bound, $w(\pi)$ has to be maximized. In their paper of 1970, Held and Karp gave two methods for the maximization of $w(\pi)$, a LP with a column generating technique and an ascent method. In Held and Karp (1971) an improved ascent method was given together with a detailed description of the used branch and bound method. The above described method can be used to compute bounds for any subproblem, since fixed included and excluded edges can easily be taken into account.

Improvements of this method are reported by Helbig Hansen and Krarup (1974) and Thompson (1975), both reporting that their version is about 25 times faster than the original one. The improvements of Helbig Hansen and Krarup were achieved by a proper choice of the parameters of the ascent method for the maximization of $w(\pi)$, by using different methods for the computation of 1-trees in the start phase of the implicit enumeration and afterwards and by improving the branching strategy. The improvements of Thompson were achieved by a time-sparing method of changing 1-trees and again by improving the branching strategy.

11. Further results on TSP

Though the TSP is NP-complete three special cases are reported in the literature which can be solved by polynomial algorithms. The first solvable case was described by Gilmore and Gomory (1964) and is related to a machine scheduling problem. Let (a_i, b_i) be given pairs of real numbers and let f, g be real valued, integrable functions with $f(x)+g(x) \geq 0$ for all $x \in \mathbf{R}$. Then the cost coefficients of the Gilmore–Gomory problem are defined by

$$c_{ij} := \int_{b_i}^{a_j} f(x)\,dx, \quad \text{if} \quad a_j \geq b_i$$

and

$$c_{ij} := \int_{a_j}^{b_i} g(x)\,dx, \quad \text{if} \quad a_j \leq b_i.$$

Lawler (1971) pointed out that a TSP with upper triangular cost matrix can be solved in $O(n^3)$ steps. Sysło (1973) shows that the TSP on a Hamiltonian digraph with special costs can be solved by a polynomial algorithm.

At present there is a considerable interest in a worst case analysis of heuristic methods. Christofides (1976) describes an $O(n^3)$ heuristic for the symmetric TSP, the cost matrix of which satisfies the triangularity condition. He shows that the ratio of the heuristic solution to the optimum TSP solution is strictly less than $\frac{3}{2}$ and improves thereby a result of Rosenkrantz et al. (1974). Golden (1976) investigates the sequential savings heuristic (cf. Clarke–Wright (1964), Webb (1972), Yellow (1970)) applied to the same class of TSP and obtain the following result:

For each $m \geq 2$ there exists a graph with $n = 2^m$ nodes such that the ratio of the sequential Clarke–Wright solution to the optimal solution is strictly larger than $[6 \ln (n)+5]/21$.

No polynomially bounded algorithms which guarantee a certain percentage of the optimal objective value are known for asymmetric TSP. Korte and Hausmann (1976) as well as Fisher, Nemhauser and Wolsey (1977) consider heuristics for *maximizing* the objectives of TSP. For symmetric TSP they derive a bound of $\frac{2}{3}$ and for the asymmetric case the bound $\frac{1}{2}$.

Since these results are probably not acceptable in practice, Karp (1976) originated a probabilistic analysis of combinatorial search algorithms. He derives for the Euclidean TSP, i.e. a TSP with given points in the plane as its vertices, the following result.

Theorem (Karp, 1976). *If n points are drawn independently from a uniform distribution in the unit cube, there is an algorithm for the corresponding TSP with the following property: the algorithm finds almost everywhere an ε-approximate solution within a time of $O(n \log n)$.*

Garey, Graham and Johnson (1976) showed that even the Euclidean TSP is NP-complete.

One of the most effective suboptimal algorithms for symmetric TSP is the method described by Lin and Kernighan (1973), which uses interchanges to improve feasible solutions. Wiorkowski and McElvain (1975) report on a rapid heuristic algorithm which is useful in certain circumstances (e.g. intraurban delivery), when the costs of computer time are incorporated in the costs of the tour.

References

E. Balas and N. Christofides (1976), A new penalty method for the travelling salesman problem, presented at the IX. Symposium on Mathematical Programming, Budapest, Hungary (August 1976).
M.L. Balinski and R.E. Gomory (1964), A primal method for the assignment and transportation problem, **Management** Sci. 10 (1964) 578–593.
M.L. Balinski and A. Russakoff (1974), On the assignment polytope, SIAM Rev. 16 (1974) 516–525.
R.S. Barr, F. Glover and D. Klingman (1977), The alternating basis algorithm for assignment problems. Math. Programming 13 (1977) 1–13.
M. Bellmore and J.C. Malone (1971), Pathology of travelling salesman subtour elimination algorithms, Operations Res. 19 (1971) 278–307.
C. Berge (1973), Graphs and Hypergraphs (North Holland, Amsterdam, 1973).
G. Birkhoff (1944), Tres observaciones sobre el algebra lineal, Rev. univ. nac. Tucumán (ser. A) 5 (1940) 147–151.
R.A. Brualdi and P.M. Gibson (1976), The assignment polytope. Math. Programming 11 (1976) 97–101.
R.E. Burkard (1973), Die Strörungsmethode zur Lösung quadratischer Zuordnungsprobleme, Operations Res. Verfahren 16 (1973) 89–108.
R.E. Burkard (1974), Quadratische Bottleneckprobleme, Operations Res. Verfahren 18 (1974) 26–41.
R.E. Burkard (1975a), Numerische Erfahrungen mit Summen und Bottleneck Zuordnungsproblemen, in: L. Collatz, G. Meinardus and H. Werner, eds., Numerische Methoden bei graphentheoretischen und kombinatorischen Problemen (Birkhäuser, Basel, 1975) 9–25.
R.E. Burkard (1975b), Heuristische verfahren zur lösung quadratischer zuordnungsprobleme, Z. Operations Res. 19 (1975) 183–193.
R.E. Burkard, W. Hahn and U. Zimmermann (1977), An algebraic approach to assignment problems. Math. Programming 12 (1977) 318–327.
R.E. Burkard and J. Offermann (1977), Entwurf von Schreibmaschinentastaturen mittels quadratischer Zuordnungsprobleme, Z. Operations Res. 21 (1977), B121–B132.
R.E. Burkard and K.H. Stratmann (1976), Numerical investigations on quadratic assignment problems, Report 76-3, Mathematical Institute, University of Cologne (1976), Naval Res. Logist. Quart. 25 (1) (1978) 129–148.
R.E. Burkard and U. Zimmermann (1977), Weakly admissible transformations for solving algebraic assignment and transportation problems, Report 77-3, Mathematical Institute, University of Cologne, W. Germany (1977), forthcoming in Math. Programming Studies.
M.I. Celebiler (1969), A probabilistic approach to solving assignment problems, Operations Res. 17 (1969) 993–1004.
N. Christofides (1972), Bounds for the travelling salesman problem, Operations Res. 20 (1972) 1044–1056; 22 (1974) 1121.
N. Christofides (1975), Graph Theory: An Algorithmic Approach (New York, Academic Press, 1975).
N. Christofides (1976), Worst case analysis of a new heuristic for the travelling salesman problem, Working paper, Carnegie–Mellon University, Pittsburgh (Feb. 1976).
N. Christofides (1977), Graphical analysis of bounds for the QAP, Communicated at the Summer School in Combinatorial Optimization, SOGESTA, Urbino, Italy (June 1977).
N. Christofides and S. Eilon (1972), Algorithms for large-scale travelling salesman problems. Operational Res. Quart. 23 (1972) 511–518.

N. Christofides and M. Gerrard (1976), Special cases of the quadratic assignment problem, Carnegie–Mellon University, Management Sciences Research Report No. 391, Pittsburgh, PA (April 1976).

V. Chvátal (1973), Edmonds polytopes and weakly Hamiltonian graphs, Math. Programming 5 (1973) 29–40.

G. Clarke and J. Wright (1964), Scheduling of vehicles from a central depot to a number of delivery points, Operations Res. 12 (1964) 568–581.

A. Cruse (1975), A note on symmetric doubly-stochastic matrices, Discrete Math. 13 (1975) 109–119.

G. Dantzig, D.R. Fulkerson and S. Johnson (1954), Solution of a large scale-travelling salesman problem, Operations Res. 2 (1954) 393–410.

G. deGhellink (1977), A primal-dual method for linear assignment problems; presorting of matrix entries, Communication at ARIDOSA, Banff, Canada (August 1977).

U. Derigs and U. Zimmermann (1977); An augmenting path method for solving linear bottleneck assignment problems, Report 77–4, Mathematical Institute, University of Cologne (July 1977), Computing 19 (1978) 285–295.

M. Devine and F. Glover (1972), Computational study of the symmetric assignment problem, presented at the 41st National ORSA Meeting, New Orleans (April 1972).

E.A. Dinic and M.A. Kronrad, An algorithm for the solution of the assignment problem (translated from Russian), Soviet Math. Dokl. 10 (1969) 1324–1326.

B. Dorhout, Het lineaire toewijzingsproblem — Vergelijking van algoritmen, Report BN 21/73, Mathematisch Centrum, Amsterdam (1973).

P.J. Eberlein, Remarks on the van der Waerden conjecture II, Linear Algebra Appl. 2 (1969) 311–320.

P.J. Eberlein and G.S. Mudholkar, Some remarks on the van der Waerden conjecture, J. Combinatorial Theory 5 (1968) 386–396.

J. Edmonds, Optimum branchings, J. Res. Nat. Bur. Standards 71B (1967) 233–240.

M.L. Fisher, G.L. Nemhauser and L.A. Wolsey, An analysis for finding a maximum weight Hamiltonian circuit (submitted for publication).

A.M. Frieze, A bilinear programming formulation of the 3-dimensional assignment problem, Math. Programming 7 (1974) 376–379.

E.Ja. Gabovič, The travelling salesman problem I, Trudy Vyčisl. Centra Tartu. Gos. Univ. 19 (1970) 52–96 (in Russian).

E.Ja. Gabovič, The little travelling salesman problem, Trudy Vyčisl. Centra Tartu. Gos. Univ. 19 (1970) 27–51 (in Russian).

E.Ja. Gabovič, A. Ciz and A. Jalas (1971), The bottleneck travelling salesman problem. Trudy Vyčisl. Centra Tartu. Gos. Univ. 22 (1971) 3–24 (in Russian).

D. Gale, Optimal assignments in an ordered set, An application of matroid theory. J. Combinatorial Theory 4 (1968) 176–180.

M.R. Garey, R.L. Graham and P.S. Johnson, Some NP-complete geometric problems, Proc. Eighth ACM Symp. on Theory of Computing (1976).

R.S. Garfinkel (1971), An improved algorithm for the bottleneck assignment problem, Operations Res. 19 (1971) 1747–1751.

R.S. Garfinkel (1973), On partitioning the feasible set in a branch and bound algorithm for the asymmetric travelling salesman problem, Operations Res. 21 (1973) 340–343.

G.K. Gaschütz and J.H. Ahrens (1968), Suboptimal algorithms for the quadratic assignment problem, Naval Res. Logist. Quart. 15 (1968) 49–62.

W. Gaul and A. Heinecke, Einige Aspekte in der Zuordnungstheorie. Z. Operations Res. 19 (1975) 89–99.

J.W. Gavett and N.V. Plyter, The optimal assignment of facilities to locations by branch and bound, Operations Res. 14 (1966) 210–232.

B. Gavish, P. Schweitzer and E. Shlifer, The zero pivot phenomenon in transportation and assignment problems and its computational implications, Math. Programming 12 (1977) 226–240.

R. Gianessi, A decomposition method for the travelling salesman problem, Report A–40, Dipartimento di recerca operativa e scienze statistiche, Universitá di Pisa, Pisa, Italy (November 1976).

P.C. Gilmore, Optimal and suboptimal algorithms for the quadratic assignment problem, J. SIAM 10 (1962), 305–313.

P.C. Gilmore and R.E. Gomory (1964), A solvable case of the travelling salesman problem. Proc. Nat. Acad. Sci. 51 (1964) 178–181.

F. Glover, D. Karney, D. Klingman and A. Napier, A computation study on start procedures, basic change criteria and solution algorithms for transportation problems, Management Sci. 20 (1974) 793–813.

B.L. Golden (1976), Evaluating the Clarke-Wright vehicle routing algorithm, Working paper MS/S76-003, College of Business and Management, University of Maryland, College Park, Maryland (November 1976).

M. Grötschel, An optimal tour through 120 Cities in the federal republic of germany (private communication).

M. Grötschel and M.W. Padberg (1974a), Lineare Charakterisierung des Travelling Salesman Polytopen, Report 7416, Institut für Ökonometrie und Operations Research, University of Bonn (June 1974).

M. Grötschel and M.W. Padberg (1974b), Linear characterization of the symmetric travelling salesman polytope. Report 7417, Institut für Ökonometrie und Operations Research, University of Bonn (June 1974).

M. Grötschel and M.W. Padberg (1975), Partial characterization of the asymmetric travelling salesman polytope. Math. Programming 8 (1975) 378–381.

M. Grötschel and M.W. Padberg (1977), On the symmetric travelling salesman problem I: Inequalities, II: Lifting theorems and facets, Research Reports RC6820, 6821 IBM Thomas J. Watson Res. Center, Yorktown Heights (1977).

O. Gross, The bottleneck assignment problem. RAND paper P-1630, The RAND Corporation. Santa Monica, Calif. (March 1959).

P. Hansen and L. Kaufman (1973), A primal-dual algorithm for the three dimensional assignment problem, Cahiers Centre Études Recherche Opér. 15 (1963) 327–336.

L.H. Harper and J.E. Savage, On the complexity of the marriage problem, Advances in Math. 9 (1972) 299–312.

K. Helbig Hansen and J. Krarup, Improvements of the Held–Karp algorithm for the symmetric travelling salesman problem, Math. Programming (1974) 87–96.

M. Held and R.M. Karp, The travelling salesman problem and minimum spanning trees. Operations Res. 18 (1970) 1138–1162.

M. Held and R.M. Karp, The travelling salesman problem and minimum spanning trees: Part II, Math. Programming 1 (1971) 6–25.

I. Heller (1953), On the problem of shortest path between points I, Bull. Amer. Math. Soc. 59 (1953) 551.

I. Heller (1955), Geometric characterization of cyclic permutations. Bull. Amer. Math. Soc. 61 (1955) 227.

S. Hong (1971), A linear programming approach for the travelling salesman problem, unpublished Ph.D. Thesis, The Johns Hopkins University (1971).

S. Hong, The asymmetric travelling salesman problem and spanning arborescences of directed graphs, forthcoming in Proc. of Symposium on Operations Research, Heidelberg (Nov. 1976).

R.M. Karp, The fast approximate solution of hard combinatorial problems, Proc. Sixth Southeastern Conf. on Combinatorics, Graph Theory and Computing, Winnipeg (Utilitas Mathematical Publ., 1975).

R.M. Karp, The probabilistic analysis of some combinatorial search algorithms, forthcoming in: Proc. of the Symposium on Algorithms and Complexity: New Directions and Recent Results, Pittsburgh (1976).

R.M. Karp and S.Y. Li (1965), Two special cases of the assignment problem, Discrete Maths. 13 (1975) 129–142.

C. Kastning (Ed.), Integer programming and related areas, a classified bibliography, Lecture Notes in Economics and Math. Systems 128 (Springer–Verlag, Berlin–Heidelberg–New York, 1976).

M. Katz, On the extreme points of a certain convex polytope, J. Combinatorial Theory 8 (1970) 417–423.

B. Korte and D. Hausmann, An analysis of the greedy heuristic for independence systems, Report No. 7645, Institut für Ökonometrie und Operations Research, Universität Bonn (May 1976).

J. Krarup, The peripatetic salesman and some related unsolved problems, in: B. Roy, ed., Combinatorial Programming—Methods and Applications (Reidel, Dordrecht–Boston, 1974) 173–178.

H.W. Kuhn, The Hungarian method for assignment problems, Nav. Res. Logist. Quart. 2 (1955) 83–97.

A.H. Land, A problem of assignment with interrelated costs, Operational Res. Quart. 14 (1963) 185–198.

E.L. Lawler, The quadratic assignment problem, Management Sci. 9 (1963) 586–599.

E.L. Lawler, A solvable case of the travelling salesman problem, Math. Programming 1 (1971) 267–269.

E.L. Lawler (1974), The quadratic assignment problem, A brief review, in: B. Roy, ed., Combinatorial Programming Methods and Applications (Reidel, Dordrecht–Boston, 1974).

J.K. Lenstra and A.H.G. Rinnoy Kan, Some simple applications of the travelling salesman problem, Operational Res. Quart. 26 (1975) 717–733.

O. Leue, Methoden zur Lösung dreidimensionaler Zuordnungsprobleme, Angewandte Informatik 4 (1972) 154–162.

S. Lin and B.W. Kernighan, An effective heuristic algorithm for the travelling salesman problem, Op. Research 21 (1973) 498–516.

M. Markus and H. Minc, Permanents, Am. Math. Monthly 72 (1965) 577–591.

M. Markus and M. Newman, On the minimum of the permanent of a doubly stochastic matrix, Duke Math. J. 26 (1959) 61–72.

J.-F. Maurras, Some results on the convex hull of the Hamiltonian cycles of symmetric complete graphs, in: B. Roy, ed., Combinatorial Programming, Methods and Applications (Reidel, Dordrecht–Boston, 1974).

P. Miliotis, Integer programming approaches to the travelling salesman problem, Math. Programming 10 (1976) 367–378.

P. Miliotis, Using cutting planes to solve the travelling salesman problem, Working paper, London School of Economics (1977).

H. Müller–Merbach, Optimale Reihenfolgen (Springer-Verlag, Berlin–Heidelberg–New York, 1970).

K.G. Murty, The symmetric assignment problem, Report ORC 67–12, University of California, Berkeley (1967).

K.G. Murty, On the tours of a travelling salesman, SIAM J. Control 7 (1969) 122–131.

C.St.J.A. Nash-Williams, Hamiltonian Circuits, in: D.R. Fulkerson, ed., Studies in Graph Theory, Part II (The Math. Association of America, 1975).

R.Z. Norman, On the convex polyhedra of the symmetric travelling salesman problem, Bull. Amer. Math. Soc. (1955) 559.

M.W. Padberg and S. Hong, On the symmetric travelling salesman problem: A computational study, presented at ARIDOSA, Banff, Canada (August 1977).

W, Padberg and M.R. Rao, The travelling salesman problem and a class of polyhedra of diameter two, Math. Programming 7 (1974) 32–45.

U. Pape and B. Schön, Verfahren zur Lösung von Summer und Engpaßzuordnungsproblemen, Elektronische Datenverarbeitung 12 (1970) 149–163.

J.F. Pierce and W.B. Crowston, Tree search algorithms for quadratic assignment problems, Naval Res. Logist. Quart. 18 (1971) 1–36.

W.P. Pierskalla, The tri substitution method for the three-dimensional assignment problem, CORS Journal 5 (1967) 71–81.

W.P. Pierskalla, The multidimensional assignment problem, Op. Research. 16 (1968) 422–431.

T.C. Raymond, Heuristic algorithm for the travelling salesman problem, IBM J. Res. Develop. 13 (1969) 400–407.

D.J. Rosenkrantz, R.E. Stearns and P.M. Lewis, Approximate algorithms for the travelling salesperson problem, Proc. 15th IEEE Symposium on switching and automata theory (1974) 33–42.

G.T. Ross and R.M. Soland, A branch and bound algorithm for the generalized assignment problem, Math. Programming 8 (1975) 91–103.

S. Sahni and T. Gonzalez, P-complete approximation problems, J. Assoc. Comput. Mach. 23 (1976) 555–565.

W. Schiebel, G. Unger and J. Terno, A cutting plane algorithm for the travelling salesman problem, Report 07–15–76, Sektion Mathematik, Technische Universität Dresden, GDR (July 1975).

Th.H.C. Smith, A computational comparison of an improved pair assignment algorithm and a pair exclusion algorithm for the quadratic assignment problem, Carnegie–Mellon University, Man. Sci. Res. Report No. 383, Pittsburgh, PA (November 1975).

V. Srinivasan and G.L. Thompson, Benefit-cost analysis of coding techniques for the primal transportation algorithm, J. Assoc. Comput. Mach. 20 (1973) 194–213.

H. Steckhan and R. Thome, Vereinfachungen der Eastmanschen Branch and Bound Lösung für symmetrische Travelling Salesman Probleme, Operations Res. Verfahren 14 (1972) 360–389.

M.M. Sysło, A new solvable case of the travelling salesman problem, Math. Programming 4 (1973) 347–348.

M.M. Sysło, Generalizations of the standard travelling salesman problem, Paper presented at the IX Intern. Symposium on Math. Progr., Budapest (Sept. 1976).

G.L. Thompson, Algorithmic and computational methods for solving symmetric and asymmetric travelling salesman problems, Working paper, presented at the Workshop on Integer Programming (WIP), Bonn (September 1975).

N. Tomizawa, On some techniques useful for solution of transportation network problems, Networks 1 (1971) 173–194.

B.L. van der Waerden, Aufgabe 45. Jahresbericht der Deutschen Mathematischen Vereinigung 35 (1926) 117.

M. Vlach, Branch and bound method for the three-index assignment problem, Ekonom.–Mat. Obzor 3 (1967) 181–191.

M. Webb, Relative performance of some sequential methods of planning multiple delivery journeys, Operational Res. Quart. 23 (1972) 361–372.

J.J. Wiorkowski and K. McElvain, A rapid heuristic algorithm for the approximate solution of the travelling salesman problem, Transp. Res. 9 (1975) 181–185.

P. Yellow, A computational modification of the savings method of vehicle scheduling, Operational Res. Quart. 21 (1970) 281–283.

CUTTING STOCK, LINEAR PROGRAMMING, KNAPSACKING, DYNAMIC PROGRAMMING AND INTEGER PROGRAMMING, SOME INTERCONNECTIONS

Paul C. GILMORE

Mathematical Sciences Department, IBM Thomas J. Watson Research Center, Yorktown Heights, N.Y. 10598, U.S.A.

Cutting stock problems have many connections with Linear, Dynamic and Integer Programming, many of the connections being through the Knapsack Problem. This paper relates some of the connections arising from the one dimensional cutting stock. Little background in mathematical programming is presumed.

1. Introduction

Cutting stock problems have many connections with Linear, Dynamic and Integer Programming, many of the connections being through the Knapsack Problem. This paper relates some of the connections arising from the one dimensional cutting stock problem. Little background in mathematical programming is presumed.

In Section 2 the one dimensional fractional cutting stock problem is used to introduce linear programming, and then the essentials of the primal simplex algorithm are sketched in Section 3. How the knapsack problem arises naturally as a column generation problem in the solution of a fractional cutting problem by linear programming is described in Section 4.

Section 5 provides a description of an algorithm for solving the knapsack problem. The algorithm is a modified dynamic programming calculation arising directly from the recursive definition of a knapsack function. The description of the algorithm is given as a series of recursive definitions of functions that describe transformations of the state vector of the algorithm.

Studies of the algorithms for the solution of the knapsack problem have led to new insights into integer programming. Some of these insights are described in Section 6. Finally in Section 7 the linear programming formulation of the cutting stock problem is contrasted with the linear programming formulation of the integer programming problem.

This paper is really a history of my involvement with cutting stock, linear programming, knapsacking, dynamic programming and integer programming. I was first exposed to a cutting stock problem through a friend's description of a problem of U.S. Steel. But the insights of this paper were obtained through a long and rewarding collaboration with Ralph Gomory, and a study of his work and work inspired by him. I want to express my gratitude for his collaboration and inspiration. My thanks also to Ellis Johnson and Manfred Padberg for many helpful conversations during the preparation of this paper.

2. The cutting stock problem

A supplier is assumed to have a stock of steel rods of several fixed lengths $L_1, \ldots, L_k$, called the stock lengths. These lengths could be determined, for example, by the rod making machinery, or by limitations on warehouse space. The cost per unit length of the stock lengths can differ from stock length to stock length because of different manufacturing, shipping or storage costs. Let C_i be the cost of one piece of stock of length L_i.

Generally the stock lengths of the rod are not the lengths which users of the rods require. A user orders from the supplier the number of pieces of each length of rod that he requires. The supplier then amalgamates these orders into a single demand for d_i pieces of length l_i, for $i = 1, \ldots, n$, and fills the orders by cutting the stock lengths into the ordered lengths. Any pieces of rod cut from the stock lengths that do not contribute to the filling of an order are taken to be scrap. Therefore the supplier's cost of filling the orders is the sum of the costs of all the stock lengths that he had to cut to produce the ordered lengths in the demanded numbers. The supplier's problem, to fill the orders at least cost, is called the one dimensional integer cutting stock problem.

The supplier's problem is *one* dimensional because lengths, rather than say rectangles, are ordered and cut. It is *integer* because the demand d_i, say 7, for an ordered length l_i, say 3, must be met by supplying 7 whole pieces of rod of length 3, and not by supplying say one bundle of $1\frac{1}{2}$ pieces, another bundle of $1\frac{5}{6}$ pieces, and a third bundle of $3\frac{2}{3}$ pieces for a total of 7 pieces.

A meaningful *fractional* one dimensional cutting stock problem arises when the demand d_i is for a sheet of material of length d_i and width l_i, and the sheet can be supplied in any number of pieces as long as all the lengths add up to d_i. For example, as illustrated in Fig. 1 a demand for a piece of material of length 7 and width 3 can be met by supplying one piece of length $1\frac{1}{2}$, a second piece of length

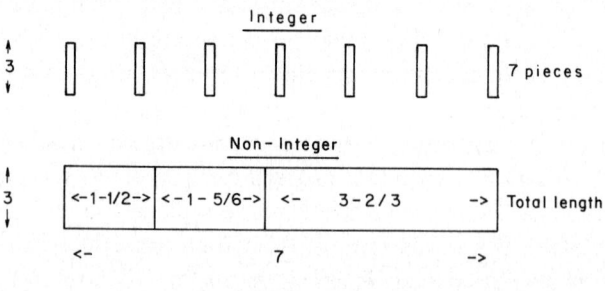

Fig. 1.

$1\frac{5}{6}$, and a third piece of length $3\frac{2}{3}$. The difference between the formulations of the fractional and the integer cutting stock problems may not be profound but, as will be seen later, it is of great mathematical and practical significance. The fractional problem is so much more amenable to solution than the integer that frequently approximate solutions to the integer problem are obtained by rounding solutions to the fractional form of the problem.

To our knowledge Kantorovich, in the remarkable paper [12], was the first to formulate and propose a solution to the cutting stock problem. Kantorovich described one and two dimensional cutting stock problems as but one of several applications of a method he had developed for solving what are now called linear programming problems, so called because of Dantzig's independent work [3]. Kantorovich did not distinguish between the integer and fractional cutting stock problems. However it is only necessary to make the distinction after they have been formulated as a linear programming problem since the integer condition translates into an integer condition on the variables of the linear programming problem.

To see how a cutting stock problem can be formulated as a linear programming problem and at the same time to give an illustration of a linear programming problem, a simple cutting stock problem will be formulated as a linear programming problem. The problem is adapted from Example 3, page 380 of [12].

Let there be two stock lengths of lengths 7.4 and 7.0 with costs respectively of 7.3 and 7.0. Let there be three ordered lengths of lengths 2.9, 2.1 and 1.5, each with a demand for 100. Consider first all the different ways that the lengths 2.9, 2.1 and 1.5 can be cut from the lengths 7.4 and 7.0 with a resulting scrap piece smaller than 1.5. These ways are enumerated as the columns of the matrix of cutting patterns of Fig. 2. Each column of the matrix consists of a triple of integers a_1, a_2 and a_3 for which

$$2.9 \times a_1 + 2.1 \times a_2 + 1.5 \times a_3 \leq L \tag{2.1}$$

where L is 7.4 or 7, and for which the scrap piece, the difference between the right and left hand sides of 2.1 is less than 1.5. Each such triple appears as a column or cutting pattern in Fig. 2. The demands for the ordered lengths also appears as a column in Fig. 2, a column with each entry 100.

Ordered Length	Cutting Patterns							Demands
2.9	2	1	1	1	0	0	0	100
2.1	0	2	1	0	3	2	1	100
1.5	1	0	1	3	0	2	3	100
Costs	7.3	7.3	7.0	7.3	7.0	7.3	7.0	

Fig. 2

An additional cost row appears below the matrix of cutting patterns in Fig. 2. The entry in the cost row below a given column is the cost of the smallest stock length from which the cutting pattern can be cut. For example, the second cutting pattern cuts 1 piece of length 2.9, and 2 pieces of length 2.1 from the stock length 7.4 of cost 7.3. If the problem was simply to minimize the scrap, rather than the cost, then instead of 7.3 would appear 7.4 in the cost row.

Associate now a variable with each of the cutting patterns of Fig. 2. A value taken by such a variable indicates the number of times the cutting pattern is to be used. For example, let $x_1, x_2, \ldots, x_7$ be associated with the 7 cutting patterns in order, and let $x_1 = 2$, $x_2 = 5$, and $x_3 = 7$, while all the other variables are 0. Then there will be $2 \times 2 + 5 \times 1 + 7 \times 1$ or 16 pieces of length 2.9 cut, $2 \times 0 + 5 \times 2 + 7 \times 1$ or 17 pieces of length 2.1 cut, and $2 \times 1 + 5 \times 0 + 7 \times 1$ or 9 pieces of length 1.5 cut. The cost of supplying the pieces will be $2 \times 7.3 + 5 \times 7.3 + 7 \times 7.0$ or 100.1.

To meet the demands there must be at least 100 pieces of each ordered length produced. Thus an inequality is associated with each row of the matrix A of cutting patterns. Let d be the column vector of demands and x the row vector of variables associated with the columns of A. Then the demands are satisfied when non-negative values of the variables x exist for which

$$Ax \geq d. \tag{2.2}$$

Actually the inequalities 2.2 may be assumed to be equalities, since all the elements of A and d are non-negative and A is non-singular. m linearly independent columns of A can be simply constructed by taking each ordered length l_i in turn and cutting it as often as possible from the largest stock length. If c is the row vector of costs then the fractional cutting stock problem can be formulated as the following linear programming problem:

L.P.: Find x satisfying

$$x \geq 0, \tag{2.3}$$

and

$$Ax = d, \tag{2.4}$$

for which

$$c \cdot x \tag{2.5}$$

is as small as possible.

A solution x to L.P. with the elements of x integer is a solution to the integer cutting stock problem. The jth element of x indicates how often the cutting pattern represented by the jth column of A should be used to supply the demands. A fractional solution x to L.P. is a solution to the fractional cutting stock problem. For most of this paper attention will be focused only on the fractional cutting stock problem; only in the last section will the integer problem be discussed.

There now exist several different methods for solving a linear program and many good introductory and reference works for the methods; see for example Dantzig [5] or Simmonard [15]. The efficiency of the methods is reflected in the extensive use made of linear programming in the solution of scientific, engineering and business problems. Nevertheless linear programming formulations of cutting stock problems result in linear programs with special computational problems.

The columns of the matrix A in 2.4 are all the possible cutting patterns that can be used to cut ordered lengths from stock lengths. Matrices with very large numbers of columns can therefore result. To get some idea of how large, consider a cutting stock problem with 20 ordered lengths, a modest sized problem. Assume that 5 pieces of the largest ordered length can be cut from the largest stock length. Then a lower bound on the number of cutting patterns is $20^5/5!$ or approximately 24,000. When larger problems are considered, for example, one with 30 ordered lengths, then the comparable lower bound is 190,000. The storage requirements for such matrices make it impossible for the corresponding linear programming problem to be solved by ordinary methods on a modest sized computer.

A proposal for overcoming the difficulty of the large number of columns was made by Eiseman [6]. The proposal was to generate a library of only good columns, by some measure of "good", and use only these columns in the matrix. But unfortunately it is difficult to find measures of "good" that sufficiently restrict the size of the matrix without distorting the solution. In Section 4 another proposal will be described after the outlines of a method for solving linear programming problems is described in the next section.

3. A primal simplex algorithm

A digression is necessary to describe the primal simplex method of solution of a linear programming problem like L.P. of the last section.

A basis for the m equations 2.4 is a submatrix B of A consisting of m linearly independent columns of A. Let x_B be the variables associated with the columns of B. They are called the basic variables for B. The values $B^{-1} \cdot d$ of x_B form the unique solution to the equations

$$B \cdot x_B = d. \qquad (3.1)$$

If $B^{-1} \cdot d \geq 0$ then B is said to be a feasible basis for the equations 2.4.

A feasible solution for the equations 2.4 is a solution x satisfying 2.3. From a feasible basis a unique feasible solution can be determined: Let the values of the basic variables be determined by (3.1) and let the values of the other variables be zero. Although it is not immediately obvious it is possible in solving an L.P. problem to restrict attention to only feasible solutions determined by bases, or what are called basic feasible solutions. For any given feasible solution a basic feasible solution of no greater cost can be found.

The cost of the basic feasible solution determined by a feasible basis B can be easily calculated. Let c_B be the cost coefficients corresponding to the columns of B. Let x be the feasible solution determined by B. Then the cost of x is $c_B \cdot x_B$, which is $c_B \cdot (B^{-1} \cdot d)$.

The L.P. problem can now be reformulated:

B.L.P.: Find a submatrix B of m linearly independent columns of A which satisfies

$$B^{-1} \cdot d \geq 0, \qquad (3.2)$$

and for which

$$c_B \cdot (B^{-1} \cdot d) \qquad (3.3)$$

is as small as possible.

In other words, find a feasible basis B of 2.4 for which 3.3 is as small as possible. Since the dot product is associative the cost 3.3 can equally well be calculated as $(c_B \cdot B^{-1}) \cdot d$. This trivial transformation has profound significance.

Let Π_B be $(c_B \cdot B^{-1})$ so that 3.3 is $\Pi_B \cdot d$. The elements of Π_B can be interpreted as unit prices for the demands d. That is, the ith element of Π_B can be interpreted as the price for one unit of material of the size corresponding to the ith equation of 2.4. For to obtain the cost $\Pi_B \cdot d$ each such unit price is multiplied by the amount of the material being produced and all such products are summed. The elements of Π_B are often called the shadow prices for the basis B.

The shadow prices Π_B for a feasible basis B figure prominently in the fundamental result upon which the primal simplex algorithm depends:

Theorem. *Let B be any feasible basis of 2.4, and let A_j be any column of A satisfying*

$$\Pi_B \cdot A_j > c_j. \qquad (3.4)$$

Then, in general, a feasible basis of lower cost can be obtained from B by replacing a column of B by A_j. Further if no column of A satisfies 3.4 then no feasible basis of cost smaller than $\Pi_B \cdot d$ can be found.

The hedge "in general" covers a possibility, usually referred to as degeneracy, when the cost of the new basis is only equal to that of B, not lower. An understanding of how degeneracy is handled in the primal simplex method is, however, not important for this paper, as is neither the method of determining the new basis from B and a column A_j satisfying 3.4.

A primal simplex algorithm for solving B.L.P. proceeds then as follows: An initial feasible basis B_0 is obtained in some fashion. For the B.L.P. formulation of the cutting stock it is easy to find a B_0: The jth column of B_0 is $[L/l_j]$ where L is the largest stock length. For any feasible basis B_k, the columns of A not in B_k are "priced out" to see if one can be found satisfying 3.4 with B taken to be B_k. If one is found then B_k is "updated" to B_{k+1} by replacing some column of B_k by A_j, and the pricing out of the columns of A is repeated for B_{k+1}. If one cannot be found then the algorithm halts with the basis B_k as a solution to B.L.P. Thus should

$$\Pi_B \cdot A_j \leq c_j, \quad \text{for all } j \tag{3.5}$$

then the basis B defines a feasible solution of least cost.

An L.P. problem in which the objective $c \cdot x$ is to be maximized instead of minimized can be solved similarly. The problem can be changed to a minimization problem by taking $(-c) \cdot x$ as the objective. Equivalently, the dual inequalities 3.4 and 3.5 can be reversed; that is, the inequality 3.5, for example, becomes

$$\Pi_B \cdot A_j \geq c_j, \quad \text{for all } j. \tag{3.6}$$

4. Cutting stock and knapsacking

As remarked earlier solving a cutting stock problem as a B.L.P. presents some computational difficulties because of the size of the matrix A of cutting patterns. However a way around the difficulties exists.

Now let B be a feasible basis for a B.L.P. formulation of a cutting stock problem. To determine if a feasible basis of lower cost can be found it is necessary to price out each column A_j of A to determine if one can be found satisfying 3.4. The cost c_j of a column A_j is the cost c of the stock length L from which A_j is being cut. Thus to determine if there exists a column A_j cut from L satisfying 3.4 it is sufficient to find the maximum of $\Pi_B \cdot A_j$, where A_j is cut from L, and compare this maximum with c. If there is more than one stock length, then a calculation of the maximum can be carried out separately for each stock length.

The problem for given L of determining the maximum of $\Pi_B \cdot A_j$, subject to L being cut from L, is called a knapsack problem. More generally it is formulated:

K.P.: Find non-negative integers $x_1, x_2, \ldots, x_m$ satisfying

$$l_1 x_1 + l_2 x_2 + \cdots + l_m x_m \leq L \tag{4.1}$$

for which

$$\pi_1 x_1 + \pi_2 x_2 + \cdots + \pi_m x_m \tag{4.2}$$

is as large as possible.

The name was given to a simpler form of the problem by Dantzig [4]. For the simpler problem the x's must be 0 or 1. It can be said to be the problem facing a hiker in filling his knapsack. For each i let l_i be the weight of an object the hiker might place in his knapsack, and let L be the maximum weight he wishes to carry. Let π_i be the worth of the ith object to the hiker. Then the **knapsack** problem is to fill the knapsack with objects of greatest total worth subject to the weight limitation. The 0/1 Knapsack Problem has received considerable attention lately from those interested in its computational complexity; see for example Sahni [13].

The more general knapsack problem arises if the hiker is allowed to select any number of a given object, not just 0 or 1.

The knapsack problem is an integer L.P. problem with $m = 1$. To see this it is only necessary to replace 4.1 with an equality with one additional non-negative integer variable:

$$l_1 x_1 + l_2 x_2 + \cdots + l_m x_m + x_0 = L \tag{4.3}$$

and to replace 4.2 with

$$(-\pi_1) x_1 + (-\pi_2) x_2 + \cdots + (-\pi_m) x_m.$$

Then like L.P. it becomes a problem of minimizing a linear function subject to a single equation and subject to the variables being non negative-integers. This transformation is possible provided $l_1, l_2, \ldots, l_m$, and L are integer. But that may safely be assumed to be the case. If they are instead fractional then they can all be brought to a common denominator and then replaced by their numerator. For example, a knapsack problem with three ordered lengths 2.9, 2.1, and 1.5, and a stock length of 7.4, is converted into one with ordered lengths 29, 21 and 15, and a stock length of 74.

The method proposed in [7] for avoiding the difficulty of the large number of columns in A is to replace the pricing out step of the primal simplex algorithm by the solution of a K.P. With this method it is only necessary to know the current basis at any given time; if a column A_j exists satisfying 3.4, it will be found as the solution of a knapsack problem. The method depends heavily, of course, on being able to rapidly solve knapsack problems. Thus although the original integer cutting stock problem was relaxed to a fractional cutting stock problem to make it easier to solve, an integer programming problem, in the form of a knapsack problem, must still be solved.

5. Computing the knapsack function by dynamic programming

The maximum $F(L)$ of 4.2 subject to 4.1 is, for given $l_1, l_2, \ldots, l_m$, a function of L called the knapsack function. To solve the knapsack problem K.P. is to determine $F(L)$ and non-negative integers $x_1, \ldots, x_m$ for which

$$F(L) = l_1 x_1 + \cdots + l_m x_m. \tag{5.1}$$

The method proposed for solving the fractional cutting stock problem requires an algorithm for solving the knapsack problem that is efficient in time and space. Some of the results of an effort to develop such an algorithm are recorded in [8]. Although finally the branch and bound algorithm described in the second of the papers [7] was found to be most practical, interesting insights into the knapsack function were obtained from an algorithm suggested by a recursive definition of F.

As noted in the last section the maximum of 4.2 subject to 4.1 is the same as the maximum of 4.2 subject to 4.3, provided the variables $x_0, x_1, \ldots, x_m$ are all non-negative integer. Define l_0 to be 1 and π_0 to be 0. Then x_0 can be treated exactly as the other varaibles of 4.3 and F can be recursively defined by the following two equations:

Define $F(0) = 0$, and

$$F(L+1) = \max\{F(L+1-l_i) + \pi_i : L+1 \geq l_i \text{ and } 0 \leq i \leq m\}. \tag{5.2}$$

The first equation is clear enough. The second equation must hold because the best that can be done with a piece of length $L+1$ is the best that can be done by cutting some piece l_i from it with a value π_i and receiving an additional value $F(L+1-l_i)$ from the remaining piece.

The order in which the lengths $l_i, \ldots, l_m$ are cut from L has no effect upon the value $F(L)$. This follows from the definition 5.2 of F:

Theorem Let $F(L) = F(L - l_i) + \pi_i$ and $F(L - l_i) = F(L - l_i - l_j) + \pi_j$. Then $F(L) = F(L - l_j) + \pi_j$.

(5.3)

Proof. From 5.2 follows

$$F(L) \geq F(L - l_j) + \pi_j \quad \text{and} \quad F(L - l_j) \geq F(L - l_j - l_i) + \pi_i.$$

Consequently

$$F(L) \geq F(L - l_j) + \pi_j \geq F(L - l_j - l_i) + \pi_i + \pi_j.$$

But

$$F(L - l_j - l_i) + \pi_i + \pi_j = F(L - l_i) + \pi_i = F(L). \quad \square$$

It is not enough to know $F(L)$. It is also necessary to know how the value $F(L)$ is achieved, that is, the values of $x_1, \ldots, x_m$ for which 5.1 holds. To know this it is sufficient to know the value of the following Function I for certain positive arguments:

Define: $I(0) = m$, and for $L > 0$,

$$I(L) = \min\{i : F(L) = F(L - l_i) + \pi_i \text{ and } L - l_i \geq 0 \text{ and } 0 \leq i \leq m\}. \tag{5.4}$$

To determine from I the values $x_i, \ldots, x_m$ for which 5.1 holds, proceed as follows: Let L_1 be L and k_1 be $I(L_1)$. For $j > 1$ and $L_j > 0$ let L_{j+1} be $L_j - l_{k_j}$, and let k_{j+1} be $I(L_{j+1})$. Let p be the largest j for which $L_j > 0$. Then $F(L)$ is $\pi_{k_1} + \pi_{k_2} + \cdots + \pi_{k_p}$ and is achieved by cutting from L one piece each of the lengths $l_{k_1}, l_{k_2}, \ldots, l_{k_p}$. Since the k's need not be distinct neither need these lengths.

Note that necessarily

$$k_1 \geq k_2 \geq \cdots \geq k_p. \tag{5.5}$$

For let $I(L)=i$ and $I(L-l_i)=j$ so that $F(L)=F(L-l_i)+\pi_i$ and $F(L-l_i)=F(L-l_i-l_j)+\pi_j$. From Theorem 5.3 follows $F(L)=F(L-l_j)+\pi_j$ so that from 5.4 follows $j \geqslant i$. For the same reason the definition 5.2 of F can be modified by replacing m in the condition of minimization by $I(L+1-l_i)$.

The function F is a monotone increasing step function. Its breakpoints, that is points L for which $F(L+1)>F(L)$, are the points $B(K)$ for $K \geqslant 1$, where B is defined: Define:

$$B(0)=0,$$

and

$$B(K+1)=\min\{L: B(K)<L \text{ and } F(L)>F(B(K))\}. \tag{5.6}$$

To know F it is sufficient to know $F(B(K))$ for $K \geqslant 0$, since:

$$F(L)=F(B(K)), \quad \text{when} \quad B(K) \leqslant L < B(K+1). \tag{5.7}$$

Similarly to know I it is sufficient to know $I(B(K))$ for $K \geqslant 0$, since:

$$I(L)=0, \quad \text{when} \quad B(K)<L<B(K+1). \tag{5.8}$$

Simultaneous recursive definitions of $F(B(K))$, $I(B(K))$, and $B(K)$ are possible: Define:

$$B(0)=F(B(0))=0 \quad \text{and} \quad I(B(0))=m. \tag{5.9}$$

Define:

$$B(K+1)=\min\{B(J)+l_i: 0 \leqslant J \leqslant K \text{ and } 1 \leqslant i \leqslant I(B(J))$$
$$\text{and } B(K)<B(J)+l_i \text{ and } F(B(J))+\pi_i>F(B(K))\},$$
$$F(B(K+1))=\max\{F(B(J))+\pi_i: 0 \leqslant J \leqslant K \text{ and } B(J)+l_i=B(K+1)\},$$

and

$$I(B(K+1))=\min\{i: F(B(J))+\pi_i=F(B(K+1)) \text{ and } 0 \leqslant J \leqslant K\}.$$

Because of 5.7 $F(B(K+1))$ must be some value $F(B(J))+\pi_i$, where $0 \leqslant J \leqslant K$ and $B(K)<B(J)+l_i$. Because of 5.5 it may be assumed that $1 \leqslant i \leqslant I(B(J))$. Thus the definitions of $B(K+1)$ and $F(B(K+1))$ are justified. The definition of $I(B(K+1))$ then follows immediately.

These definitions suggest the step-off calculation of $F(L)$ that is described in [8]. It is called a "step-off" because, from each breakpoint $B(K)$ in turn, one steps-off a distance l_i, where $1 \leqslant i \leqslant I(B(K))$, and compares $F(B(K))+\pi_i$ with the best possible value previously obtained for the point $B(K)+l_i$. This calculation can be described by the simultaneous recursive definitions of $B(K)$ and two new functions $F(K, X)$ and $I(K, X)$. It will become apparent that $F(B(K))$ is $F(K, B(K))$ and $I(B(K))$ is $I(K, B(K))$. The variable K on which the recursion is based is a counter which keeps track of the number of times a loop has been traversed. During the $K+1$st execution of the loop, all step-offs are taking place from the breakpoint $B(K)$. The results of the Kth execution of the loop are recorded by $F(K, X)$ and $I(K, X)$ for $0 \leq X$.

For $0 \leq X$: define

$$F(0, X) = 0 \quad \text{and} \quad I(0, X+1) = 0.$$

Define

$$I(0, 0) = m \quad \text{and} \quad B(0) = 0.$$

For $1 \leq i \leq I(K, B(K))$ and $0 < K$, let

$$V_i = F(K, B(K) + l_i) \quad V'_i = F(K, B(K)) + \pi_i,$$

and

$$k_i = I(K, B(K) + l_i),$$

and define

$$F(K+1, B(K) + l_i) = \max \{V_i, V'_i\}$$

$$I(K+1, B(K) + l_i) = \begin{cases} k_i & \text{if } V_i > V'_i, \\ \min \{k_i, i\} & \text{if } V_i = V'_i, \\ i & \text{if } V_i < V'_i. \end{cases}$$

For $0 \leq X$ and $X \neq B(K) + l_i$, where $1 \leq i \leq I(K, B(K))$: define

$$F(K+1, X) = F(K, X), \quad \text{and} \quad I(K+1, X) = I(K, X).$$

Define

$$B(K+1) = \min \{L : B(K) < L \quad \text{and} \quad F(K+1, L) > F(K+1, B(K))\}.$$

One observation is important for the design of algorithms to compute B, F and I: To compute $F(K+1, X)$ and $I(K+1, X)$ only $B(K)$ and $F(K, X)$ and $I(K, X)$ need be known. Further, the only values of $F(K+1, X)$ and $I(K+1, X)$ that can differ from $F(K, X)$ and $I(K, X)$ respectively are those with $X = B(K) + l_i$, when $1 \leq i \leq I(K, B(K))$.

The method of computation described played a significant role in the work described in the next two sections. A natural implementation of the method is by means of two vectors of memory, one for the values of $F(K, X)$ and one for the values of $I(K, X)$ for $X = 0, 1, \ldots, L$. For successive K's new vectors need not be created since those for $K+1$ can be overlayed on those for K. It was during some experimental runs with the method that it was discovered that for sufficiently large K, $I(K, B(K))$ is always 1. The explanation of this empirical fact led to new insights into the relationship between linear and integer programming.

6. The knapsack function and integer programming

A one constraint linear programming problem results when the variables of a knapsack problem, KP of Section 4, are not constrained to be integer. Any one of the 1×1 matrices $l_1, l_2, \ldots, l_m$ forms a feasible basis since $L/l_i \geq 0$. l_i is optimal feasible if it satisfies condition 3.6, that is, if $(\pi_j/l_i)l_i \geq c_j$, for all j.

The method of computation described in the last section is not dependent upon the order in which the ordered lengths are listed. Consequently the following ordering may be assumed:

$$\pi_1/l_1 \geq \pi_2/l_2 \geq \cdots \geq \pi_m/l_m. \tag{6.1}$$

The ordering implies that l_1 is an optimal feasible basis for the linear programming form of KP. The ordering will be assumed for this section.

Theorem. *Let L_k be the sum of the lowest common multiples of l_k and l_i for $k < i \leq m$. Then for all L, $L \geq L_k$, $I(L) \leq k$.* (6.2)

Proof. Let $L \geq L_k$ and let $F(L) = \pi_1 x_1 + \cdots + \pi_m x_m$. Let i be the smallest index for which $x_i \neq 0$ so that $I(L) \leq i$. Hence if $i \leq k$ then $I(L) \leq k$. Assume therefore that $k < i \leq m$.

Since $L \geq L_k$ it follows that for some i, $k < i \leq m$, $x_i \geq lcm_i/l_i$, where lcm_i is the lowest common multiple of l_k and l_i. Consider now

(a) $\pi_k(lcm_i/l_k) + \pi_i(x_i - lcm_i/l_i) + \pi_{i+1}x_{i+1} + \cdots + \pi_m x_m$, and

(b) $l_k(lcm_i/l_k) + l_i(x_i - lcm_i/l_i) + l_{i+1}x_{i+1} + \cdots + l_m x_m$.

Necessarily (b) is L. Further (a) $\geq \pi_1 x_1 + \cdots + \pi_m x_m$, for

$$\pi_k(lcm_i/l_k) + \pi_i(x_i - lcm_i/l_i) = \pi_i x_i + (\pi_k/l_k - \pi_i/l_i)lcm_i \geq \pi_i x_i,$$

from $k < i$ and 6.1. Consequently (a) is $F(L)$ and therefore $I(L) \leq k$. □ (6.3)

Corollary. *For $L \geq L_1$, $F(L) = F(L - l_1) + \pi_1$.*

Proof. $I(L) \leq 1$. If $I(L) = 1$ then the conclusion is immediate. Assume therefore that $I(L) = 0$. Let L_0 be the smallest integer for which $I(L') = 0$ for all L', $L_0 < L' \leq L$. Necessarily $L_1 \leq L_0$ since $F(L_1) = (\pi_1/l_1)L_1$. Hence $I(L_0) = 1$ and $F(L_0) = F(L_0 - l_1) + \pi_1$. But $F(L) = F(L_0) = F(L_0 - l_1) + \pi_1 \leq F(L - l_1) + \pi_1 \leq F(L)$. □

To compute $F(L)$ for $L \geq L_1$ when it has already been computed for $0 \leq L < L_1$ is easy; for by the corollary

$$F(L) = F(L - l_1\langle(L - L_1)/l_1\rangle) + \pi_1\langle(L - L_1)/l_1\rangle.$$

Here $\langle x \rangle$ is the least integer y for which $x \leq y$. It is therefore important to find an integer smaller than L_1 with the same property. One can be found, while calculating F by the method of the last section, by applying the following theorem.

Theorem. *Let L' be such that $I(L) \leq k$ for all L for which $L' \leq L < L' + l_{max}$, where l_{max} is $\max\{l_1, \ldots, l_m\}$. Then $I(L) \leq k$ for all $L, L' \leq L$.* (6.4)

Proof. Let $L' + l_{max} \leq L$ and let $I(L) = i$. Let $L_0 = L$ and $i_0 = i_1$. Let $L_{j+1} = L_j - l_{i_j}$ and let $i_{j+1} = I(L_{j+1})$. For some $j, L' \leq L_{i_j} < L' + l_{max}$. Consequently $i_j \leq k$ and therefore $i_0 \leq k$. □

Let a be the least integer for which $I(L) \leq 1$ when $L > l_1 a$. Such an a is found, for example, by letting it be the least for which $I(L) \leq 1$ is true for all $L, l_1 a < L \leq l_1 a + l_{max}$. L will be said to be in the periodic range of F if $l_1 a < L$ since then $F(L) - (\pi_1/l_1)L$ is periodic. For

$$F(L) - (\pi_1/l_1)L = F(L - l_1) + \pi_1 - (\pi_1/l_1)L = F(L - l_1) - (\pi_1/l_1)(L - l_1).$$

To compute $F(L)$ for any L, therefore, it is only necessary to have computed it for the range $0 \leq L \leq l_1 a + l_{max}$. However, a computation of it strictly in its periodic range can be made with possibly less effort.

Consider the cyclic group with elements the integers a for which $0 \leq a < l_1$, and with addition $\oplus$ modulo l_1. Any integer L has a representative L^* in the group, where $L^* = L - l_1[L/l_1]$. Here $[x]$ is the greatest integer y for which $y \leq x$. A knapsack problem over the group can be defined as follows. Let

$$\pi_i^* = \pi_i - (\pi_1/l_1)l_i \quad \text{and} \quad \pi_0^* = -\pi_1/l_1. \tag{6.5}$$

KP*: Let $0 \leq a < l_1$. Find non-negative integers $x_0, x_2, \ldots, x_m$ satisfying

$$l_2^* x_2 \oplus \cdots \oplus l_m^* x_m \oplus x_0 = a, \tag{6.6}$$

for which the following is as large as possible

$$\pi_2^* x_2 + \cdots + \pi_m^* x_m + \pi_0^* x_0. \tag{6.7}$$

There is no bound on the integers satisfying 6.6. For if $x_2, \ldots, x_m, x_0$ satisfies 6.6 so does also $x_2 + l_1 y_2, \ldots, x_m + l_1 y_m, x_0 + l_1 y_0$ for any integers $y_2, \ldots, y_m, y_0$. Consequently if some $\pi_i^* > 0$ then 6.7 could be made as large as you please. However this is not the case because of the assumption 6.1, and hence 6.7 is bounded above by 0.

Let now F^* be the knapsack function defined by KP*; that is, $F^*(a)$ is the maximum of 6.7 subject to 6.6. A simple relationship exists between F and F^*.

Integers $x_2, \ldots, x_m, x_0$ satisfy 6.6 if and only if there is an integer b_0 for which

$$l_2 x_2 + \cdots + l_m x_m + x_0 = a + l_1 b_0. \tag{6.8}$$

The objective 6.7 can be rewritten using the definitions of π_i^* as

$$\pi_1 x_1 + \pi_2 x_2 + \cdots + \pi_m x_m - (\pi_1/l_1)(l_1 x_1 + l_2 x_2 + \cdots + l_m x_m + x_0), \tag{6.9}$$

where x_1 is any arbitrary integer. Consider now any length L. It can be written as $a + l_1 b$, where a is L^* and b is $[L/l_1]$. Any solution of

$$l_1 x_1 + l_2 x_2 + \cdots + l_m x_m + x_0 = a + l_1 b \tag{6.10}$$

is necessarily a solution of 6.8 with $b_0 = b - x_1$. Consequently
$$F(a+l_1b) - (\pi_1/l_1)(a+l_1b) \leqslant F^*(a).$$
Conversely, given a solution of 6.8, and given $b \geqslant b_0$, x_1 can be defined to be $b - b_0$ so that 6.10 is satisfied. Thus for such b the following is also true
$$F(a+l_1b) - (\pi_1/l_1)(a+l_1b) \geqslant F^*(a)$$
Consequently the relationship between F and F^* can be expressed as follows:

Theorem. *Given any* $a, 0 \leqslant a < l_1$, *there exists some* b_0 *such that for all* $b, b \geqslant b_0$,
$$F(a+l_1b) - (\pi_1/l_1)(a+l_1b) = F^*(a). \tag{6.11}$$

Thus if $F^*(a)$ has been computed for all $a, 0 \leqslant a < l_1$, then it may be possible to compute $F(L)$ by a simple calculation. Given L let $a = L^*$ and b be $[L/l_1]$. Let $x_2, \ldots, x_m, x_0$ be the values of the variables satisfying 6.6 for which 6.7 is $F^*(a)$; then b_0 is found from 6.8. Should $b \geqslant b_0$ then $F(L)$ can be computed simply from 6.11.

Theorem 6.11 relates the integer and linear programming solutions of a one constraint L.P. problem: $F(a+l_1b)$ is the value of the integer solution while $(\pi_1/l_1)(a+l_1b)$ is the value of the continuous solution.

A computation of F^* can be made by a method almost identical with the method described for computing F, only instead of computing on a memory array of L locations, only l_1 locations are needed. Stepping-off from a point $B(K)$, where $0 \leqslant B(K) < l_1$, with a length l_1 is done molulo l_1 so that the value to be updated is at the location $(B(K)+l_1)^*$. To stop the calculation a stopping condition analogous to the condition of Theorem 6.4 can be used.

The disadvantage of trying to compute $F(L)$ by first computing $F^*(L^*)$ is that the computation may be of no value if it turns out that $[L/l_1] < b_0$. The advantage may be a smaller computation. We will return to this point shortly but in a more general context.

The importance of Theorem 6.11 lies in the insight it brings to the relationship between a continuous and an integer solution to an L.P. problem. It was first proved in [8] and in a more general form in [14]. Its importance is also enhanced by the fact that it generalizes immediately to L.P. problems with any number of constraints. This fact was first proved in [10].

Consider a general linear programming problem L.P. with all elements of A and d integer. Let B be an optimal feasible basis and let A now be redefined to be A after the columns of B have been removed from it. Let c_B be the vector of cost coefficients for the columns of B and c_A for those of A. Then the L.P. problem can be described as follows:

$$\text{Maximize } c_B x + c_A y$$
subject to
$$Bx + Ay = d, \tag{6.12}$$

where as before x and y are non-negative, and are integer for the integer programming problem.

Define as before π_B to be $c_B \cdot B^{-1}$. Then since B is an optimal basis

$$c_A - \pi_B \cdot A \leq 0. \tag{6.13}$$

Also $c_B \cdot (B^{-1} \cdot d)$ is the largest value of $c_B \cdot x + c_A \cdot y$, when the variables need not be integer. Define $F(d)$ to be its largest value when the variables must be integer.

B defines a group on the integer vectors of m elements, just as l_1 defined a group on the integer vectors of 1 element. The elements of the group are all vectors a with elements $a_1, \ldots, a_m$ satisfying $0 \leq a_i \leq e_i$; here $e_1, \ldots, e_m$ are positive integers that can be computed from B and whose product is the absolute value of the determinant of B. Addition in the group is molulo the vector e. Any vector d of m integers, positive, zero, or negative, has a representative d^* in the group defined to be $d - B[B^{-1}d]$. The matrix A^* is the matrix of representatives of the columns of A so that it is $A - B[B^{-1}A]$. The reduced cost coefficients c_A^* for the columns of A are defined to be $c_B - \pi_B \cdot A$, which by 6.13 are all less than 0.

Let $\odot$ be dot product taken modulo e. The group maximization problem corresponding to 6.12 is then defined for $0 \leq a < e$:

Maximize $c_{A*}y$

subject to

$$A^* \odot y = a \tag{6.14}$$

with y non-negative integer.

Let $F^*(a)$ be the maximum of $c_{A*}y$. The relationship between F and F^* is expressed in the following theorem:

Theorem. *Given any $a, 0 \leq a < e$ there exists some b_0 such that for all $b, b \geq b_0$,*

$$F(a + Bb) - \pi_B(a + Bb) = F^*(a). \tag{6.15}$$

Proof. The proof of the theorem is an almost word-for-word translation of the proof of Theorem 6.11. As the group equation 6.6 became 6.8, so the group equation of 6.14 becomes

$$Ay = a + Bb_0. \tag{6.16}$$

Similarly the objective function of 6.14 can be written

$$c_B \cdot x + c_A \cdot y - \pi_B \cdot d, \tag{6.17}$$

where x is any arbitrary vector.

Any solution to the equation of 6.12 is the solution of 6.16 with $b = [B^{-1}d] - x$. Consequently

$$F(a + Bb) - \pi_B(a + Bb) \leq F^*(a).$$

Conversely given a solution of the equations 6.16 and given b, $b \geq b_0$, x can be defined to be $b - b_0$ so that the equations of 6.12 are satisfied with $d = a + Bb$. Thus for such b,

$$F(a+Bb) - \pi_B(a+Bb) \geq F^*(a)$$

and the theorem follows immediately. $\square$

The theorem reduces the computation of $F(d)$, for some d, to that of $F^*(d^*)$. For let $d = a + Bb$, where a is d^* and b is $[B^{-1}d]$. From the y which achieve $F^*(a)$ the vector b_0 can be simply calculated. Then if $b \geq b_0$, $F(d)$ can be computed from 6.15. For a discussion of methods for calculating F^* see Chen and Zionts [1] and Johnson [11].

7. Cutting stock and integer programming

Another connection between cutting stock and integer programming problems is a more superficial one although an historically interesting one. The matrix A of a linear programming formulation of a cutting stock problem has generally many more columns than it does rows, and generally the only efficient methods for solving the problem as a linear programming problem require that the columns be generated as they are needed. The matrix A of a linear programming formulation of an integer programming problem has generally many more rows than it does columns and it is also the case that generally the only efficient methods for solving the problem as a linear programming problem require that the rows be generated as they are needed. Indeed it was the method of Gomory [9] for solving integer programming problems that suggested the method of Gilmore and Gomory [7] for solving the cutting stock problem.

Consider the set of constraints

$$Ax \geq d \qquad (7.1)$$

for an integer programming problem, where now the non-negativity constraints $x \geq 0$ have been included in the matrix A. Consequently if A is $m \times n$ then $m \geq n$.

Let now λ be a vector of m non-negative elements. Then necessarily any x satisfying 7.1 also satisfies

$$(\lambda A)x \geq \lambda d. \qquad (7.2)$$

Assume further that λA is all integer and consider the inequality

$$(\lambda A)x \geq \langle \lambda d \rangle, \qquad (7.3)$$

where as before $\langle x \rangle$ is the smallest integer y for which $x \leq y$. Necessarily any integer x satisfying 7.2 satisfies 7.3 also since $(\lambda A)x$ is an integer. Therefore any integer x satisfying 7.1 also satisfies 7.3, although there may be non-integer x

satisfying 7.1 which does not satisfy 7.1. An inequality like 7.3 is called a cut; it cuts away a portion of the space of x satisfying 7.1 without cutting away any integer x satisfying 7.1.

For example, consider the following inequalities, adapted from an example attributed in Chvátal [2] to Professor Bondy:

$$\begin{align} -x &\geq -1 \\ 3x-y &\geq -1 \\ -3x-y &\geq -4 \\ x &\geq 0 \\ y &\geq 0 \end{align} \tag{7.4}$$

The space of (x, y) satisfying 7.4 is the shaded area in Fig. 3.

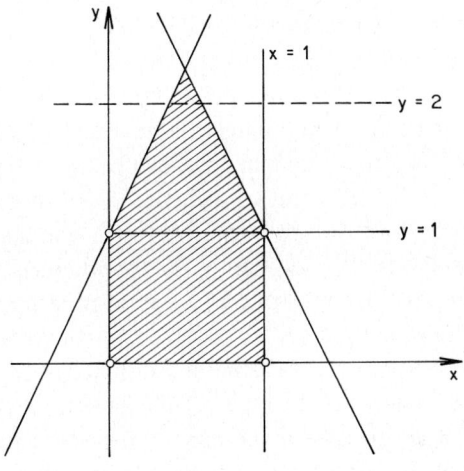

Fig. 3.

A cut of the form 7.3 is obtained by adding together the second and third inequalities of 7.4 to produce $-2y \geq -5$, dividing by 2 and noting that $\langle -\frac{5}{2} \rangle$ is -2. The cut produced is $-y \geq -2$ or $y \leq 2$. It cuts off the shaded area above the dotted line.

There may of course exist integers d_0, $d_0 > \langle \lambda d \rangle$ for which

$$(\lambda A)x \geq d_0 \tag{7.5}$$

is also a cut for 7.1. Such a cut would be deeper than 7.3, in the sense that it cuts away a larger portion of the space of x satisfying 7.1. The deepest possible cut 7.5 can be obtained by taking d_0 to be the minimum of $(\lambda A)x$ subject to 7.1 and x being integer; there necessarily is such a minimum since $\langle \lambda d \rangle$ is a lower bound for it. Thus finding the deepest possible cut 7.5 requires solving an integer programming problem with the same inequalities 7.1, as the original problem.

For the example 7.4 the deepest possible cut obtained by adding together the second and third inequalities is

$$-y \geq -1 \tag{7.6}$$

When this is added to the inequalities 7.4 the space of points (x, y) satisfying them is reduced to the shaded square below the line $y = 1$. In this case each corner of the square, or each vertex of the polyhedra determined by 7.4 and 7.5 as they are called, has all integer components. As a consequence any integer programming problem with 7.4 as its set of inequalities can be reduced to a linear programming problem with 7.4 and 7.6 as its set of inequalities.

In the case of 7.4 only one additional constraint 7.6 is needed to have a polyhedra with all integer vertices. It can be shown that for any set 7.1 of inequalities, where all components of A and d are rational, that finitely many additional constraints of the form 7.5 can be found for which the resulting polyhedra has all integer vertices; see, for example, Gomory [9]. Thus any integer linear programming problem can be converted to a linear programming problem with additional constraints. However the number of such constraints needed can be very very large, making it impractical in general to add all such constraints.

The calculation of the cut 7.3 from the constraints 7.1 is very easy, assuming that λ is given. The choice of λ for a good cut 7.3 is however, anything but easy. It should be noted too that not all cuts 7.5 can be obtained *directly* from 7.1 as cuts 7.3. For example the cut 7.6 cannot be so obtained from 7.4. Rather it is necessary to add the cuts 7.3 generated to the original inequalities 7.1 and form new cuts 7.3 from the enlarged set. Of course the coefficients of the variables of cuts obtained in this fashion can be obtained directly from A by some λ, but the right hand side d_0 is not of the form $\langle \lambda A \rangle$. For example taking 2 times 7.6 and adding it to $3x - y \geq -1$ results in the cut $x - y \geq -1$, yet the best cut with the same coefficients that can be obtained from 7.4 directly is $x - y \geq -2$. However Chvátal shows in [2] that when the x satisfying 7.1 can be bounded then any cut can be obtained by the process of forming cuts 7.3, adding them to the set 7.1 and repeating.

A linear programming formulation of a cutting stock problem results in a matrix with many columns. A linear programming formulation of an integer programming problem results in a matrix with many rows. Clearly a linear programming formulation of an integer cutting stock problem results in a matrix that has many columns and many rows. It is not surprising that it is difficult to find exact solutions to the integer cutting stock problem.

References

[1] D.-S. Chen and S. Zionts, Comparison of some algorithms for solving the group theoretic integer programming problem, Operations Res. 24 (1976) 1120–1128.
[2] V. Chvátal, Edmonds polytopes and a hierarchy of combinatorial problems, Discrete Math. 4 (1973) 305–337.

[3] G.B. Dantzig, Programming in a linear structure, Econometrica 17 (1949) abstract.
[4] G.B. Dantzig, Discrete variable extremum problems, Operations Res. 5 (1957) 266–277.
[5] G.B. Dantzig. Linear Programming and Extensions (Princeton University Press, Princeton, N.J. 1963).
[6] K. Eisemann, The trim problem, Management Sci. 3 (1957) 279–284.
[7] P.C. Gilmore and R.E. Gomory, A linear programming approach to the cutting-stock problem, I and II, Operations Res. 9 (61) 849–859, and 11(63) 863–888. Multistage cutting stock problems of two and more dimensions, Operations Res. 13 (1965) 94–120.
[8] P.C. Gilmore and R.E. Gomory, The theory and computation of knapsack functions, Operations Res. 14 (66) 1045–1074.
[9] R.E. Gomory, Outline of an algorithm for integer solutions to linear programs, Bull. Amer. Math. Soc. 64 (1958) 275–278.
[10] R.E. Gomory, On the relation between integer and non-integer solutions to linear programs, Proc. Nat. Acad. Sci. 53 (1965) 260–265.
[11] E. Johnson, Group-theoretic methods in integer programming, Annals Discrete Math. 5 (1979) next volume.
[12] L.V. Kantorovich, Mathematical methods of organizing and planning production, Management Sci. 6 (1960) 363–422. An English translation of a Russian paper published in 1939.
[13] S. Sahni, Approximate algorithms for the 0/1 knapsack problem, J. ACM 22 (1975) 115–124.
[14] J.F. Shapiro and H.M. Wagner, A finite retrieval algorithm for the knapsack and turnpike models, Operations Res. 15 (1967) 319–341.
[15] M. Simmonard, Linear Programming (Prentice-Hall, Englewood Cliffs, NJ, 1966).

USE OF FLOYD'S ALGORITHM TO FIND SHORTEST RESTRICTED PATHS*

Victor KLEE

University of Washington, Seattle, WA, 98195, U.S.A.

David LARMAN

University College, London, U.K.

In a directed network with no negative circuit, Floyd's algorithm finds, for each pair of nodes x and y, a shortest path from x to y. Here the procedure is extended to minimize more general length-functions over sets of paths that are restricted in various ways.

1. Introduction

Throughout this paper, G denotes a complete directed graph with n nodes and n^2 edges. Each edge is an ordered pair (i, j) of nodes and has as its *length* a number $\lambda(i, j) \in \mathbf{R}^* =]-\infty, \infty]$. For notational convenience G's node-set is assumed to be the set $N = \{1, \ldots, n\}$. In applying the algorithms developed here to an incomplete graph $G = (N, E)$ and real-valued edge-weighting $\lambda : E \to R$, simply extend λ by setting $\lambda(i, j) = \infty$ for all $(i, j) \notin E$.

A *walk from x to y* is a node-sequence $(x_0, \ldots, x_t)$ such that $x_0 = x$, $x_t = y$, and $t > 0$. It is a *chain* if no node is repeated and a *circuit* if $x_t = x_0$ but there is otherwise no repetition. Both chains and circuits are called *paths*, a practice that is unusual but is convenient for our purposes.

Floyd's algorithm [12, 13, 5, 8, 11] initializes $S[i, j] \leftarrow \lambda(i, j)$ for all $i, j \in N$ and then proceeds as follows:

    for $k \leftarrow 1$ **until** n **do**
        for $i \leftarrow 1$ **until** n **do**
            for $j \leftarrow 1$ **until** n **do**
                $S[i, j] \leftarrow \min \{S[i, j], S[i, k] + S[k, j]\}$.

If there are no circuits of negative length then $S[x, y]$ emerges as the length of a shortest path from x to y. The computation is easily modified to find shortest paths in addition to their lengths.

In the present paper the procedure is extended to deal with a family $\mathfrak{F}$ of sets of walks and with a walk-length function L more general than the usual one. Under suitable assumptions the extended procedure finds, for each choice of $x, y \in N$ and

---

* This research was supported in part by the U.S. Office of Naval Research.

$\mathcal{L} \in \mathfrak{F}$, a shortest $\mathcal{L}$-path from x to y. That is, L is minimized over the set of all paths from x to y that belong to $\mathcal{L}$. In the "classical" case, $\mathcal{L}$ is the set of all paths (or walks) in G, $\mathfrak{F} = \{\mathcal{L}\}$, and the length of a walk is the sum of the length of its edges.

2. The assumptions

The function L is used to measure the length of a walk in terms of the lengths of its edges. It is assumed the range $\mathrm{rng}\, L$ of L is contained in $\mathbf{R}^*$, the domain $\mathrm{dmn}\, L$ of L is the set of all finite sequences in $\mathbf{R}^*$, and the following two conditions are satisfied:

(1) if $\alpha_1, \ldots, \alpha_t \in \mathbf{R}^*$ and $0 < s t$, then
$$L(\alpha_1, \ldots, \alpha_t) = L(L(\alpha_1, \ldots, \alpha_s), L(\alpha_{s+1}, \ldots, \alpha_t));$$

(2) if $\alpha_1, \alpha_2, \beta_1, \beta_2 \in \mathrm{rng}\, L$ with $\alpha_1 \leq \beta_1$ and $\alpha_2 \leq \beta_2$, then $L(\alpha_1, \alpha_2) \leq L(\beta_1, \beta_2)$.

The *length* of a walk $W = (x_0, \ldots, x_t)$ is defined as
$$L_\lambda(W) = L(\lambda(x_0, x_1), \ldots, \lambda(x_{t-1}, x_t)).$$

By (1), $L_\lambda(UV) = L_\lambda(L_\lambda(U), L_\lambda(V))$ when U is a walk from x to y and V is a walk from y to z. Here UV denotes the walk that follows U from x to y and then follows V from y to z.

Among the admissible functions L are

$L^p(\alpha_1, \ldots, \alpha_t) = (\alpha_1^p + \cdots + \alpha_t^p)^{1/p}$ for an integer $p > 0$,

$L^\infty(\alpha_1, \ldots, \alpha_t) = \max(\alpha_1, \ldots, \alpha_t)$,

$L^p(\alpha_1, \ldots, \alpha_t) = (|\alpha_1|^p + \cdots + |\alpha_t|^p)^{1/p}$ for a real $p > 0$.

The usual L is L_1. The function L_∞ is also of practical interest, for if G is initially equipped with nonnegative real edge-weights $\gamma(i, j)$ representing flow capacities and if $\lambda(i, j) = -\gamma(i, j)$ for all i and j, then the shortest paths with respect to L_∞ are those of maximum flow capacity for specified initial and terminal nodes [7, 8, 11].

When $\mathcal{U}$ and $\mathcal{V}$ are sets of walks let
$$\mathcal{U}\mathcal{V} = \{UV : U \in \mathcal{U}, V \in \mathcal{V}, U \text{ ends where } V \text{ starts}\}.$$

Thus a walk $(x_0, \ldots, x_t)$ belongs to $\mathcal{U}\mathcal{V}$ if and only if there exists s such that $0 < s < t$, $(x_0, \ldots, x_s) \in \mathcal{U}$ and $(x_s, \ldots, x_t) \in \mathcal{V}$. The first assumption about $\mathfrak{F}$ is:

(3) if $0 < s < t$ and $(x_0, \ldots, x_t) \in \mathcal{W} \in \mathfrak{F}$, then there exist $\mathcal{U}, \mathcal{V}$ such that
$$(x_0, \ldots, x_s) \in \mathcal{U} \in \mathfrak{F}, \quad (x_s, \ldots, x_t) \in \mathcal{V} \in \mathfrak{F}, \quad \text{and} \quad \mathcal{U}\mathcal{V} \subset \mathcal{W}.$$

For a simple but interesting example, suppose that a set of special edges of G is given, and for $0 < k < l$ let $\mathcal{W}(k)$ denote the set of all walks $(x_0, \ldots, x_t)$ such that (x_{i-1}, x_i) is special for at most k values of i. Let
$$\mathfrak{F} = \{\mathcal{W}(k) : 0 \leq k \leq l\}.$$

Note that $\mathcal{W}(i)\mathcal{W}(j) \subset \mathcal{W}(k)$ when $i+j \le k$. This is example (A) of the last section of the paper. The reader may wish, before continuing in the present discussion, to turn to examples (B)–(G) of that section. They will provide an idea of the sort of shortest restricted paths that can be found efficiently by applying the extended version of Floyd's algorithm (EVFA) to a length-function L and a family $\mathfrak{F}$ of sets of walks satisfying conditions (1)–(4).

The *edges* of a walk $W = (x_0, \ldots, x_t)$ are $(x_0, x_1), \ldots, (x_{t-1}, x_t)$. A path P is *associated* with W if P also starts at x_0 and ends at x_t, P is a subsequence of W, and each edge of P is an edge of W. It is assumed $\mathfrak{F}$, L and λ are interrelated as follows:

(4) *if* $W \in \mathcal{W} \in \mathfrak{F}$ *and P is a path associated with W, then* $P \in \mathcal{W}$ *and* $L_\lambda(P) \le L_\lambda(W)$.

Since there are only finitely many paths in G, a consequence of (4) is:

(5) *For each* $x, y \in N$ *and* $\mathcal{L} \in \mathfrak{F}$, *either there is no $\mathcal{L}$-walk from x to y or there is a $\mathcal{L}$-path which is a shortest $\mathcal{L}$-walk from x to y.*

3. The algorithms

The extended version of Floyd's algorithm (EVFA) starts with the $n \times n$ matrix λ of edge-lengths of G, procedures for computing $L(\alpha)$ and $L(\alpha, \beta)$ for all $\alpha, \beta \in \mathbf{R}^*$, and a suitable representation of the family $\mathfrak{F}$ of sets of walks. Also required is a set $\mathfrak{T}$ of triples $(\mathcal{U}, \mathcal{V}, \mathcal{W})$ of members of $\mathfrak{F}$ such that:

(6) $\mathcal{U}\mathcal{V} \subset \mathcal{W}$ *for each* $(\mathcal{U}, \mathcal{V}, \mathcal{W}) \in \mathfrak{T}$;

(7) *if* $0 < s < t$ *and* $(x_0, \ldots, x_t) \in \mathcal{W} \in \mathfrak{F}$, *then there exists $\mathcal{U}$ and $\mathcal{V}$ such that* $(x_0, \ldots, x_s) \in \mathcal{U}$, $(x_s, \ldots, x_t) \in \mathcal{V}$, *and* $(\mathcal{U}, \mathcal{V}, \mathcal{W}) \in \mathfrak{T}$.

Of course $\mathfrak{T}$ may be taken as the set of all triples $(\mathcal{U}, \mathcal{V}, \mathcal{W})$ of members of $\mathfrak{F}$ such that $\mathcal{U}\mathcal{V} \subset \mathcal{W}$, but it is most efficient to have $\mathfrak{T}$ as small as possible subject to (6) and (7). (In the example following (3) in the preceding section, it would be best to let $\mathfrak{T} = \{(\mathcal{W}(i), \mathcal{W}(j), \mathcal{W}(k)): i+j = k \le l\}$ rather than using $i+j \le k$.)

For each $\mathcal{L} \in \mathfrak{F}$ the algorithm outputs four $n \times n$ matrices: an $\mathbf{R}^*$-valued $S_\mathcal{L}$, an integer-valued $M_\mathcal{L}$, an $\mathfrak{F}$-valued $U_\mathcal{L}$ and an $\mathfrak{F}$-valued $V_\mathcal{L}$. At the time of output these satisfy the following conditions for all $x, y \in N$:

(8) *if there is no $\mathcal{L}$-path from x to y, then* $S_\mathcal{L}[x, y] = \infty$ *and* $M_\mathcal{L}[x, y] = -1$.

(9) *if there is a $\mathcal{L}$-path from x to y, then*
   (a) $S_\mathcal{L}[x, y]$ *is the (possibly ∞) length of a shortest $\mathcal{L}$-path from x to y;*
   (b) $M_\mathcal{L}[x, y] = 0$ *if* (x, y) *is such a shortest path; otherwise,* $M_\mathcal{L}[x, y]$ *is the index m of an intermediate node on such a shortest path, and the path itself is formed from a shortest $U_\mathcal{L}[x, y]$-path from x to m followed by a shortest $V_\mathcal{L}[x, y]$-path from m to y.*

Using the output M, U, and V of EVFA the path-tracing algorithm (PTA) actually finds the shortest paths.

EVFA: EXTENDED VERSION OF FLOYD'S ALGORITHM

begin
  for $i \leftarrow 1$ **until** n **do**
    for $j \leftarrow 1$ **until** n **do**
      for each $\mathcal{W} \in \mathfrak{F}$ **do**
        if $(i, j) \in \mathcal{W}\lambda$ **then begin**
$$S_\mathcal{W}[i, j] \leftarrow L(\lambda(i, j));$$
$$M_\mathcal{W}[i, j] \leftarrow 0$$
        end
          else   **begin**
$$S_\mathcal{W}[i, j] \leftarrow \infty;$$
$$M_\mathcal{W}[i, j] \leftarrow -1$$
        end of initialization;
  for $k \leftarrow 1$ **until** n **do**
    for $i \leftarrow 1$ **until** n **do**
      for $j \leftarrow 1$ **until** n **do**
        for each $(\mathcal{U}, \mathcal{V}, \mathcal{W}) \in \mathfrak{T}$ **do**
          if $(L(S_\mathcal{U}[i, k], S_\mathcal{V}[k, j]) < S_\mathcal{W}[i, j])$
            or $(M_\mathcal{W}[i, j] = -1$ **and** $M_\mathcal{U}[i, k] \neq -1$ **and** $M_\mathcal{V}[k, j] \neq -1)$
          then begin
$$S_\mathcal{W}[i, j] \leftarrow L(S_\mathcal{U}[i, k], S_\mathcal{V}[k, j]);$$
$$M_\mathcal{W}[i, j] \leftarrow k;$$
$$U_\mathcal{W}[i, j] \leftarrow \mathcal{U};$$
$$V_\mathcal{W}[i, j] \leftarrow \mathcal{V}$$
        end of main loop;
  for each $\mathcal{W} \in \mathfrak{F}$ **do begin**
        print $S_\mathcal{W}, M_\mathcal{W}, U_\mathcal{W}$ and $V_\mathcal{W}$
    end
end

In the path-tracing algorithm, STACK's members are alternately node-indices and members of $\mathfrak{F}$. (It is often convenient in practice to represent the members of $\mathfrak{F}$ by negative integers.) When a shortest $\mathfrak{L}$-path from x to y is desired, STACK is initialized as $(y, \mathfrak{L}, x)$. As STACK is processed, node-indices are added to PATH, which emerges as a shortest $\mathfrak{L}$-path from x to y. (PTA has no output when there is no $\mathfrak{L}$-path from x to y.)

PTA: PATH-TRACING ALGORITHM

begin
  if $M_\mathfrak{L}[x, y] = -1$ **then goto** NONE
  STACK[1] $\leftarrow y$; STACK[2] $\leftarrow \mathfrak{L}$; STACK[3] $\leftarrow x$;
  $s \leftarrow 3$; $p \leftarrow 0$;
  while $s \geq 3$ **do**

```
begin
    i ← STACK[s];
    𝒲 ← STACK[s−1];
    j ← STACK[s−2];
    m ← M_𝒲[i, j];
    if m = 0 then begin
                p ← p+1;
                PATH[p] ← i;
                s ← s−2
            end
        else begin
                STACK[s−1] ← V_𝒲[i, j];
                STACK[s] ← m;
                STACK[s+1] ← U_𝒲[i, j];
                s ← s+2;
                STACK[s] ← i
            end
    end of loop;
    PATH[p+1] ← STACK[1];
    print PATH;
    NONE:
end
```

In describing the efficiency of EVFA and PTA we use the uniform cost criterion in the RAM model of random access computation [1]. Initialization of EVFA requires time $O(|\mathfrak{F}| n^2)$ and each passage through the main loop requires time $O(|\mathfrak{T}| n^2)$, so the overall time-complexity of EVFA is

$$O(|\mathfrak{F}| n^2 + |\mathfrak{T}| n^3).$$

This counts each evaluation of $L(\alpha)$ or $L(\alpha, \beta)$ as a single step.

For each choice of $x, y, \mathcal{L}$, PTA requires time $O(n)$ to find a shortest $\mathcal{L}$-path from x to y. Thus, starting from the output of EVFA, time $O(|\mathfrak{F}| n^2)$ is required to find, for all $x, y \in N$ and $\mathcal{L} \in \mathfrak{F}$, a shortest $\mathcal{L}$-path from x to y.

It remains to show that EVFA and PTA compute what is claimed for them. In the case of PTA, this follows from a routine inductive argument showing that at the end of the initialization and also at the end of each passage through the loop there exists a shortest $\mathcal{L}$-path P from x to y such that the following three conditions are satisfied:

(a) s is odd; STACK[1] = y; if $p > 0$ then PATH[1] = x;

(b) PATH[1], ..., PATH[p], STACK[s], STACK[$s-2$], ..., STACK[1] is a subsequence of P;

(c) for each odd k with $3 \leq k \leq s$, the segment of P that joins $i = $ STACK[k] to $j = $ STACK[$k-2$] is a shortest STACK[$k-1$]-path from i to j.

We turn now to EVFA. For $W \in \mathfrak{F}$ and $0 \leq k \leq n$, let W_k denote the set of all walks $(x_0, \ldots, x_t) \in W$ such that $x_s \leq k$ when $0 < s < t$. Thus the end nodes of walks in W_k are unrestricted but all intermediate nodes are in $\{1, \ldots, k\}$. Let $\mathfrak{F}_k = \{W_k : W \in \mathfrak{F}\}$. Since $W_n = W$ for all $W \in \mathfrak{F}$, it suffices to prove the following for $0 \leq k \leq n$:

(10_k) *After the kth passage through the main loop of* EVFA, *conditions* (8) *and* (9) *are satisfied for all* $x, y \in N$ *and* $\mathscr{L} \in \mathfrak{F}_k$.

The proof is by induction on k. Here initialization is regarded as the 0th passage through the main loop and assertion (10_0) is obvious because $\mathscr{L}_0$ is merely the set of all edges (paths (x_0, x_1)) in $\mathscr{L}$.

From the argument below it follows that (10_k) holds for all k regardless of the order in which i, j and $(\mathcal{U}, \mathcal{V}, W)$ appear in the main loop. In particular, the main loop could be written as

for $k \leftarrow 1$ **until** n **do**
  for each $(\mathcal{U}, \mathcal{V}, W) \in \mathfrak{T}$ **do**
    for each $(i, j) \in \{1, \ldots, n\} \times \{1, \ldots, n\}$ **do** . . . .

That is convenient for programming in some languages with special array-handling capabilities, such as APL.

Now suppose, with $0 < k \leq n$, that (10_{k-1}) holds, and consider the kth passage through the main loop. We note first that if i or j is k, then there do not exist $\mathcal{U}, \mathcal{V} \in \mathfrak{F}$ such that $\mathcal{U}\mathcal{V} \subset W$ and

(d) $M_W[i, j] = -1$ but $M_\mathcal{U}[i, k] \neq -1 \neq M_\mathcal{V}[k, j]$, or
(e) $L(S_\mathcal{U}[i, k], S_\mathcal{V}[k, j]) < S_W[i, j]$.

Suppose, for example, that $j = k$. If (d) holds there is no W_{k-1}-path from i to k but there is a $\mathcal{U}_{k-1}$-path U from i to k and there is a V_{k-1}-path V from k to k. But then UV is a W_k-walk from i to k and by condition (4) there is an associated W_k-path P from i to k. Plainly P is, in fact, a W_{k-1}-path, and that is a contradiction. A similar contradiction is derived from (e), using conditions (1)–(4) and the inductive hypothesis. It follows that the kth rows and kth columns of S_W, M_W, U_W and V_W are unchanged by the kth passage through the main loop. Hence (10_k) holds for all $\mathscr{L} \in \mathfrak{F}_k$ and $x, y \in N$ with $x = k$ or $y = k$. The case in which $x \neq k \neq y$ remains.

Supposing, still, that $0 < k \leq n$ and (10_{k-1}) holds, consider $W \in \mathfrak{F}$ and $i, j \in N$ with $i \neq k \neq j$. We discuss only the case in which (e) holds at some time during the kth passage, for the other cases (described in terms of (d) and (e)) are similar. Let

$$\mu = \min\{L(S_\mathcal{U}[i, k], S_\mathcal{V}[k, j]) : (\mathcal{U}, \mathcal{V}, W) \in \mathfrak{T}\},$$

the minimum during the kth passage, and let $(\mathcal{U}', \mathcal{V}')$ be the first pair $(\mathcal{U}, \mathcal{V})$ for which the minimum is attained. Then at the end of the kth passage,

$$S_W[i, j] = \mu, \quad M_W[i, j] = k, \quad U_W[i, j] = \mathcal{U}', \quad V_W[i, j] = \mathcal{V}'.$$

Let U be a shortest $\mathcal{U}'_{k-1}$-path from i to k and let V be a shortest $\mathcal{V}'_{k-1}$-path from k to j, whence $L_\lambda(UV) = \mu$. Then $UV \in \mathcal{U}'\mathcal{V}' \subset \mathcal{W}$ and hence UV is a $\mathcal{W}_k$-walk from i to j. For (10_k) it suffices to show UV is a shortest $\mathcal{W}_k$-path from i to j. Consider an arbitrary shortest $\mathcal{W}_k$-walk W from i to j and an arbitrary associated path $P = (x_0, \ldots, x_t)$. Then $P \in \mathcal{W}_k$ and $L_\lambda(P) \leq L_\lambda(W)$, whence of course

$$L_\lambda(P) = L_\lambda(W) \leq L_\lambda(UV).$$

The node k appears in P for otherwise it is true at the end of the $(k-1)$th passage that $S_W[i, j] = L_\lambda(P)$ and then (e) never holds during the kth passage, contrary to hypothesis. With $x_s = k$ there exist $\mathcal{U}$ and $\mathcal{V}$ such that $(x_0, \ldots, x_s) \in \mathcal{U}$, $(x_s, \ldots, x_t) \in \mathcal{V}$ and $(\mathcal{U}, \mathcal{V}, \mathcal{W}) \in \mathfrak{T}$. But then

$$L_\lambda(UV) \leq L(S_\mathcal{U}[i, k], S_\mathcal{V}[k, j]) \leq L(L_\lambda(x_0, \ldots, x_s), L_\lambda(x_s, \ldots, x_t)) = L(P),$$

whence $L_\lambda(UV) = L_\lambda(W)$ and UV is a shortest $\mathcal{W}_k$-walk from i to j. If UV is not a path it has an associated path that misses k, and that was shown to be impossible.

4. Applications

Though conditions (1)–(4) suffice for the validity of EVFA, some additional conditions aid in verifying condition (4) for specific applications. The function L is said to be *nice* if in addition to (1) and (2) it satisfies the following two conditions:

(11) *each point of rng L is fixed under L; that is,* $L(\alpha_1, \ldots, \alpha_t) = L(L(\alpha_1, \ldots, \alpha_t))$;

(12) *if $\beta \geq 0$, then $L(\alpha, \beta) \geq L(\alpha) \leq L(\beta, \alpha)$ for all $\alpha \in \text{rng } L$.*

Note that each of L_p, L_∞ and L^p is nice.

If W is a walk $(x_0, \ldots, x_t)$ and the proper segment $(x_r, \ldots, x_s)$ of W is a circuit C then C is called an *intermediate circuit* of W and W_{rs} denotes the walk that remains when all of C but x_r or x_s is removed from W. More precisely, when $0 < r$ W_{rs} is the walk $(x_0, \ldots, x_r, x_{s+1}, \ldots, x_t)$, $(x_0, \ldots, x_r, x_{s+1})$ or $(x_0, \ldots, x_r)$ according as $s+1 < t$, $s+1 = t$ or $s = t$, and when $s < t$ W_{rs} is the walk $(x_0, \ldots, x_{r-1}, x_s, \ldots, x_t)$, $(x_{r-1}, x_s, \ldots, x_t)$ or $(x_s, \ldots, x_t)$ according as $r-1 > 0$, $r-1 = 0$ or $r = 0$.

Note that a walk is a path if and only if it has no intermediate circuit. Hence condition (4) can be deduced from repeated application of the following condition:

(13) *if $W = (x_0, \ldots, x_t) \in \mathcal{W} \in \mathfrak{F}$ and $(x_r, \ldots, x_s)$ is an intermediate circuit of W, then $W_{rs} \in \mathcal{W}$ and $L_\lambda(W_{rs}) \leq L_\lambda(W)$.*

Note that the inequality of (13) always holds when $L = L_\infty$. In other cases it can often be deduced from the following result.

(14) If L is nice, $C = (x_r, \ldots, x_s)$ is an intermediate circuit of a walk $W = (x_0, \ldots, x_t)$ and $L_\lambda(C) \geq 0$, then $L_\lambda(W_{rs}) \leq L_\lambda(W)$.

To prove (14), note that $W = CW_{rs}$ if $r = 0$, $W = W_{rs}C$ if $s = t$, and if $0 < r < s < t$ there are walks U and V such that $W = UCV$ and $W_{rs} = UV$. We consider only the third case for the others are similar to it. With $L_\lambda(C) \geq 0$, it follows from (11), (12) and (1) that

$$L_\lambda(U) = L(L_\lambda(U)) \leq L(L_\lambda(U), L_\lambda(C)) = L_\lambda(UC),$$

then from (1) and (2) that

$$L_\lambda(W_{rs}) = L_\lambda(UV) = L(L_\lambda(U), L_\lambda(V)) \leq L(L_\lambda(UC), L_\lambda(V))$$
$$= L_\lambda(UCV) = L_\lambda(W).$$

Below are some illustrative problems on shortest restricted paths. In each case, "find shortest paths" means that for each $x, y \in N$, either a shortest path from x to y (among those satisfying the indicated restrictions) must be found or it must be concluded that no path from x to y satisfies the restrictions.

(A) A set of edges is given. For each $k \leq l$, find shortest paths that use at most k of the special edges.

(B) A set of nodes is given. For each $k \leq l$, find shortest paths that use at most k of the special nodes.

(C) A sequence of s sets is given, each consisting of nodes or edges or a mixture. For each choice of $(k_1, \ldots, k_s)$ with $k_r \leq l_r$ for all r, find shortest paths that use (for all r) at most k_r of the elements of the rth set.

(D) In addition to the $\mathbf{R}^*$-valued edge-lengths $\lambda(x, y)$, integer edge-lengths $\pi(x, y) \geq 0$ are given. Each walk has its usual length L_λ and also a length I_π where I is L_1. An integer $l \geq 0$ is given. For each $k \leq l$, find L_λ-shortest paths P subject to the restriction that $I_\pi^-(P) \leq k$.

(E) The nodes of G are partitioned into two disjoint sets A and B, and an integer $l \geq 0$ is given. For each $k \leq l$, find shortest paths that oscillate at most k times between A and B.

(F) A subgraph H of G and an integer $l \geq 0$ are given. For each $k \leq l$, find shortest paths P for which $P \cap H$ has at most k components.

(G) A set M of edges of G is given. Find shortest M-alternating paths.

As can be seen by reference to conditions (1)–(4) and (13)–(14), the discussions of (A)–(G) below are valid (that is, EVFA can be applied for the stated purpose) if L is L_∞ and also if L is nice and $L_\lambda(C) \geq 0$ for each circuit C intermediate to a walk belonging to a member of $\mathfrak{F}$, where $\mathfrak{F}$ is the family of sets of walks used for the particular problem. (Problem (G) requires an additional condition, stated later.)

(A) This problem, which was mentioned earlier, is straightforward. For $0 \leq k \leq l$, let $\mathcal{W}(k)$ denote the set of all walks $(x_0, \ldots, x_t)$ such that the edge (x_{i-1}, x_i) is special for at most k values of i. Let $\mathfrak{F} = \{\mathcal{W}(k): 0 \leq k \leq l\}$. Let $\mathfrak{T} = \bigcup_{0=k}^{l} \mathfrak{T}_k$ where

$$\mathfrak{T}_k = \{(\mathcal{W}(i), \mathcal{W}(j), \mathcal{W}(k)) : i + j = k\}.$$

Then $|\mathfrak{T}_k| = k+1$ and $|\mathfrak{T}| = (l+1)(l+2)/2$. The overall time-complexity of EVFA for this problem is $O(l^2 n^3)$.

(B) This problem is similar to (A), but it is included to illustrate the way in which the end behaviour of walks must sometimes be considered in constructing $\mathfrak{F}$ and $\mathfrak{T}$ for the application of EVFA. For $0 \leq k \leq l$ let $\mathcal{W}_{--}(k)$ [resp. $\mathcal{W}_{-+}(k)$, $\mathcal{W}_{+-}(k)$, $\mathcal{W}_{++}(k)$] denote the set of all walks $(x_0, \ldots, x_t)$ such that the node x_i is special for at most k values of i with $0 < i < t$ and, in addition, neither x_0 nor x_t is special [resp. x_t is special but x_0 is not, x_0 is special but x_t is not, both x_0 and x_t are special]. Let $\mathfrak{F}$ consist of the sets $\mathcal{W}_{--}(k)$ for $k \leq l$, the sets $\mathcal{W}_{-+}(k)$ and $\mathcal{W}_{+-}(k)$ for $k \leq l-1$, and the sets $\mathcal{W}_{++}(k)$ for $k \leq l-2$. Let $\mathfrak{T} = \bigcup_{0=k}^{l} \mathfrak{T}_k$, where $\mathfrak{T}_k$ consists of the triples

$$(\mathcal{W}_{--}(i), \mathcal{W}_{--}(j), \mathcal{W}_{--}(k)) \quad \text{for } i+j=k,$$
$$(\mathcal{W}_{--}(i), \mathcal{W}_{-+}(j), \mathcal{W}_{-+}(k-1)), \quad (\mathcal{W}_{+-}(i), \mathcal{W}_{--}(j), \mathcal{W}_{+-}(k-1))$$

and

$$(\mathcal{W}_{-+}(i), \mathcal{W}_{+-}(j), \mathcal{W}_{--}(k)) \quad \text{for } i+j=k-1,$$
$$(\mathcal{W}_{-+}(i), \mathcal{W}_{++}(j), \mathcal{W}_{-+}(k-1)), \quad (\mathcal{W}_{++}(i), \mathcal{W}_{+-}(j), \mathcal{W}_{+-}(k-1))$$

and

$$(\mathcal{W}_{+-}(i), \mathcal{W}_{-+}(j), \mathcal{W}_{++}(k-2)) \quad \text{for } i+j=k-2,$$

and

$$(\mathcal{W}_{++}(i), \mathcal{W}_{++}(j), \mathcal{W}_{++}(k-2)) \quad \text{for } i+j=k-3.$$

Again, $|\mathfrak{T}|$ is $O(l^2)$ and the complexity of EVFA is $O(l^2 n^3)$.

(C) This is included to illustrate the application of EVFA when the desired paths are subject to several restrictions. In order to avoid a notational morass, only the case of sets of edges is discussed. For $k_1 \leq l_1, \ldots, k_s \leq l_s$, let $\mathcal{W}(k_1, \ldots, k_s)$ denote the set of all walks $(x_0, \ldots, x_t)$ such that, for $1 \leq r \leq s$, (x_{i-1}, x_i) belongs to the rth set of edges for at most k_r values of i. Let $\mathfrak{T}$ consist of all triples

$$(\mathcal{W}(i_1, \ldots, i_s), \mathcal{W}(j_1, \ldots, j_s), \mathcal{W}(k_1, \ldots, k_s))$$

such that for $1 \leq r \leq s$, $i_r + j_r = k_r \leq l_r$. Then

$$|\mathfrak{T}| = \sum_{k_1=0}^{l_1} \sum_{k_2=0}^{l_2} \cdots \sum_{k_s=0}^{l_s} ((k_1+1)(k_2+1) \cdots (k_s+1))$$
$$= 2^{-s} \prod_{r=1}^{s} (l_r+1)(l_r+2)$$

and the complexity of EVFA is

$$O(2^{-s}(l_1+2)^2 \cdots (l_s+2)^2 n^3).$$

A theorem of Gabow, Maheshwari and Osterweil [6] implies that this problem is

NP-complete even for the special case in which all edge-lengths are 0 or 1, the special sets are pairwise disjoint and each one consists of two nodes (or of two edges), and the l_r's are all 1.

(D) This may be regarded as the integer-weighted version of a problem of which (A) is the cardinality-weighted version. Similar extensions are available for the other problems considered here. For $0 \leq k \leq l$, let $\mathcal{W}(k)$ be the set of all walks W for which $L_\pi(W) \leq k$. Define $\mathfrak{F}$ and $\mathfrak{T}$ in the obvious ways. The complexity of EVFA is $O(l^2 n^3)$. For a closely related treatment of this problem and of (A), see the discussion of the Bellman–Ford method in [11, pp. 74–75, 92–93]. EVFA is similar to the Bellman–Ford method but is more general. Roughly speaking, it amounts to replacing the additive semigroup $\{0, 1, 2, \ldots\}$ of Bellman–Ford by an arbitrary semigroup.

(E) This is a special case of a more general problem, which may be formulated as follows: A function ϕ is defined on a set of nodes and edges of G, with rng $\phi \subset \{1, \ldots, m\}$, and an integer $l \geq 0$ is given. For each $k \leq l$, find shortest paths along which ϕ has at most k relative extrema.

As the term is used here, a *relative extremum* of a real sequence $(\alpha_0, \ldots, \alpha_u)$ is an ordered pair (r, s) such that $0 < r \leq s < u$ and

$$\alpha_{r-1} < \alpha_r = \cdots = \alpha_s > \alpha_{s+1} \quad \text{or} \quad \alpha_{r-1} > \alpha_r = \cdots = \alpha_s < \alpha_{s+1}.$$

For a walk $W = (x_0, \ldots, x_t)$, let

$$\exp W = (x_0, (x_0, x_1), x_1, \ldots, (x_{t-1}, x_t), x_t),$$

the expanded version of W in which nodes and edges alternate. Let W_ϕ denote the sequence of ϕ-values corresponding to the elements of $\exp W$ that belong to dmn ϕ, and let $\rho_\phi(W)$ denote the number of relative extrema of W_ϕ. The general problem is to find shortest paths P for which $\rho_\phi(P) \leq k$. Problem (E) is the special case in which

(*) $m = 2$ and dmn $\phi = N = A \cup B$, with $\phi = 1$ on A and $\phi = 2$ on B.

As is shown below, this can be handled by EVFA. However, we do not know how to use EVFA efficiently for the general problem, or even for the following special cases:

$m = 3$ and dmn $\phi = N$;

$m = 2$ and dmn ϕ is a proper subset of N;

$m = 2$ and dmn ϕ is the set E of all edges of G.

In each case there is difficulty, even when L is L_1 and all values of the edge-length λ are positive, in constructing a suitable family $\mathfrak{F}$ satisfying conditions (3) and (4).

Now let us return to (E) in the formulation provided by (*), except that the condition dmn $\phi = N$ may be replaced by dmn $\phi \supset N$. For $0 \leq k \leq l$ and $u, v \in \{1, 2\}$, let $\mathcal{W}_{uv}(k)$ denote the set of all walks $W = (x_0, \ldots, x_t)$ such that $\phi(x_0) =$

u, $\phi(k_t) = v$, and W_ϕ has at most k relative extrema (equivalently, W oscillates at most k times between A and B). Let

$$\mathfrak{F} = \{W_{u,v}(k) : u, v \in \{1, 2\}, k \leq l\}$$

and let $\mathfrak{T}$ consist of all triples.

($W_{uu}(i)$, $W_{uu}(j)$, $W_{uu}(k)$) for $u \in \{1, 2\}$ and ($\{i, j\} = \{0, k\}$ or ($i > 0 < j$ and $i + j = k - 1$)),

($W_{uv}(i)$, $W_{uu}(j)$, $W_{uu}(k)$) for $\{u, v\} = \{1, 2\}$ and $i + j = k - 1$,

($W_{uv}(i)$, $W_{vv}(j)$, $W_{uv}(k)$) for $\{u, v\} = \{1, 2\}$ and ((i, j) = ($k, 0$) or ($j > 0$ and $i + j = k - 1$)),

($W_{uu}(i)$, $W_{uv}(j)$, $W_{uv}(k)$) for $\{u, v\} = \{1, 2\}$ and ((i, j) = ($0, k$) or ($i > 0$ and $i + j = k - 1$)).

Then EVFA can be applied, solving problem (E) in time $O(l^2 n^3)$.

(F) Define ϕ on all nodes and edges of G, with $\phi = 1$ on nodes and edges of the graph H and $\phi = 2$ otherwise. With the W_{uv} as in the preceding paragraph, the paths P for which $P \cap H$ has at most k components are precisely the paths in

$$W_{11}(2k - 3) \cup W_{12}(2k - 2) \cup W_{21}(2k - 2) \cup W_{22}(2k - 3).$$

Hence (F) can also be handled by EVFA in time $O(l^2 n^3)$.

(G) This problem is also discussed in a more general setting. With ϕ as in the discussion of (E) and with $m = 2$, let a walk W be called ϕ-*alternating* if the sequence W_ϕ alternates between 1 and 2. How can shortest ϕ-alternating paths be found? Problem (G) is the special case in which dmn $\phi = E$, $\phi = 1$ on M, and $\phi = 2$ on $E \sim M$.

When dmn $\phi \supset N$, EVFA can be applied by taking $\mathfrak{F} = \{W_{11}, W_{12}, W_{21}, W_{22}\}$, where W_{uv} is the set of all ϕ-alternating walks W such that the sequence W_ϕ starts with u and ends with v. Then let $\mathfrak{T}$ consist of all triples (W_{uv}, W_{vu}, W_{uu}) and (W_{uv}, W_{vv}, W_{uv}) for all $u, v \in \{1, 2\}$.

Now consider the case in which dmn $\phi = E$. Then each pair (i, j) forms a ϕ-alternating path (recall the standing hypothesis that G is the complete graph on N) but of course we are interested only in paths of finite length. Define $\mathfrak{F}$ by restricting the W_{uv}'s of the preceding paragraph to include only walks of finite length, and assume

(+)   each alternating circuit of finite length has an even number of edges.

Then problem (F) can be handled by EVFA with $\mathfrak{T}$ consisting of all triples (W_{up}, W_{qv}, W_{uv}) for $u, v \in \{0, 1\}$ and $\{p, q\} = \{0, 1\}$. However, this approach may fail when (+) fails for then a path associated with an alternating walk need not be an alternating path and thus condition (4) may fail. For example, consider Fig. 1 and note that (p_2, p_3, p_4, p_5) is an alternating circuit according to our definition, where the solid edges are those in M.

Brown [2] suggests a method for finding shortest M-alternating paths in a

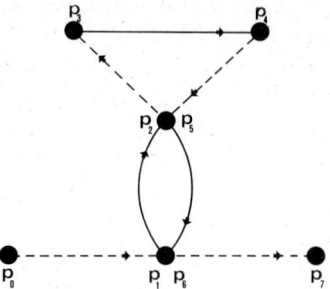

Fig. 1. The walk $(p_0, \ldots, p_7)$ is alternating but the associated path (p_0, p_1, p_7) is not.

directed graph $D = (N, E)$ (no longer assumed complete). Another directed graph D^* is constructed, having two nodes x' and x'' for each node x of D, and the edges of D^* are obtained as follows for each edge (x, y) of D with length $\lambda(x, y) < \infty$:

when $(x, y) \in M$, (x', x'') is an edge of D^* with length $\lambda(x, y)$;
when $(x, y) \notin M$, (x'', y') is an edge of D^* with length $\lambda(x, y)$.

It is claimed [2] there is a natural one-to-one correspondence between alternating paths in D and ordinary paths in D^*. Thus the problem of finding shortest alternating paths in D is equivalent to the problem of finding shortest ordinary paths in D^*. The claim is correct when (+) holds but not in general, as can be seen from Fig. 1. For example, if D is the graph of Fig. 1, then the path $(p_0'', p_1', p_2'', p_3', p_4'', p_2', p_1'', p_4')$ in D^* corresponds to the walk $(p_0, p_1, p_2, p_3, p_4, p_2, p_1, p_7)$ in D.

In general, Brown's construction does produce a one-to-one correspondence between the walks in D^* and the alternating walks in D. If D has no negative alternating circuit then D^* has no negative circuit and Floyd's algorithm can be applied to find shortest paths (= shortest walks) in D^* and hence alternating walks in D. The latter may or may not be paths. For general graphs, even when $L = L_1$ and $\lambda > 0$, we do not know how to apply EFVA directly to find shortest alternating paths. However, the more complicated "blossom" methods of Edmonds [3, 4, 11] will apparently apply to this problem.

We close with a query. For $0 \leq l \leq m < n$ and for $\rho \in \{\leq, =, \geq\}$ let $\mathfrak{P}_0(l, m, n, \rho)$ (resp. $\mathfrak{P}_1(l, m, n, \rho)$) denote the following problem:

A complete graph G is given, with n nodes and positive edge-lengths. In addition, a set of m special nodes (resp. edges) of G is given. Find shortest paths P in G such that the number of special nodes (resp. edges) used by P is in the relation ρ to l.

We have seen here that the extended version of Floyd's algorithm solves $\mathfrak{P}_0(l, m, n, \leq)$ and $\mathfrak{P}_1(l, m, n, \leq)$ in time $O(l^2 n^3)$. By contrast, the problem $\mathfrak{P}_0(n, n, n, =)$ is essentially the traveling salesman problem and hence is NP-hard [9, 10, 1]. What else of interest can be said about the computational complexity of $\mathfrak{P}_0(l, m, n, \rho)$ and $\mathfrak{P}_1(l, m, n, \rho)$ for $\rho \in \{=, \geq\}$?

Note added in proof

We are indebted to R. Ladner and N. Megiddo for calling our attention to the following paper: S. Fortune, J. Hopcroft and J. Wyllie, The directed subgraph homeomorphism problem, Technical Report TR 78-342, Department of Computer Science, Cornell University, Ithaca, NY, 1978.

This paper shows that the following problem is NP-complete. Given a directed graph G and nodes s, x and t of G, is there a simple path from s to t that passes through x? Since this problem is reducible to $\mathfrak{P}_0(l, m, n, \rho)$, and also to $\mathfrak{P}_1(l, m, n, \rho)$, for $\rho \in \{=, \geq\}$ and for any given l and m with $1 \leq l \leq m$, these problems are also NP-hard.

References

[1] A.V. Aho, J.E. Hopcroft and J.D. Ullman, The Design and Analysis of Computer Algorithms (Addison–Wesley, Reading, MA, 1974).
[2] J.R. Brown, Shortest alternating path algorithms, Networks 4 (1974) 311–334.
[3] J. Edmonds, Paths, trees and flowers, Canad. J. Math. 17 (1965) 449–467.
[4] J. Edmonds, Maximum matching and a polyhedron with 0, 1 vertices, J. Res. Nat. Bur. Standards 69B (1965) 125–130.
[5] R.W. Floyd, Algorithm 97, Shortest path. Comm. ACM 5 (1962) 345.
[6] H.N. Gabow, S.N. Maheshwari and L. Osterweil, On two problems in the generation of program test paths, IEEE Trans. Software Eng. SE-2 (1976) 227–231.
[7] T.C. Hu, The maximum capacity route problem. Operations Res. 9 (1961) 898–900.
[8] T.C. Hu, Integer Programming and Network Flows (Addison–Wesley, Reading, MA, 1969).
[9] R.M. Karp, Reducibility among combinatorial problems, in: Complexity of Computer Computations (Proc. Sympos. IBM Thomas J. Watson Res. Center, Yorktown Heights, N.Y., 1972), Plenum, New York (1972) 85–103.
[10] R.M. Karp, On the computational complexity of combinatorial problems, Networks 5 (1975) 45–68.
[11] E.L. Lawler, Combinatorial Optimization: Networks and Matroids (Holt, Rinehart and Winston, New York, 1976).
[12] B. Roy, Transitivité et connexité, C.R. Acad. Sci. Paris 249 (1959) 216–218.
[13] S. Marshall, A theorem on Boolean matrices, J. Assoc. Comput. Mach. 9 (1962) 11–12.

SHORTEST PATH AND NETWORK FLOW ALGORITHMS*

Eugene L. LAWLER

Department of Electrical Engineering and Computer Sciences and the Electronics Research Laboratory, University of California at Berkeley, CA 94720, U.S.A.

1. Introduction

A truly remarkable variety of discrete optimization problems can be formulated and solved as shortest path or network flow problems, or can be solved by procedures which employ shortest path or network flow algorithms as subroutines. It follows that these network computations are among the most fundamental and important in the entire area of discrete optimization.

We shall not attempt to make a comprehensive survey of the sizeable literature on shortest paths and network flows. Instead, we shall present a selective summary of some of the more "classic" results for some of the more "standard" problem formulations, most of which are contained in [25]. In each case, we shall give an estimate of the worst-case running time of the algorithm in question. The reader can find a more extensive discussion of computation experience in the surveys of Bradley [4] and Bradley, Brown and Graves [5]. A very recent survey of "dynamic" network flow problems is given by Bookbinder and Sethi [3].

2. Shortest paths

2.1. Definitions and notation

It is appropriate to state a few conventions we shall adhere to, unless we explicitly state otherwise. We shall deal with a directed graph $G = (N, A)$, where $n = |N|$ is the number of *nodes* and $m = |A|$, $m \geq n$, is the number of *arcs*. Each arc (i, j) has a specified *length* a_{ij}. Generally, these arc lengths may be either positive or negative. However, with few exceptions, as in Sections 2.11–2.13, we assume that G contains no directed cycle for which the sum of the arc lengths is strictly negative.

For the purpose of obtaining running time bounds, we shall usually assume that the input data are presented in the form of *arc lists*. That is, for each node i, there is given a list of those arcs (i, j) which are incident to node i, together with their

---

* Preparation of this paper was supported by National Science Foundation grant MCS76-17605.

lengths a_{ij}. When the graph G is sparse, it is usually more advantageous to work with such a data structure than with, say, a matrix of arc lengths.

By a *path*, we mean a "simple", "loop-free" directed path, i.e. one with no repeated nodes (except in Section 2.15). The *length* of a path is the sum of the lengths of the arcs contained within it. We shall be wont to describe algorithms for computing shortest path lengths without indicating explicitly how one obtains the paths yielding these lengths. We do so assuming that the reader can fill in the necessary details.

2.2. Bellman's equations

Virtually all methods for finding the length of a shortest path between two specified nodes embed the problem in the larger problem of finding the lengths of shortest paths from an origin to each of the other nodes in the network.

Let

u_j = the length of a shortest path from the origin to node j.

Let the origin be numbered 1 and the other nodes be numbered $2, 3, \ldots, n$. Then the shortest path lengths must satisfy the following equations, which we refer to as Bellman's equations:

$$u_1 = 0,$$
$$u_j = \min_{(i,j) \in A} \{u_i + a_{ij}\}, \quad j = 2, 3, \ldots, n. \tag{1}$$

There exists a directed tree rooted from the origin, such that the length of the unique path from the origin to mode j in this tree is equal to u_j, if any such path exists. Such a tree is called a *shortest path tree*. Each of the algorithms described in the next several sections can be viewed as a technique for solving Bellman's equations and for constructing such a shortest path tree, under certain specified conditions.

2.3. Dijkstra's algorithm

Consider the special case in which all arc lengths are nonnegative. The well-known algorithm of Dijkstra [9] can be summarized as follows.

Step 0 Set $u_1 = 0$

Set $u_j = \begin{cases} a_{1j}, & \text{if } (1, j) \in A \\ +\infty, & \text{otherwise} \end{cases}$

Set $S = \{2, 3, \ldots, n\}$.

Step 1 Find $k \in S$, where $u_k = \min_{j \in S} \{u_j\}$.
If $u_k = +\infty$, stop; there are no paths to the nodes remaining in S.
Set $S = S - \{k\}$.
If $S = \emptyset$, stop; the computation is completed.

Step 2 Set $u_j = \min\{u_j, u_k + a_{kj}\}$, for all $(k, j) \in A$.
Go to Step 1.

Note that Step 2 calls for $O(m)$ additions and comparisons overall. Step 1 calls for $O(n^2)$ comparisons overall, if the u_j values are maintained in an array or unordered list. This yields an overall running time of $O(n^2)$.

It is possible to maintain the u_j values in a priority queue. This queue requires $O(n)$ time to establish initially as part of Step 0. Each execution of Step 1 then requires constant time, or $O(n)$ time overall. Each time a u_j value is changed, in Step 2, the new value must be entered into the queue in place of the old value, which requires $O(\log n)$ time or $O(m \log n)$ time overall. Thus the algorithm can also be implemented in $O(m \log n)$ time.

Note that a planar network has only $O(n)$ arcs. It follows that Dijkstra's method can compute shortest paths from a fixed origin to all other nodes in a planar network in $O(n \log n)$ time (provided, of course, the arc lengths are nonnegative).

D.B. Johnston [21] has suggested an implementation for priority queues which yields the following result. For any fixed $\varepsilon > 0$, Dijkstra's algorithm can be implemented to run in time $O(\min\{n^{1+\varepsilon} + m, m \log n\})$. Thus, for "dense" graphs with $m \geq n^{1+\varepsilon}$, $O(m)$ running time is attainable.

2.4. Acyclic networks

Suppose the network is acyclic, i.e. it contains no directed cycles. It is well known that the nodes of such a network can be numbered in such a way that there exists an arc from i to j only if $i < j$. Moreover, such a numbering can be obtained in $O(m)$ time.

It is evident that Bellman's equations can be solved by simple substitution: $u_1 = 0$, u_2 depends on u_1 only, u_3 depends on u_1 and $u_2, \ldots, u_j$ depends on $u_1, u_2, \ldots, u_{j-1}$, and so on. Moreover, the equations can be solved in $O(m)$ time. We attribute this result to folklore.

2.5. Bellman–Ford method

Now consider the general case in which the network is neither acyclic nor are all the arc lengths nonnegative. A method of successive approximations, which we attribute to Bellman and Ford is as follows:

Let

$u_j^{(k)}$ = the length of a shortest path from the origin to node j, subject to the condition that the path contains no more than k arcs.

Initially, set

$$u_1^{(1)} = 0,$$

$$u_j^{(1)} = \begin{cases} a_{1j}, & \text{if } (1, j) \in A, \\ +\infty, & \text{otherwise.} \end{cases}$$

Then

$$u_j^{(k+1)} = \min\{u_j^{(k)}, \min_{(i,j) \in A}\{u_i^{(k)} + a_{ij}\}\}. \qquad (2)$$

and

$$u_j = u_j^{(n-1)}.$$

There are $O(n)$ iterations, for $k = 2, 3, \ldots, n-1$. Each iteration requires $O(m)$ time. Hence the procedure requires $O(mn)$ time overall. Note that this implies $O(n^2)$ time for a planar network.

2.6. Yen's modifications

Jin Yen [31] has suggested modifications of the Bellman–Ford procedure which together reduce the total number of additions and comparisons by a factor of approximately four in the case of a complete network, and by a factor of two, in all cases.

Let us call an arc (i, j) *upward* if $i < j$ and *downward* if $i > j$. A path is said to contain a *change in direction* whenever a downward arc is followed by an upward arc, or vice versa. Let

$u_j^{(k)} =$ the length of a shortest path from the origin to node j, subject to the condition that it contains no more than $k-1$ changes in direction.

Let

$$u_1^{(0)} = 0,$$

$$u_j^{(0)} = \begin{cases} a_{1j}, & \text{if } (1, j) \in A, \\ +\infty, & \text{otherwise.} \end{cases}$$

Then we have

$$u_j^{(k+1)} = \min\{u_j^{(k)}, \min_{\substack{(i,j) \in A \\ i<j}}\{u_i^{(k+1)} + a_{ij}\}\}, \quad k \text{ even,}$$

$$u_j^{(k+1)} = \min\{u_j^{(k)}, \min_{\substack{(i,j) \in A \\ i>j}}\{u_i^{(k+1)} + a_{ij}\}\}, \quad k \text{ odd;}$$

and

$$u_j = u_j^{(n-1)},$$

as before.

The above modification reduces the running time by a factor of approximately two. A second modification is as follows. At each iteration, at least one additional $u_j^{(k)}$ value becomes a correct shortest path length and thereafter ceases to affect

the calculations. Accordingly, let
$$I_1 = \{2, 3, \ldots, n\},$$
and
$$I_{k+1} = \{i \mid u_i^{(k+1)} < u_i^{(k-1)}\}, \quad k \geq 1.$$

Then instead of minimizing over arcs (i, j) such that $i < j$ (or $i > j$), we minimize only over arcs (i, j) such that $i < j$ and $i \in I_{k+1}$. Since $|I_{k+1}| \leq n - (k+1)$, it follows that the number of additions and comparisons is reduced by a factor of about two, in the case of a complete network.

2.7. All-pairs problems

One way to find shortest paths between *all pairs* of nodes in a directed network is to carry out n distinct single-origin computations. There are sometimes advantages to this seemingly redundant approach.

For example, suppose we wish to solve the all pairs problem for a sparse network with positive and negative arc lengths. Suppose each of the other nodes can be reached from node 1. As we have seen, an $O(mn)$ computation suffices to find shortest paths from node 1 to each of the other nodes.

We can then replace each arc length a_{ij} by

$$\bar{a}_{ij} = a_{ij} + u_i - u_j \geq 0. \tag{3}$$

The length of a path from i to j, with respect to the $\bar{a}_{ij}$ values, differs from the true path length by a constant, $u_i - u_j$. It follows that the solution of the all-pairs problem for the $\bar{a}_{ij}$ lengths solves the original problem.

Since all arc lengths are now nonnegative, we can apply Dijkstra's method to each of the remaining $n-1$ single-origin problems. It follows that the all-pairs problem can be solved in $O(mn \log n)$ time. In the case of a planar network, this implies $O(n^2 \log n)$ time.

2.8. Spira's algorithm

Spira [29] has proposed an algorithm for the all-pairs problem in the case that all arc lengths are nonnegative. Separate Dijkstra type computations are carried out, as in the previous section, after an initial sorting of the arcs. The worst-case bound is $O(n^3 \log n)$. However, Spira shows that the *expected* number of operations is $O(n^2 (\log n)^2)$ if arc lengths are independently and identically distributed random variables from any distribution. Fredman [16] has shown that the expected number of comparisons can be reduced to $O(n^2 \log n)$, but at the expense of a considerable increase in the number of housekeeping operations.

2.9. Floyd–Warshall algorithm

A very elegant, but not necessarily the most efficient, method for solving the all-pairs problem can be attributed to Floyd [14] and Warshall [30].

Let

$u_{ij}^{(k)}$ = the length of a shortest path from i to j, subject to the condition that the path does not pass through any of the nodes $k, k+1, \ldots, n$ (i and j excepted).

Then we have

$$u_{ij}^{(1)} = \begin{cases} a_{ij}, & \text{if } (i,j) \in A \\ +\infty, & \text{otherwise.} \end{cases}$$

$$u_{ij}^{(k+1)} = \min \{u_{ij}^{(k)}, u_{ik}^{(k)} + u_{kj}^{(k)}\}$$

(4)

and $u_{ij}^{(n+1)}$ is the length of a shortest path from i to j.

It is evident that this computation requires $O(n^3)$ time, regardless of the density or sparsity of the network. Other algorithms, e.g. [7] achieve comparable time bounds by carrying out operations like (4) in a different order.

2.10. Fredman's procedure

Fredman [16] has been able to show that, *for fixed n*, the all-pairs shortest path problem for a nonnegative network can be solved in $O(n^{2.5})$ time. His procedure utilizes a precompiled table that is used for all problems of a given size n. The best available method for compiling this table requires $O(n^3 \log \log n/\log n)$ time. Because of the complexity of the algorithm, and the apparent lack of testing, Fredman's result is probably of only theoretical significance at this time.

2.11. Detection of negative cycles

Most shortest path algorithms can be easily adapted to detect the existence of negative cycles. For example, the Bellman–Ford procedure can be continued for one additional iteration. If the network is strongly connected, a negative cycle exists if and only if $u_j^{(n)} < u_j^{(n-1)}$ for some node j.

It is thus evident that a network can be tested for the existence of negative cycles in $O(mn)$ time, and this is the best time bound known for the problem.

2.12. Minimizing mean cycle arc length

The *mean arc length* of a directed cycle C is

$$\lambda(C) = \frac{1}{|C|} \sum_{(i,j) \in C} a_{ij},$$

where $|C|$ denotes the number of arcs in the cycle. The problem of finding a cycle for which the mean arc length is minimum generalizes the problem of detecting a negative cycle, inasmuch as a negative cycle exists if and only if

$$\lambda_{\min} = \min_{C} \{\lambda(C)\} < 0.$$

A procedure of Karp [23] solves the mean arc length problem in $O(mn)$ time. Suppose the network is strongly connected. Let

$u_j^{(k)}$ = the length of the shortest sequence ("walk", "nonsimple directed path") of exactly k arcs from node 1 to node j.

The $u_j^{(k)}$ values can be computed by a modification of equations (2):

$$u_j^{(k+1)} = \min_{(i,j) \in A} \{u_i^{(k)} + a_{ij}\}, \quad k = 1, 2, \ldots, n.$$

It can be shown that

$$\lambda_{\min} = \min_{j \in N} \min_{0 \leq k \leq n-1} \left\{ \frac{u_j^{(n)} - u_j^{(k)}}{n-k} \right\},$$

and the value of $\lambda_{\min}$ can be found in $O(mn)$ time.

2.13. Minimizing weighted mean cycle arc length

Suppose in addition to a_{ij}, each arc (i, j) is assigned a weight b_{ij}. The *weighted mean arc length* of a cycle C is

$$\lambda(C) = \frac{\sum_{(i,j) \in C} a_{ij}}{\sum_{(i,j) \in C} b_{ij}}.$$

The problem discussed in the previous section clearly corresponds to the special case that $b_{ij} = 1$, for all (i, j).

The value of $\lambda_{\min}$ can be determined by testing various trial values. For a given trial value λ, a negative cycle is sought in the network with arc lengths

$$\bar{a}_{ij} = a_{ij} - \lambda b_{ij}.$$

If a strictly negative cycle exists, the trial value λ is too large. If the minimum cycle length is strictly positive λ is too small. If the minimum cycle length is exactly zero, then the trial value is correct.

It can be shown that the number of trial values which must be tested is roughly $O(\log n)$. (More precisely, $O(\log n + \log a + \log b)$, assuming all a_{ij} and b_{ij} values are integers, $|a_{ij}| \leq a$, $|b_{ij}| \leq b$, and there is no directed cycle for which the sum of the b_{ij}'s is negative.) Thus, this problem can be solved in roughly $O(mn \log n)$ time [25, 27].

2.14. Undirected shortest paths

Suppose the network is undirected and we wish to find a shortest path between two specified nodes. If all arc lengths are nonnegative, it is possible to replace each undirected arc $\{i, j\}$ by a symmetric pair of directed arcs (i, j) and (j, i), with

$a_{ij} = a_{ji}$, and apply any of the *directed* shortest path procedures to solve the problem. However, if the length of arc (i, j) is negative, this approach is invalid, inasmuch as it will introduce a negative directed cycle: (i, j), (j, i).

This problem can be reduced to a weighted nonbipartite matching problem, provided the network contains no (undirected) negative cycle. There are weighted matching algorithms with running time $O(n^3)$, and this establishes an upper bound on the complexity of the problem [25].

2.15. Ranking shortest paths

It is sometimes useful to be able to compute the shortest path, the second, third, ..., Kth shortest paths between a specified pair of nodes. There are two algorithms, for different versions of the problem, which seem to dominate all others.

The K shortest paths between two specified nodes of a directed network can be found in $O(C_1(m, n) + KnC_2(m, n))$ time and $O(m + Kn)$ space, where $C_1(m, n)$ is the time required to solve a single-origin problem and $C_2(m, n)$ is the time required to apply Dijkstra's procedure when arc lengths are made positive by a transformation like (3) [25, 26]. Thus, for example, the K shortest paths can be found in a planar network in $O(Kn^2 \log n)$ time.

The procedure can be adapted to undirected networks to obtain the K shortest paths between two specified nodes in $O(Kn^4)$ time.

If *nonsimple* directed paths are to be ranked, a very efficient procedure due to Dreyfus [11] can be applied. The K shortest paths from a single origin to each of the other nodes can be found in $O(Kn \log n)$ time and $O(Kn + m)$ space, once an initial single-origin problem is solved.

3. Maximal flows

3.1. Definitions and notation

We suppose that a directed graph $G = (N, A)$ is given, with a specified positive *capacity* c_{ij} for each arc $(i, j) \in A$. A *source* s and a *sink* t are designated. A *flow* is a set of numbers x_{ij}, such that

$$\sum_{(i,j) \in A} x_{ij} - \sum_{(i,j) \in A} x_{ji} = 0, \quad j \neq s, t$$

$$0 \leq x_{ij} \leq c_{ij}, \quad (i, j) \in A$$

The *value* v of a flow is the net amount of flow out of the source or into the sink:

$$v = \sum_{(s,i) \in A} x_{si} - \sum_{(i,s) \in A} x_{is} = \sum_{(i,t) \in A} x_{it} - \sum_{(t,i) \in A} x_{ti}.$$

By the "maximal flow problem" we mean the problem of finding a flow for which the value v is maximum.

It is well known that many problems can be formulated and solved as maximal flow problems. This includes many problems which at first sight do not necessarily appear to have any relation to the physical notion of flows. See [25].

3.2. Ford–Fulkerson procedure

The "classical" procedure for solving maximal flow problems is due to Ford and Fulkerson and is based on the notion of augmenting paths.

An undirected path between the source and the sink is said to be an *augmenting path* if $x_{ij} < c_{ij}$ for each "forward" arc in the path and $x_{ij} > 0$ for each "backward" arc. (An arc is said to be *forward* if it is directed from s to t, and *backward* otherwise.) One can *augment* the flow along an augmenting path, increasing the flow through each forward arc and decreasing the flow through each backward arc, thereby obtaining a larger-value flow. Conversely, a flow is maximal only if no augmenting path exists with respect to that flow.

Ford and Fulkerson proposed a labelling procedure, requiring $O(m)$ time, for finding an augmenting path. (More precisely, the time bound is $O(m+n)$, but we assume $m \geq n$.) However, they did not offer a bound on the number of augmentations required to find a maximal flow, except to note that: (*a*) If all arc capacities are integer, the number of augmentations does not exceed the maximal flow value v, indicating a bound of $O(mv)$ on running time. (*b*) If arc capacities are irrational, there may not be finite termination of the algorithm.

3.3. Edmonds–Karp analysis

In 1972, Edmonds and Karp [12] showed that a very simple restriction on the Ford–Fulkerson algorithm causes the running time to be polynomial bounded in m and n. In particular, they showed that if each successive augmenting path is chosen to contain as few arcs as possible, no more than $O(mn)$ augmentations are necessary, *regardless of whether or not arc capacities are rational numbers.*

Thus an $O(m^2 n)$ time bound was established for the restricted version of the Ford–Fulkerson algorithm. Perhaps somewhat ironically, the majority of computer codes which had been written for the Ford–Fulkerson algorithm had observed the restriction of choosing a shortest augmenting path, since this results from one of the more natural implementations of the labelling procedure.

N. Zadeh [33] has been able to characterize a class of networks for which $O(n^3)$ augmentations are required by the restricted Ford–Fulkerson algorithm.

3.4. Dinic's algorithm

In 1970, the Russian mathematician Dinic [10] published an algorithm for the maximal flow problem which involves a labelling procedure that constructs an "auxiliary" graph from which a maximal set of disjoint shortest augmenting paths can be constructed. Augmentations are made simultaneously through all augmenting paths in such a set. Then the labelling procedure is repeated. The overall time bound is $O(mn^2)$.

Dinic's algorithm was at first unknown or unappreciated in the West, as can be inferred from its date of publication prior to the Edmonds–Karp paper. Moreover, many of Dinic's ideas anticipated the $O(mn^{1/2})$ procedure of Hopcroft and Karp [19] for unweighted bipartite matching.

3.5. Karzanov's algorithm

More recently, A.V. Karzanov [24] improved Dinic's algorithm to run in $O(n^3)$ time. We shall not discuss Karzanov's algorithm, except to comment that it is unique in that it does *not* make use of the idea of augmenting paths. An excellent exposition of Karzanov's algorithm has been given by Even [13].

3.6. Planar networks

The Edmonds–Karp, Dinic and Karzanov algorithms all yield running time bounds of $O(n^3)$ in the case of planar networks, for which m is $O(n)$. Recently, Itai and Shiloach [20] have obtained some better results for this special case. If s and t are on the same face, then an algorithm of Berge [2] can be improved and implemented to run in $O(n \log n)$ time. Otherwise, a flow of value v, for any given v, can be found in $O(n^2 \log n)$ time. If the network is undirected, a minimum cut can be found in $O(n^2 \log n)$ time. All these algorithms require only $O(n)$ space.

4. Minimum cost flows

4.1. Definitions and notation

Suppose each arc of a flow network has assigned to it a *cost* a_{ij}, and it is desired to find a flow of given value v from s to t, such that

$$\sum_{(i,j)\in A} a_{ij} x_{ij}$$

is minimized. A variety of algorithms have been proposed for this problem and for its generalizations. Among these are adaptations of the primal and dual simplex algorithms of linear programming [6, 8], and algorithms using shortest path and negative cycle procedures as subroutines [12]. Possibly the most aesthetically pleasing algorithm is the out-of-kilter method, reviewed below.

4.2. Out-of-kilter method

The "out-of-kilter" algorithm was developed independently by Minty and by Fulkerson [15]. It appears that many of the ideas were anticipated by Yakovleva.

The out-of-kilter method accommodates a very general class of optimal flow problems. We may assume that each arc (i, j) has assigned to it a lower bound l_{ij} on flow as well as a capacity c_{ij}, so that we require

$$0 \leq l_{ij} \leq x_{ij} \leq c_{ij}.$$

We wish to find a circulation (i.e. a flow satisfying the conservation law at each node), for which

$$\sum_{(i,j)\in A} a_{ij} x_{ij}$$

is minimized.

There is, of course, a linear programming dual to this minimization problem. The essential variables of the dual problem are *node numbers* or *potentials* u_j. If a given set of arc flows x_{ij} and node numbers u_j, are all "in kilter", then the primal and dual solutions are both optimal. The degree to which a given arc (i, j) is "out-of-kilter" is given by the *kilter number* K_{ij}, which indicates the absolute value of the change in x_{ij} necessary to bring arc (i, j) into kilter. Specifically,

$$K_{ij} = \begin{cases} |x_{ij} - l_{ij}|, & \text{if } u_j - u_i < a_{ij} \\ \max\{l_{ij} - x_{ij}, x_{ij} - c_{ij}, 0\}, & \text{if } u_j - u_i = a_{ij} \\ |x_{ij} - c_{ij}|, & \text{if } u_j - u_i > a_{ij} \end{cases}$$

The sum of the kilter numbers,

$$K = \sum_{(i,j)\in A} K_{ij}$$

is a measure of the degree to which a given pair of primal and dual solutions is out of kilter.

At each iteration of the out-of-kilter algorithm a change is made in either the primal or the dual solution. The kilter number of no arc is increased and, at most iterations, at least one kilter number is reduced. A bound on the overall running time is $O(n^2 K_0)$. where K_0 is the initial sum of the kilter numbers. (Note: In [25], $O(mK_0)$ is erroneously given as a bound.)

4.3. Scaling technique

Suppose the out-of-kilter algorithm is initiated with all $x_{ij} = 0$, $u_j = 0$. Then the initial sum of the kilter numbers may be as large as $mc_{\max}$, where

$$c_{\max} = \max_{(i,j)\in A} \{c_{ij}\}.$$

It is thus seen that $O(n^2 K_0)$ becomes $O(mn^2 c_{\max})$, and this is not a polynomial bound, in the commonly accepted use of the term.

Nevertheless, it is possible to modify the out-of-kilter algorithm by the scaling technique of Edmonds and Karp [12] so as to obtain a bound of $O(mn^2(\log c_{\max} + 1))$. See [25]. Since the out-of-kilter method solves transportation and transhipment problems, this result establishes that these problems are polynomial bounded, contrary to a statement in [4].

4.4. Computational experience

Minimum-cost flow algorithms are commonly applied to very large networks. The primal algorithms of Mulvey [28] and Karney and Klingman [22] have reportedly been used to solve problems with over 20,000 nodes or over 450,000 arcs. Bradley, Brown and Graves [5] report solving a problem with 3,000 nodes, 35,000 arcs in 97 seconds on the IBM 360/67 and another with 5,000 nodes and 15,000 arcs in 113 seconds.

Several empirical studies [1, 17, 18, 22] of the relative efficiency of various algorithms seem to favor primal methods for transshipment and transportation problems. Time savings of about 30–40%, compared with the out-of-kilter method are typical. These results are persuasive, but not entirely conclusive.

References

[1] R.S. Barr, F. Glover and D. Klingman, An improved version of the out-of-kilter method and a comparative study of cumputer codes, Math. Programming 7 (1974) 60–87.
[2] C. Berge and A. Ghouila–Houri, Programming, Games and Transportation Networks (Methuen, London, 1976).
[3] J.H. Bookbinder and S.P. Sethi, The dynamic transportation problem: A survey, Working paper, Faculty of Management Studies, University of Toronto (1977).
[4] G.K. Bradley, Survey of deterministic networks, AIIE Trans. 7 (1975) 222–234.
[5] G.H. Bradley, G.G. Brown and G.L. Graves, Design and implementation of large scale primal transshipment algorithms, Management Sci., 24 (1977) 1–34.
[6] A Charnes and M. Kirby, The dual method and the method of Balas and Ivanescu (Hammer) for the transportation model, Cahiers Centre Etudes Recherche Opér. 6 (1964) 5–18.
[7] G.B. Dantzig, All shortest routes in a graph, Operations Research House, Stanford University, Technical Report 66-3 (Nov. 1966).
[8] G.B. Dantzig, Linear Programming and Extensions (Princeton Univ. Press, Princeton, NJ, 1963).
[9] E.W. Dijkstra, A note on two problems in connexion with graphs, Numer. Math. 1 (1959) 269–271.
[10] E.A. Dinic, Algorithm for solution of a problem of maximum flow in a network with power estimation, Soviet Math. Dokl. 11 (1970) 1277–1280.
[11] S.E. Dreyfus, An appraisal of some shortest-path algorithms, Operations Res. 17 (1969) 395–412.
[12] J. Edmonds and R.M. Karp, Theoretical improvements in Algorithm Efficiency for network flow problems, Assoc. Comput. Mach. 19 (1972) 248–264.
[13] S. Even, The max flow algorithm of Dinic and Karzanov, unpublished manuscript (1976).
[14] R.W. Floyd, Algorithm 97, shortest path, Comm. Assoc. Comput. Mach. 5 (1962) 345.
[15] L.R. Ford and D.R. Fulkerson, Flows in Networks (Princeton University, Press, Princeton, N.J., 1962).
[16] M.L. Fredman, New bounds on the complexity of the shortest path problem, SIAM J. Comput. 5 (1976) 83–89.
[17] F. Glover, D. Karney and D. Klingman, Implementation and computational study on start procedures and basis change criteria for a primal network code, Networks 4 (1974) 191–212.
[18] F. Glover, D. Karney, D. Klingman and A. Napier, A computational study on state procedures, basis change criteria and solution algorithms for transportation problems, Management Sci. 20 (1974) 793–819.
[19] J.E. Hopcroft and R.M. Karp, A $n^{5/2}$ algorithm for maximum matchings in bipartite graphs, SIAM J. Comput. 2 (1973) 225–231.
[20] A. Itai and Y. Shiloach, Maximum flow in planar graphs, Technical Report No. 88, Technion, Israel (1976).

[21] D.B. Johnson, Shortest paths in dense networks, J. Assoc. Comput. Mach. (1976).
[22] D. Karney and D. Klingman, Implementation and computational study of an in-core out-of-core primal network code, Report CS 158, University of Texas (1973).
[23] R.M. Karp, unpublished manuscript (1977).
[24] A.V. Karzanov, Determining the maximal flow in a network by the method of preflows, Soviet Math. Dokl. 15 (1974) 434–437.
[25] E.L. Lawler, Combinatorial optimization: Networks and Matroids (Holt, Rinehart and Winston, New York, 1976).
[26] E.L. Lawler A procedure for computing the K best solutions to discrete optimization problems and its application to the shortest path problem, Management Sci. 18 (1972) 401–405.
[27] E.L. Lawler, Optimal cycles in doubly weighted directed linear graphs in: P. Rosenstiehl, ed., Theory of Graphs, (Dunod, Paris, and Gordon and Breach, New York, 1967) 209–214.
[28] J. Mulvey, Special structure in network models and associated applications, Ph.D. dissertation, UCLA, (August 1975).
[29] P.M. Spira, A new algorithm for finding all shortest paths in a graph of positive arcs in average time $O(n^2 \log^2 n)$, SIAM J. Comput. 2 (1973) 28–32.
[30] S. Warshall, A theorem on boolean matrices, J. Assoc. Comput. Mach. 9 (1962) 11–12.
[31] J.Y. Yen, An algorithm for finding shortest routes from all source nodes to a given destination in general network, Quart. Appl. Math. 27 (1970) 526–530.
[32] J.Y. Yen, Finding the K shortest loopless paths in a network, Management Sci. 17 (1971) 712–716.
[33] N. Zadeh, Theoretical efficiency of the Edmonds–Karp algorithm for computing maximal flows, J. Assoc. Comput. Mach. 19 (1972) 184–192.

COVERING, PACKING AND KNAPSACK PROBLEMS

Manfred W. PADBERG

IBM T. J. Watson Research Center, P.O. Box 218, Yorktown Heights, NY 10598, U.S.A.

New York University, Graduate School of Business Administration, 100 Trinity Place, New York, NY 10006, U.S.A.

We survey some of the recent results that have been obtained in connection with covering, packing, and knapsack problems formulated as linear programming problems in zero-one variables.

0. Introduction

Covering, packing and knapsack problems are specially structured integer programming problems that along with the travelling salesman problem have been extensively investigated during the past 25 years. The focus of interest — originally tending exclusively to algorithmic aspects of these problems — has shifted in the recent past years to an investigation of the facial structure of polyhedra associated with the respective problems. This research is motivated by the desire to obtain in a systematic way *tighter* formulations of the respective problems in terms of linear inequalities than is provided for by an ad-hoc formulation of such integer programming problems, and it is necessitated by the fact that existing algorithms — cutting plane approaches as well as enumerative procedures such as branch-and-bound-exhibit erratic performance on large-scale pure integer programming problems: The same procedure may perform reasonably well on some problems, while other problems just cannot be solved in reasonable computing times, see e.g. [15] for relatively small-sized set covering problems that appear to be quite intractable. A systematic way to obtain a tighter formulation of an integer programming problem is to investigate classes of linear inequalities that constitute *facets* of the convex hull of the feasible integer points of the respective problem. Facets, i.e. their respective linear inequalities, belong to the system of linear inequalities that uniquely define the convexified solution set of an integer programming problem. By contrast, cutting planes do in general not have this property and are not even guaranteed to *intersect* the convex hull of integer solutions, see e.g. [32] for a related example. Of course, a *complete* description of hard integer programming problems by way of linear inequalities will probably prove to be elusive, but even *partial* results in this direction are bound to be of considerable practical help in the solution of integer programming problems. This follows from the straightforward observation that e.g. a branch-and-bound procedure will tend to generate fewer "live" nodes and thus can be expected to terminate faster the better the bound is that has been obtained by solving the underlying linear programming problem. A recent computational study concerning the usefulness of *facetial* inequalities in the solution of symmetric travelling salesman problems [38] has substantiated the fact that excellent lower bounds on the integer optimum can be obtained this way in reasonable computing times. The largest problem attempted in this study was a 318-city problem with data as published in [28]. The associated linear programming problem min$\{cx \mid Ax = 2e_m, 0 \le x \le e_n\}$ has $m = 318$ equality-constraints and $n = 50{,}403$ variables with upper bounds of one. The optimal objective function value of this linear program is 38.766 inches and the best tour found has 41.349 inches. Thus a gap of 2.583 inches between the two values must be "mended" and faced with a gap of such relatively large magnitude, one would be most reluctant to use e.g. a branch-and-bound procedure in an attempt to prove optimality or even near-optimality of the tour of length 41.349 inches. (Note that in this case not just an integer solution to $Ax = 2e_m$, $0 \le x \le e_n$, must be found, but an integer vector that corresponds to a *tour*.) With the procedure described in [38], adding 183 automatically generated facetial inequalities and solving the augmented, *tighter* linear program to optimality took 37 minutes of CPU-time on the IBM 370/168 MVS of the IBM T. J. Watson Research Center. The optimal objective function value of the augmented linear program was 41.237 inches, i.e. the gap was now reduced to 0.112 inches and the above solution is within $\frac{1}{4}$% of the optimum. If the economics of that particular application demanded a true optimal solution, one would at this point not hesitate to recommend use of any commercial branch-and-bound code such as e.g. the IBM-package MPSX-MIP for the solution of the *augmented* linear program. At the optimum of the augmented linear program 48,871 of the 50,403 variables could be eliminated by a simple bounding argument. Thus a problem with 318 equations, 183 inequalities and $n = 1{,}715$ variables, all but 183 of them zero-one variables, would have to be solved by branch-and-bound and that seems entirely realistic for a commercial package *if* the "gap" is as small as indicated above. The conclusions of [38] are not based on a singular problem, but rather on similar experience with a total of 74 problems of varying sizes that were used in the study.

There are two issues involved in the use of facetial inequalities in actual large-scale pure-integer computation: First of all a *descriptive* identification of facetial inequalities and secondly, an *algorithmic* identification that finds a particular facetial inequality, are needed. The general objective of the descriptive identification of facetial inequalities is to provide a clear-cut combinatorial description of it, such as provided e.g. by cliques, odd cycles, etc. in a graph or more generally, combinatorial configurations in the non-negative constraint matrix A that are "easily verified". Once a descriptive identification is obtained, it is then reasonable to look for an algorithmic identification of one or several members of a particular class of inequalities that is precisely described by its combinatorial attributes and which, if it is possible, will cut off a given point not belonging to the convex hull of integer solutions. See [38] for a more detailed discussion of this point.

In this paper we first survey the formulation of covering, packing and knapsack problems along with some applications and the more interesting specializations of these general, structured pure integer programming problems. Next we discuss commonalities among the three different types of problems. In Section 2 we pay attention to conditions under which the respective problems can be solved as linear programs, i.e. without an integrality stipulation enforced explicitly on the variables. In Section 3 we survey the results concerning the descriptive identification of facetial inequalities which have been obtained to date. In fact, research done along these lines has focused so far on assembling descriptive information about facetial inequalities, rather than on their algorithmic identification, since the second step obviously cannot be done before the first one is reasonably resolved.

The paper presumes some knowledge about linear programming and the more elementary concepts in graph theory such as presented in the books by Berge [6], Dantzig [10], Garfinkel and Nemhauser [17], Harary [24], Salkin [41] and Simmonard [43]. Furthermore, we refer the reader to the (excellent, naturally!) survey [3] on set-packing and set partitioning which includes a survey of the state-of-the-art of algorithms in this area, and to [8] for a recent comparative study of various algorithms for set-covering problems.

1. Packing, covering and knapsack problems

1.1. *Problem formulations and their uses*

A great variety of scheduling problems can be formulated as follows. Given

(i) a finite set of M of distinguishable elements, e.g. $M = \{1, 2, \ldots, m\}$,
(ii) a set of rules defining a family F of "acceptable" subsets of M, and
(iii) a utility or profit (disutility or cost) associated with each member of F,

find a maximum-profit (minimum-cost) collection of members of F such that every element of M is contained

(a) in at most one of the selected members of F,
(b) in exactly one of the selected members of F, or
(c) in at least one of the selected members of F.

To formulate these problems as zero-one decision problems, one introduces a zero-one variable x_j which is 1 if the jth member of F is selected and zero if not and denotes c_j the profit (cost) associated with the jth member of F. (We can assume that the members of F have been indexed $1, 2, \ldots, n$). Let A denote the $m \times n$ incidence matrix of the members of F (columns of A) versus the elements of M (rows of A), i.e. if F_j is the jth member of F, then

$$a_{ij} = \begin{cases} 1 & \text{if } i \in F_j \\ 0 & \text{if not.} \end{cases}$$

The (weighted) *set packing* problem is

$$\max \sum_{j=1}^{n} c_j x_j,$$
$$\text{s.t.} \quad Ax \leq e_m,$$
$$x_j = 0 \quad \text{or} \quad 1 \quad \text{for} \quad j = 1, \ldots, n, \qquad \text{(SP)}$$

where e_m is vector of m ones. Problem (SP) captures the formulation of problem (a), while (b) calls for a partitioning of the set M and is referred to as the (weighted) *set partitioning* problem

$$\max \sum_{j=1}^{n} c_j x_j,$$
$$Ax = e_m, \qquad \text{(SPP)}$$
$$x_j = 0 \quad \text{or} \quad 1 \quad \text{for} \quad j = 1, \ldots, n.$$

Finally, problem (c) calls for a collection of members of F that cover all elements of M. The (weighted) *set covering* problem is

$$\min \sum_{j=1}^{n} c_j x_j,$$
$$Ax \geq e_m, \qquad \text{(SC)}$$
$$x_j = 0 \quad \text{or} \quad 1 \quad \text{for} \quad j = 1, \ldots, n.$$

If the weights c_j are all equal to one, one is interested in the maximum (minimum) cardinality collection of members of F with the required properties. To rule out trivial pathologies, we assume that A has neither a zero column nor a zero row, i.e. the empty set is not in F and every element of M is contained in some member F_j of F. Note that these assumptions imply e.g. that both (SP) and (SC) admit feasible solutions, while feasibility of (SPP) is a far more complicated matter. (SPP) can, however, be brought to both the form of (SP) and (SC) while ensuring that the respective optimal solutions coincide. To this end, let θ be a sufficiently large number, e.g. $\theta > \sum_{j=1}^{n} c_j$. Then

$$\min \{-cx + \theta e_m y \mid Ax - y = e_m, y \geq 0, x_j = 0 \text{ or } 1, \forall j \in N\}$$
$$= \min \{c'x - \theta m \mid Ax \geq e_m, x_j = 0 \text{ or } 1, \forall j \in N\},$$

where c is the vector with components c_j, $j \in N$, $c' = \theta e_m A - c$ and $N = \{1, \ldots, n\}$. (Transposition of vectors is not separately indicated.) On the other hand, we can write

$$\max \{cx - \theta e_m y \mid Ax + y = e_m, y \geq 0, x_j = 0 \text{ or } 1, \forall j \in N\}$$
$$= \max \{c''x - \theta m \mid Ax \leq e_m, x_j = 0 \text{ or } 1, \forall j \in N\},$$

where $c'' = c + \theta e_m A$.

A whole host of applications of these problems has been reported in the literature and a partial list includes: railroad crew scheduling, truck deliveries, airline crew scheduling, tanker-routing, information retrieval, switching circuit design, stock cutting, assembly line balancing, capital equipment decisions, location of offshore drilling platforms, other facility location problems, political districting, etc. The usefulness and wide applicability of this class of zero-one problems follows from the simple observation that in most cases a problem of the above form can be solved to a satisfactory degree of approximation by the following two-stage procedure:

Stage 1. Using the set of rules defining "acceptable" subsets of M, generate explicitly a subset $\bar{F}$ of F such that the probability of an optimal solution being contained in $\bar{F}$ is sufficiently high.

Stage 2. Replace F in the problem definition of type (a), (b) or (c) (whichever is being considered) by a list of the members of $\bar{F}$ and solve the associated set-packing, set-partitioning or set-covering problem.

The most widely used application seems to be the airline crew scheduling problem, in which M corresponds to the set of flight legs (from city A to city B, at time t) to be covered during a planning period (usually a few days or the time after which the flight-schedule repeats itself), while each subset F_j stands for a possible rotation (sequence of flight legs with the same initial and terminal point) for a crew. In order to be acceptable, a rotation must satisfy certain regulations. To set up the problem, one starts with a given set of (usually several hundred) flight legs, and one generates by computer a set of (normally several thousand) acceptable rotations with their respective cost. This produces the matrix A (of density usually ≤ 0.05) after which one attempts to solve the associated set-partitioning problem. If the attempt is successful, the solution yields a minimum-cost collection of acceptable rotations such that each flight leg is included in exactly one rotation of the collection. (Some formulations of the airline crew scheduling problem permit "deadheading" of crews, i.e. the assignment of more than one crew to a flight-leg. In this case, the problem is treated as a set-covering problem.) Airline crew scheduling problems with 300–500 constraints and 2,500–4,000 variables are sometimes solved (to optimality); but frequently much smaller problems (with several hundred variables) defy solution within reasonable time limits.

Another frequently used application is the following specially structured set-packing problem which is usually referred to as the simple *plant-location problem*: Suppose that at n possible locations plants can be built, necessitating a fixed investment f_i at location i, for producing a homogeneous good demanded by m customers. Assuming that the capacity of each plant is sufficiently large to satisfy total demand, one has to decide at which locations to build the plants so as to minimize the total cost of satisfying the customers demands. Denoting by x_{ij} the *fraction* of customer's j demand filled by plant i and letting $y_i = 1$ if a plant is located in location i, $y_i = 0$ if not, the problem can be formulated as follows:

$$\min \sum_{j=1}^{m}\sum_{i=1}^{n} c_{ij}x_{ij} + \sum_{i=1}^{n} f_i y_i,$$

$$\text{s.t.} \quad \sum_{i=1}^{n} x_{ij} = 1 \quad \text{for} \quad j=1,\ldots,m, \quad \text{(SPL)}$$

$$0 \leq x_{ij} \leq y_i \quad \text{for all } i,j,$$

$$y_i = 0 \quad \text{or} \quad 1 \quad \text{for} \quad i=1,\ldots,n.$$

Here c_{ij} are the transportation costs involved in satisfying the total demand of customer j from plant i and moreover, transportation costs are assumed to be linear. (More realistic problem formulations include economies of scale of production at different plants as well as limited production capacities.) One notes that for each choice of the variables y_i, problem (SPL) decomposes into m trivial problems whose solution is zero-one and in fact, (SPL) can be transformed into the form (SP) by first using the substitution $y'_i = 1 - y_i$ and secondly, by converting the set of m equations into less-than-or-equal inequalities. The latter can be accomplished by introducing an artificial variable for each equation having a "large enough" cost and subsequent elimination of these variables.

Other interpretations of problem (SPL) include the *lock-box location* problem faced by corporations that are interested in collecting funds due to them as quickly as possible. Alternatively, in times of tight money markets, corporations may be interested in maintaining accounts in strategically located banks so as to maximize the time it takes to clear a check, i.e. so as to "stretch" or "delay" the cashing of checks issued by the corporation as long as possible.

The *multi-dimensional knapsack problem* is

$$\max \quad cx,$$

$$\text{s.t.} \quad Ax \leq b, \quad \text{(MKP)}$$

$$x_j = 0 \quad \text{or} \quad 1 \quad \text{for} \quad j=1,\ldots,n,$$

where A is a $m \times n$ matrix of *non-negative* integers and b is a vector of positive integers. Necessarily, c is a vector of n *non-negative* integers. If $m = 1$, i.e. in the case of a single constraint, problem (MKP) is referred to as the *knapsack problem*. The name stems from the following interpretation of (MKP) for $m = 1$: A hiker is faced with the problem of including or not including from a possible list of items a subset whose total weight does not exceed the weight that he can carry while making the most desirable selection possible. Here the components of c measure the "utils" derived from selecting an item and the components of the single row of A specify the weight of the respective items while the right-hand side element of b gives the total weight the hiker can carry. Despite its simple form, knapsack problems are interesting theoretically as well as difficult to solve if one considers large n and arbitrary coefficients. In the multi-dimensional knapsack problem, additional restrictions, such as on volume, height, length, width, etc. are also imposed. In this interpretation, knapsack problems arise naturally in the context of loading a ship or a box car with the most valuable cargo.

A different interpretation of (MKP) occurs in the context of *capital rationing problems*. The general problem is that one of selecting from a set of possible investment projects that subset which maximizes a certain objective function (such as total net-present value) subject to expenditure limitations in one or more subsequent periods. Specifically, let c_j be the net-present value of project j, i.e. the discounted net cash-flow associated with project j over its life-time; let a_{ij} be the cash-outlay associated with project j in period i and let b_i be the budget-ceiling for period i. If m is the number of time periods considered, then the problem is to determine the best possible mix of investment projects while maintaining the budget-ceilings in periods $1, \ldots, m$. Additional constraints might also be included to reflect possible interdependencies among the projects such as:

(i) implications of the form "if project i is accepted, then project j must be accepted" which is translated into the condition that $x_i \le x_j$, and

(ii) exclusiveness of certain projects of the form "at most one of projects i or j can be accepted" which is expressed as $x_i + x_j \le 1$.

Of course, the interdependencies listed are to serve as examples only and it is possible to express much more complicated forms of interdependencies among projects. The form of the knapsack problem in this context is referred to as the *Lorie-Savage* problem.

1.2. Commonalities: optimization problems in undirected graphs

While seemingly different problems, packing, covering and knapsack problems can ultimately be viewed as optimization problems over undirected graphs. Though primarily of theoretical interest, i.e. not of immediate algorithmic usefulness, this commonality provides insights into these problems that historically have been conceived of as being substantially different. For set-packing problems this fact is of crucial importance. We start by transforming the knapsack problem into a set-packing problem. Similar transformations apply to the multi-dimensional knapsack problem, see [18]. Let the *knapsack problem*

$$\max \quad cx,$$
$$ax \le a_0, \quad \text{(KP)}$$
$$x_j = 0 \text{ or } 1, \quad \text{for } j = 1, \ldots, n$$

have all-integer data and suppose that $0 < a_j \leq a_0$ and $c_j \geq 0$ for all $j = 1, \ldots, n$. We can replace variable x_j by a_j new zero-one variables x_{jk} for $k = 1, \ldots, a_j$ and $j = 1, \ldots, n$. Replacing each term $a_j x_j$ by the expression $\sum_{k=1}^{a_j} x_{jk}$, we must now ensure that *all* variables x_{jk} are either equal to zero simultaneously or alternatively, equal to one simultaneously. Furthermore, in order to compute the objective function value correctly, we define coefficients $c_{jk} = c_j / a_j$ for each variable x_{jk}. Thus one introduces an indicator variable z_j and requires that $x_{jk} + z_j = 1$ for $k = 1, \ldots, a_j$ and $j = 1, \ldots, n$. This way we obtain a new problem (KP1)

$$\max \sum_{j=1}^{n} \sum_{k=1}^{a_j} c_{jk} x_{jk}, \quad \text{(KP1)}$$

$$\text{s.t.} \quad \sum_{j=1}^{n} \sum_{k=1}^{a_j} x_{jk} \leq a_0,$$

$$x_{jk} + z_j = 1 \quad \text{for} \quad k = 1, \ldots, a_j; \quad j = 1, \ldots, n,$$

$$x_{jk} = 0 \quad \text{or} \quad 1 \quad \text{and} \quad z_j = 0 \quad \text{or} \quad 1 \quad \text{for all } k \text{ and } j$$

having a binary constraint matrix. In order to obtain a set-packing problem from (KP1), one introduces for all j and k new variables x_{jkh} with objective function coefficients $c_{jkh} = c_{jk}$ for $h = 1, \ldots, a_0$. We can now write down the equivalent problem (KP2)

$$\max \sum_{j=1}^{n} \sum_{k=1}^{a_j} \sum_{h=1}^{a_0} c_{jkh} x_{jkh},$$

$$\text{s.t.} \quad \sum_{j=1}^{n} \sum_{k=1}^{a_j} x_{jkh} \leq 1 \quad \text{for} \quad h = 1, \ldots, a_0, \quad \text{(KP2)}$$

$$\sum_{h=1}^{a_0} x_{jkh} + z_j = 1 \quad \text{for} \quad k = 1, \ldots, a_j; \quad j = 1, \ldots, n,$$

$$x_{jkh} = 0 \quad \text{or} \quad 1, z_j = 0 \quad \text{or} \quad 1 \quad \text{for all } h, k, \text{ and } j.$$

To transform problem (KP2) into the general form of the set-packing problem (SP), one utilizes the "trick" mentioned in the previous section.

To transform a set-covering problem into a set-packing problem, one first uses the substitution $x'_j = 1 - x_j$ so as to get the problem into "packing" form. Then one uses the same "multiplication" device that we used to get from (KP1) to (KP2) for each row of the problem in packing form. Again, indicator variables are used to ensure a one-to-one correspondence of zero-one solutions, and constraints in which an "old" variable appears are duplicated in the new variables. Thus one obtains ultimately a set-packing problem.

To transform a set-packing problem into an optimization problem over a finite undirected graph, we associate with the zero-one matrix A of (SP) the *intersection graph* $G_A = (N, E)$ in the following manner: G_A has a node for every column of A and an edge connecting nodes k and j iff column k and column j of A satisfy

$a_{ik} = a_{ij} = 1$ for at least one row $i \in \{1, \ldots, m\}$. Let A_G be the edges (rows of A_G) versus nodes incidence matrix of G_A and associate with node j of G_A the weight c_j for $j = 1, \ldots, n$ and suppose that G_A has q edges. Then the (weighted) *node-packing* problem in the graph G_A is

$$\max \quad cx,$$
$$A_G x \leq e_q, \qquad \text{(NP)}$$
$$x_j = 0 \quad \text{or} \quad 1 \quad \text{for} \quad j = 1, \ldots, n,$$

where e_q is the vector having q ones corresponding to the edges of G_A. (More generally, we define (NP) for *any* given finite, undirected graph G with n nodes, q edges and having no loops and no multiple edges to be the node-packing problem in G). One verifies, that every feasible solution x to (NP), i.e. every independent (stable) node-set in G_A is a feasible solution to (SP) and vice versa. Moreover, the optimal solutions to (NP) and (SP) coincide. Thus any set-packing problem is equivalent to a node-packing problem on a finite undirected graph.

By definition the matrix A_G has exactly two ones per row. Thus by making the substitution $x'_j = 1 - x_j$, we obtain a closely related covering problem. The (weighted) *node-covering* problem in a graph G is

$$\min \quad cx,$$
$$A_G x \geq e_q, \qquad \text{(NC)}$$
$$x_j = 0 \quad \text{or} \quad 1 \quad \text{for} \quad j = 1, \ldots, n.$$

We note that every node-cover $\bar{x}$ of G, i.e. every subset of nodes of G that cover all edges of G, corresponds uniquely to an independent set of nodes in G $\bar{\bar{x}} = e - \bar{x}$ and vice versa, every independent node set in G yields a node-cover in G by the same transformation. (The vector e will always be used to designate the vector of all ones and is always assumed to be compatibly dimensioned.) Furthermore, whereas a set-packing problem in n variables is always equivalent to a node-packing problem in some graph G on n nodes, a set-covering problem in n variables is generally *not* equivalent to a node-covering problem in a graph on the same number of nodes.

The node-packing problem in a graph G is thus the problem of determining a subset of nodes such that no two nodes have an edge of G in common, while the total node-weight is maximum. Replacing the word "node" by "edge" and vice versa, we obtain a deceptively closely related zero-one problem: In the *edge-packing* or *edge-matching* problem we are interested in determining a subset of edges of some graph G such that no two edges have a node of G in common while the edge-weight is maximum. Denote by A_G the edge (rows of G) versus nodes incidence matrix of G and assume that G has q edges and n nodes. Furthermore, let d_j be the weight assigned to edge j. Then the edge-matching problem is

$$\max \sum_{j=1}^{q} d_j y_j,$$
$$A_G^T y \leq e_n, \tag{EP}$$
$$y_j = 0 \text{ or } 1 \text{ for } j = 1, \ldots, q.$$

Note that the linear constraint-set of (EP) has exactly two ones per column, while the number of ones in row i of A_G^T equals the degree of node i in G. Reversing the sign of the inequality in the formulation of (EP) we obtain the *edge-covering* problem in a graph G:

$$\min \sum_{j=1}^{q} d_j y_j,$$
$$A_G^T y \geq e_n, \tag{EC}$$
$$y_j = 0 \text{ or } 1 \text{ for } j = 1, \ldots, q.$$

Clearly, (NP) and (EP) are special cases of the set-packing problem (SP), while (NC) and (EC) are special cases of the set-covering problem (SC). The problem of finding a *perfect* edge-matching, i.e. the problem of finding a partitioning of a graph into mutually non-adjacent edges, is obtained if we replace the inequality in (EP) with equality and hence is a special case of (SPP).

2. Integral polyhedra

In studying integer programming problems it is most natural to ask under which conditions a given linear constraint system *implies* integrality of all *basic* feasible solutions, i.e. defines a polyhedron whose extreme points are all integer. In this section we review the results that have been obtained along these lines and pertain to packing, covering and knapsack problems.

For any $m \times n$ matrix A of non-negative integers and vector b of m positive integers let $P = P(A, b)$ and $P_I = P_I(A, b)$ denote the polyhedra

$$P(A, b) = \{x \in \mathbf{R}^n \mid Ax \leq b, 0 \leq x \leq e_n\} \tag{2.1}$$

$$P_I(A, b) = \text{conv}\{x \in P(A, b) \mid x \text{ integer}\}, \tag{2.2}$$

i.e. $P_I(A, b)$ is the convex hull of zero-one points in $P(A, b)$. It is well-known from linear programming theory, that $P(A, b)$ and $P_I(A, b)$ coincide if and only if the linear program $\max \{cx \mid x \in P(A, b)\}$ produces an integral solution for every vector c of n integers. Whenever $P(A, b)$ and $P_I(A, b)$ coincide, i.e., $P(A, b) = P_I(A, b)$, the *integer* linear programming problem $\max \{cx \mid x \in P_I(A, b)\}$ can thus be solved by solving the associated linear programming problem $\max \{cx \mid x \in P(A, b)\}$ for which — at least empirically — efficient solutions procedures exist.

Note that $P(A, b) = P_I(A, b)$ holds if and only if $\max\{cx \mid x \in P(A, b)\} \equiv 0 \bmod 1$ (i.e. is integer) holds for *all* integer vectors c. For any $m \times n$ matrix A of non-negative integers and any vector b of positive integers let $Q = Q(A, b)$ and $Q_I = Q_I(A, b)$ denote the polytopes

$$Q(A, b) = \{x \in \mathbf{R}^n \mid Ax \geq b, x \geq 0\}, \tag{2.3}$$

$$Q_I(A, b) = \text{conv}\{x \in Q(A, b) \mid x \text{ integer}\}. \tag{2.4}$$

Obviously, $Q \neq \emptyset$ if and only if A has no zero row. Furthermore, if A is a zero-one matrix and b is the vector of m ones, then every extreme point $\bar{x}$ of Q satisfies $\bar{x}_j \leq 1$ for all $j = 1, \ldots, n$. (Suppose $\bar{x}_j > 1$, then $\bar{x} = \frac{1}{2}x^1 + \frac{1}{2}x^2$ where $x_k^1 = x_k^2 = \bar{x}_k$ for all $k \neq j$ and $x_j^1 = \bar{x}_j - \epsilon$, $x_j^2 = \bar{x}_j + \epsilon$. For sufficiently small $\epsilon > 0$ both $x^1 \in Q$ and $x^2 \in Q$ and consequently $\bar{x}$ is not extreme.) The remarks that apply to the pair P and P_I apply to the polytopes Q and Q_I as well.

In the case of knapsack problems with a single constraint the following theorem answers the integrality problem completely, see [37]:

Theorem 2.1. *Let a be a vector of n positive integers and b be a positive integer satisfying $a_j \leq b$ for $j = 1, \ldots, n$ and $\sum_{j=1}^n a_j > b$. Then the following two statements are equivalent:*
  (i) $P(a, b) = P_I(a, b)$.
  (ii) $(a, b) = \lambda(e, k)$ *where λ is a positive integer, e is the vector of n ones.*

For *arbitrary* multi-dimensional knapsack problems no necessary *and* sufficient conditions are known to date that ensure integrality of the polyhedron $P(A, b)$.

We turn next to the pair of problems (NP) and (NC) defined in the previous section. All graphs that we consider are assumed to be finite, undirected, without loops and without multiple edges.

Theorem 2.2. *Let G be a graph having n nodes and q edges and let A_G be the $q \times n$ edge-node incidence matrix of G. Then the following three statements are equivalent:*
  (i) G *is a bipartite graph.*
  (ii) $P(A_G, e_q) = P_I(A_G, e_q)$.
  (iii) $Q(A_G, e_q) = Q_I(A_G, e_q)$.

To prove Theorem 2.2 we note that (ii) and (iii) are equivalent. This follows because the transformation $x'_j = 1 - x_j$ transforms every extreme point of $Q(A_G, e_q)$ into an extreme point of $P(A_G, e_q)$. On the other hand, the same transformation transforms any extreme point $\bar{x}$ of $P(A_G, e_q)$ with a *maximal* number of components equal to one into an extreme point of $Q(A_G, e_q)$. Thus $P(A_G, e_q)$ has a fractional extreme point if and only if $Q(A_G, e_q)$ has a fractional extreme point. If (i) holds, then A_G is a totally unimodular matrix, and thus (ii) follows by the Theorem of Hoffman and Kruskal [25, 46]. On the other hand, suppose that G is not bipartite, then by Theorem 2.4 of [24] G contains an odd cycle on nodes $1, \ldots, 2k+1$, say, where $k \leq [n/2]$. Consequently, the vector $\bar{x}$ with components $\bar{x}_j = \frac{1}{2}$ for $j = 1, \ldots, 2k+1$ and $\bar{x}_j = 0$ for $j \geq 2k+2$ is a fractional extreme point of $P(A_G, e_q)$, since the edge-node incidence matrix of an odd cycle is non-singular. Thus (ii) cannot hold and the theorem is proven.

While the integrality question for the pair of problems (NP) and (NC) is entirely symmetric, this is no longer true for the pair of problems (EP) and (EC). By a proof entirely analogous to the one of Theorem 2.2 one proves immediately the following theorem:

Theorem 2.3. *Let G be a graph having n nodes and q edges and let A_G be the $q \times n$ edge-node incidence matrix of G. Then the following two statements are equivalent:*
 (i) *G is a bipartite graph.*
 (ii) *$P(A_G^T, e_n) = P_I(A_G^T, e_n)$.*

To prove that (i) is not a necessary condition for $Q(A_G^T, e_n) = Q_I(A_G^T, e_n)$ to hold consider the edge-node incidence matrix of the complete graph K_4 on four nodes:

$$A_G^T = \begin{pmatrix} 1 & 1 & 1 & 0 & 0 & 0 \\ 1 & 0 & 0 & 1 & 1 & 0 \\ 0 & 1 & 0 & 1 & 0 & 1 \\ 0 & 0 & 1 & 0 & 1 & 1 \end{pmatrix}.$$

One verifies that the associated edge-covering problem (EC) has integer feasible solutions only, though K_4 certainly is not bipartite. (I am indebted to Alan J. Hoffman for this example.) On the other hand, (i) is a sufficient condition to ensure that $Q(A_G^T, e_n) = Q_I(A_G^T, e_n)$ holds. To formulate a necessary and sufficient condition, let H be any subset of the nodes of G and denote by G-H the subgraph obtained from G by deleting all nodes in H and all edges incident to the nodes in H. By A_{G-H} we denote the edge-node incidence matrix of G-H.

Theorem 2.4. *Let G be a graph having n nodes and q edges and let A_G be the $q \times n$ edge-node incidence matrix of G. Then the following two statements are equivalent:*
 (i) $Q(A_G^T, e_n) = Q_I(A_G^T, e_n)$.
 (ii) *Either, G is a bipartite graph or if not, then for every odd cycle with node set C in G $Q(A_{G-C}^T, e_r) = \emptyset$, where $r = n - |C|$.*

The proof of Theorem 2.4 again is rather straightforward: To see that (i) implies (ii), suppose that G contains an odd cycle C such that $Q(A_{G-C}^T, e_r) \neq \emptyset$. Let $\bar{z}$ be any extreme point of the latter polyhedron and define $\bar{x}_j = \frac{1}{2}$ for the edges defining the odd cycle, $\bar{x}_j = \bar{z}_j$ for the edges that are contained in G-C and $\bar{x}_j = 0$ otherwise. Then $\bar{x}$ is a fractional extreme point of $Q(A_G^T, e_n)$, hence (i) is violated.

On the other hand, suppose that (i) is violated. Then $Q(A_G^T, e_n)$ has a fractional extreme point $\bar{x}$. Let B be any nonsingular submatrix of A_G^T that defines $\bar{x}$. Then every column of B has at most two entries equal to one and since $0 < \bar{x}_j < 1$ for some j, it follows by a simple counting argument that B contains the incidence matrix of at least one odd cycle, see e.g. [32] for a related argument to this effect. Let C be the node-set of an odd cycle found this way. Then necessarily $Q(A_{G-C}^T, e_r) \neq \emptyset$ and thus (ii) is violated.

Note that a graph G is bipartite if and only if G does not contain any odd cycle. Thus we can state Theorems 2.2 and 2.3 as well by "forbidding" odd cycles in G. Equivalently, since odd cycles in G correspond uniquely to submatrices of the edge-node incidence matrix of G, we can replace the requirement that G is bipartite by the requirement that A_G does not contain submatrices of odd order having row and column sums equal to two. To state the integrality result concerning set-packing problems, we will use the latter, algebraic characterization. Let A' be any $m \times k$ zero-one matrix with $m \geq k$. A' is said to have the property $\pi_{\beta,k}$ if the following conditions are met:

(i) A' contains a $k \times k$ non-singular submatrix A'_1 whose row and column sums are all equal to β.

(ii) Each row of A' which is not a row of A'_1, either is componentwise equal to some row of A'_1, or has row sum strictly less than β.

The following theorem is proven in [34, 35]:

Theorem 2.5. *Let A be $m \times n$ matrix of zeros and ones. Then the following two statements are equivalent:*

(i) $P(A, e_n) = P_I(A, e_n)$.

(ii) *For $\beta \geq 2$ and $3 \leq k \leq n$, A does not contain any $m \times k$ submatrix A' having the property $\pi_{\beta,k}$.*

A topological characterization of zero-one matrices for which (i) holds, i.e. a characterization in terms of forbidden subgraphs of the intersection G_A associated with A, is not known to date, but would follow from a proof of C. Berge's Strong Perfect Graph Conjecture [6], see also [14]. Due to this connection, zero-one matrices A satisfying (i) of Theorem 2.5 are called *perfect* zero-one matrices.

While Theorem 2.4 answers the integrality question in the case of covering problems for zero-one matrices A having two ones per column, we turn next to the case of an arbitrary set-covering problem (SC). To this end, we call a $m \times n$ matrix A of zeroes and ones *balanced* [7, 16], if A does not contain any submatrix of odd size having row and column sum equal to two. The following theorem is from [26]:

Theorem 2.6. *If A is a balanced zero-one matrix of size $m \times n$, then $Q(A, e_m) = Q_I(A, e_m)$.*

Theorem 2.6 states a sufficient condition for $Q(A, e_m) = Q_1(A, e_m)$. (The above example of the incidence matrix of K_4 violates the condition.) We note that no decent necessary *and* sufficient characterization of arbitrary zero-one matrices A of size $m \times n$ for which $Q(A, e_n) = Q_1(A, e_n)$ is known to date.

3. Facetial inequalities

In the previous section we have been concerned with conditions under which the constraint set related to covering, packing and knapsack problem has only integer basic feasible solutions. In general, however, the polyhedra defined by (2.1) and (2.2) differ and it becomes necessary to identify additional linear inequalities that together with the linear constraints defining $P(A, b)$ linearly describe the polyhedron $P_1(A, b)$. (The *existence* of such a linear constraint system follows from the Theorem of H. Weyl [47].) In order to obtain a set of constraints that does the job and that has as few linear constraints as possible, one restricts oneself to facetial inequalities, i.e. valid linear inequalities for $P_1(A, b)$ that are *generated* by the extreme points of $P_1(A, b)$; see below for a technical definition. Such inequalities are called facetial because by definition they intersect the polyhedron $P_1(A, b)$ in a face of dimension one less than the dimension of $P_1(A, b)$, i.e. because they define a *facet* of $P_1(A, b)$. To illustrate the proceeding and to take up the point of "tighter" formulations of integer programming problems mentioned in the introduction, consider the simple *plant-location problem* of Section 1. A frequently encountered formulation of the same problem is the following, see e.g. p. 9 in [41]:

$$\min \sum_{j=1}^{m} \sum_{i=1}^{n} c_{ij} x_{ij} + \sum_{i=1}^{n} f_i y_i,$$

$$\text{s.t.} \quad \sum_{i=1}^{n} x_{ij} = 1 \quad \text{for} \quad j = 1, \ldots, m,$$

$$\sum_{j=1}^{m} x_{ij} \leq m y_i \quad \text{for} \quad i = 1, \ldots, m,$$

$$y_i = 0 \quad \text{or} \quad 1, x_{ij} \geq 0 \quad \text{for all} \quad i, j.$$

(SPL*)

While (SPL*) is a *more compact* formulation of (SPL), i.e. (SPL*) has fewer constraints than (SPL), any linear programming — based approach or e.g. an implicit enumeration algorithm using embedded linear programming subproblems of the type (SPL*), must be inferior to the same approach when applied to (SPL). This follows because many points that are feasible for the linear program associated with (SPL*) are infeasible for the linear program associated with (SPL), but not vice versa. More precisely, one can show that the constraints $x_{ij} \leq y_i$ are indeed *facets* of the (integer) polyhedron associated with (SPL*), while the inequalities involving the terms my_i in (SPL*) are obviously no facetial inequalities.

An inequality $\pi x \leq \pi_0$ is called a *facetial inequality* (or simply, a *facet*) for P_I as defined in (2.2) if

(i) $x \in P_I$ implies $\pi x \leq \pi_0$, and
(ii) there exist exactly d affinely independent points x^i of P_I satisfying $\pi x^i = \pi_0$ for $i = 1, \ldots, d$, where $d = \dim P_I$ is the dimension of P_I.

Note that it is unnecessary to require that the x^i are extreme points of P_I. Similarly, for the integer polytope Q_I defined by (2.4) we consider the reverse inequality $\pi x \geq \pi_0$ in the definition of a facet. It is not difficult to show that in the case of both P_I and Q_I the only facets having $\pi_0 = 0$ are the trivial facets $x_j \geq 0$ for $j = 1, \ldots, n$. This follows immediately from the non-negativity of the matrix A. Moreover, for the cases that are of interest to us, we can assume that every column a_j of A satisfies $a_j \leq b$ for $j = 1, \ldots, n$ and that A does not contain any zero row or zero column. It follows that the dimension of $P_I = n$ and. hence, that every inequality $x_j \geq 0$ defines a facet for $j = 1, \ldots, n$. Furthermore, every (non-trivial) facetial inequality $\pi x \leq \pi_0$ of P_I (resp. $\pi x \geq \pi_0$ of Q_I) satisfies $\pi \geq 0$ and $\pi_0 > 0$. Finally, $Ae_n > b$ implies $\dim Q_I = n$.

We start by examining the question whether a facetial inequality retains its property of defining a facet when the number of variables increases in the problem under consideration. The following theorem answers this question for set packing and knapsack problems and can be found in [30], see also [31, 32, 44]:

Theorem 3.1. *Let* $P_I^S = P_I \cap \{x \in \mathbf{R}^n \mid x_j = 0 \text{ for all } j \notin S\}$ *be the polyhedron obtained from* P_I *as given by* (2.2) *by setting all variables* x_j, $j \in N - S$, *equal to zero. If the inequality* $\sum_{j \in S} \alpha_j x_j \leq \pi_0$ *is a facet of* P_I^S, *then there exist non-negative numbers* β_j, $\beta_j \leq \pi_0$, *such that* $\sum_{j \in S} \alpha_j x_j + \sum_{j \in N-S} \beta_j x_j \leq \pi_0$ *is a facet of* P_I.

As a consequence of Theorem 3.1, one can restrict oneself to a subproblem of a set-packing or knapsack problem when one wants to identify a facetial inequality and subsequently lift (or extend) such inequality to a facetial inequality for P_I. This idea has been pursued in a number of recent papers, see e.g. [1, 2, 4, 5, 21, 22, 27, 30, 31, 32, 33, 37, 39, 44, 45, 48, 49, 50, 51, 52]. Note that in the case of the polyhedra P and P_I the *intersection* of P and P_I with a hyperplane $x_j = 0$ yields the same polyhedra P^j and P_I^j that are obtained when P and P_I are *projected* orthogonally on the hyperplane $x_j = 0$, where $j \in \{1, \ldots, n\}$. In the case of covering polytopes Q and Q_I this is not the case. (Example: Consider the problem (NC) for the complete graph K_3 on three nodes; see also [14].)

A concept that has proved useful in the descriptive identification of facetial inequalities for (multi-dimensional) knapsack problems is a *cover*: Let S be any subset of N satisfying

$$\sum_{j \in S} a_j \not\leq b \tag{3.1}$$

where the symbol $\not\leq$ is meant to say that for at least one component of the vectors involved the inequality $\leq$ is violated. If

$$\sum_{j \in Q} a_j \leq b \tag{3.2}$$

holds for all $Q \subset S$ such that $|Q| = q$, but for no $Q \subseteq S$ such that $|Q| > q$,

$$\sum_{j \in S} x_j \leq q \tag{3.3}$$

defines a facet for the lower-dimensional polyhedron $P_I^S = P_I \cap \{x \in \mathbf{R}^n \mid x_j = 0 \; \forall j \in N - S\}$. A subset S of N satisfying (3.1) and (3.2) is called a *cover* with respect to A and b and, if S does not have a proper subset S' satisfying (3.1), a *minimal cover*. Facetial inequalities derived from minimal covers have been studied in [1, 5, 22, 30, 33, 39, 49, 52] and one of the main results can be stated as follows:

Theorem 3.2. *The inequality $\sum_{j \in S} x_j \leq q$ defines a facet of $P_I(A, b)$ if and only if*

(i) *there exists $Q \subseteq S$ with $|Q| = q+1$, such that Q is a minimal cover with respect A and b, where $S \subseteq N$ and q is an integer number $q \leq |S|-1$ and*
(ii) *S is a maximal subset of N satisfying (i).*

Facetial inequalities for P_I obtained from minimal covers do, however, not always have 0—1 coefficients only. Rather due to Theorem 3.1, a variable j not in the cover may get any coefficient β_j satisfying $0 \leq \beta_j \leq q$ when the inequality given by a cover is *extended* (lifted) to a facet of P_I.

A richer class of facetial inequalities of P_I is given by the following combinatorial configuration given by A and b: Let $S \subset N$ and $t \in N - S$ satisfy

$$\sum_{j \in S} a_j \leq b, \tag{3.3}$$

$Q \cup \{t\}$ is a minimal cover for every $Q \subseteq S$ with $|Q| = k$, (3.4)

where k is any given integer satisfying $2 \leq k \leq |S|$. Due to the one-element role of the index t, we call the set $S \cup \{t\}$ a *1-configuration* with respect to A and b. If $k = |S|$ a 1-configuration is simply a minimal cover with respect to A and b. The following theorem is proven in [37]:

Theorem 3.3. *Let $T = S \cup \{t\}$ be a 1-configuration with respect to A and b. Then the inequalities $0 \leq x_j \leq 1$, $\forall j \in T$ and*

$$(r-k+1)x_t + \sum_{j \in T(r)} x_j \leq r \tag{3.5}$$

yield the complete and irredundant set of facets of the polyhedron $P_I^T(A, b) = P_I(A, b) \cap \{x \in \mathbf{R}^n \mid x_j = 0 \; \forall j \in N - T\}$, where $T(r) \subseteq S$ is any subset of cardinality r of S and r is any integer satisfying $k \leq r \leq |S|$.

It follows that a 1-configuration defines $\sum_{r=k}^{s} \binom{s}{r}$ distinct facets of P_I^T, where $s = |S|$. Using the lifting procedure [33], we obtain an even greater number of facets of the polyhedron P_I. Note that implicit in the formulation of Theorem 3.3 is the assertion that the inequalities (3.5) together with $0 \leq x_j \leq 1$ for all $j \in T$ define a polyhedron with zero-one vertices only.

We turn next to the set-covering problem (SC). Different from the knapsack and set-packing problem, the facial structure of set-covering polytopes as well as node-covering polytopes has received very little attention in the literature. Some results have been obtained by D.R. Fulkerson [14] in the more general setting of the theory of blocking polyhedra: A zero-one matrix A is called *proper* if A does not contain any row a^i which dominates (in the usual vector sense) any other row of A different from a^i. The following theorem characterizes all facetial inequalities $\pi x \geq 1$ of the covering polytope Q_I with $\pi_j \in \{0, 1\}$ for $j = 1, \ldots, n$:

Theorem 3.4. *Let A be a proper zero-one matrix. Then the facets $\pi x \geq \pi_0$ of $Q_I(A, e)$ satisfying $\pi_j \in \{0, 1\}$ for $j \in N$ and $\pi_0 = 1$ are exactly the inequalities given by $Ax \geq e$.*

The *edge-packing* or *edge-matching* polyhedron is the only case among all structured integer programming problems discussed in this paper for which a complete linear description has been obtained. The next theorem was proved *algorithmically* by J. Edmonds in 1965 [13]:

Theorem 3.5. *Let A with elements a_{ij} be a $m \times n$ matrix having exactly two ones per column, zeros elsewhere. Then*

$$P_I(A, e_m) = P(A, e_m) \cap \left\{ x \in \mathbf{R}^n \; \middle| \; \sum_{j \in E(S)} x_j \leq \frac{|S|-1}{2}, s \in \mathcal{T} \right\} \quad (3.8)$$

where $\mathcal{T}$ is the family of all $S \subset \{1, \ldots, m\}$ with $|S| \geq 3$ and odd and $E(S) = \{j \in N \mid \sum_{i \in S} a_{ij} = 2\}$.

The question of redundancy of the linear constraint set (3.8) was investigated in 1973 [40]: Inequalities that matter correspond to $k \times p$ submatrices A' of A with k odd and the following three properties:

    (i) A' has all column sums equal to two;
    (ii) $A'\xi = e_k$ does not admit a zero-one solution ξ; and
    (iii) every submatrix A'' of A' obtained by deleting a row j of A' and all columns of A' that have an entry equal to one in row j admits a zero-one solution to $A''\xi' = e_{k-1}$.

Since the matrix A of Theorem 3.7 is the node (rows of A) versus edge incidence matrix of some graph G, odd cycles in G give rise to submatrices A' with the three properties, but they are *not* the only subgraphs of G giving rise to such submatrices.

To obtain statements about types of linear inequalities that define facets of *general set packing polyhedra*, the concept of the intersection graph associated with a zero-one matrix A (see Section 1.2) has been found to be most useful. In fact, in view of the equivalence of set packing problems and node packing problems discussed in Section 1.2, one can ask the question, "what subgraphs of the intersection graph (or more generally, what kind of graphs) give rise to facets of the associated packing problem," (where the set packing problem defined with respect to some zero-one matrix A is simply viewed as a node packing problem in the appropriate graph).

A first class of graphs that gives rise to facet-defining linear inequalities are cliques, i.e., maximal complete subgraphs of a graph. More precisely, let us consider an (undirected) graph $G = (N, E)$ on n nodes having m edges where $N = \{1, \ldots, n\}$. Let A_G denote the incidence matrix of edges (rows of A_G) versus the nodes of G. The next theorem can be found in [32] and can also be deduced from Theorem 8 in [14].

Theorem 3.6. *The inequality $\sum_{j \in K} x_j \leq 1$ where $K \subseteq N$, is a facet of $P_I(A_G, e_m)$ if and only if K is a node set of a clique of G.*

Thus one knows in fact all subgraphs of a graph G that give rise to facets $\pi x \leq \pi_0$ of the associated node-packing polyhedron P_I with non-negative integer π_j, $j \in N$, and $\pi_0 = 1$. In order to study the more general question, the following concepts have been found useful; see [36]: A node-induced subgraph $G_S = (S, E_S)$ of $G = (N, E)$ with node set $S \subseteq N$ is *facet-producing* if there exists an inequality $\pi x \leq \pi_0$ with non-negative integer components π_j such that

(i) $\pi x \leq \pi_0$ is a facet of $P_I^S = P_I \cap \{x \in \mathbf{R}^n \mid x_j = 0 \text{ for all } j \notin S\}$ and
(ii) $\pi x \leq \pi_0$ is *not* a facet for $P_I^T = P_I \cap \{x \in \mathbf{R}^n \mid x_j = 0 \text{ for all } j \notin T\}$ where T is any subset of S such that $|T| = |S| - 1$.

A subgraph G_S of G is called *strongly facet-producing* if there exists an inequality $\pi x \leq \pi_0$ such that (i) holds and (ii) holds for all $T \subseteq S$ satisfying $|T| \leq |S| - 1$. A subgraph of G is *facet-defining* if there exists an inequality $\pi x \leq \pi_0$ such that (i) holds and (ii) such that $\pi_j > 0$ for $j \in S$.

Given any facet $\pi x \leq \pi_0$ of P_I, let now $S(\pi) = \{j \in N \mid \pi_j > 0\}$. It is not difficult to verify that $\pi x \leq \pi_0$ is either produced by the subgraph G_S of G having the node-set $S = S(\pi)$ or if not, that there exists a subgraph G_T of G_S with node-set $T \subseteq S$ such that the facet $\tilde{\pi} x \leq \pi_0$ of P_I^T is produced by G_T where $\tilde{\pi}_j = \pi_j$ for $j \in T$, $\tilde{\pi}_j = 0$ for $j \notin T$ and P_I^T is defined as previously. The question is, of course, given

$\bar{\pi}x \leq \pi_0$ can we retrieve the facet $\pi x \leq \pi_0$ of P_I. The answer is positive and follows easily from Theorem 3.1. This means that Theorem 3.1 permits one to restrict attention to node-induced subgraphs of G when one tries to identify a linear characterization of P_I. More precisely, we can restrict attention to graphs that are *themselves* facet-defining or facet-producing. Besides cliques, which play a "singular" role in this discussion, a large class of *regular* graphs has been proven to be strongly facet-producing. (A graph G is called regular of degree k if every node of G has k edges incident to itself.) To state the next theorem, we need the notion of a "web", see [45]. A *web*, denoted $W(n, k)$ is a graph $G = (N, E)$ such that $|N| \geq 2$ and for all $i, j \in N$ $(i, j) \in E$ iff $j = i+k, i+k+1, \ldots, i+n-k$, (where sums are taken modulo n), with $1 \leq k \leq [n/2]$. The web $W(n, k)$ is regular of degree $n - 2k + 1$, and has exactly n maximum node packings of size k. The complement $\bar{W}(n, k)$ of a web $W(n, k)$ is regular of degree $2(k-1)$ and has exactly n maximum cliques of size k. One verifies that $W(n, 1)$ is a clique on n nodes, and for integer $s \geq 2$, $W(2s+1, s)$ is an odd cycle without chords (hole), while $W(2s+1, 2)$ is an odd antihole, i.e. the complement of a hole. The next theorem is essentially from [45], with the extension appearing in [36]:

Theorem 3.7. *A web $W(n, k)$ strongly produces the facet $\sum_{j=1}^{n} x_j \leq k$ if and only if $k \geq 2$ and n and k are relatively prime. The complement $\bar{W}(n, k)$ of a facet-producing web $W(n, k)$ defines (strongly produces) the facet $\sum_{j=1}^{n} x_j \leq [n/k]$ (if and only if $n = k[n/k] + 1$).*

It would appear most natural to try to "enumerate" all possible graphs that are strongly facet-producing or facet-defining. The implementation of this idea, however, appears to be hopelessly difficult as the following construction shows which permits one to "build" arbitrarily complex strongly facet-producing graphs. Let G be any *facet-defining* graph with node set $V = \{1, \ldots, n\}$ with $n \geq 2$ and consider the graph G^* obtained by joining the ith node of G to the ith node of the "claw" $K_{1,n}$ by an edge. The claw $K_{1,n}$ — also referred to as a "cherry" or "star", see e.g. [24] — is the bipartite graph in which a single node is joined by n edges to n mutually non-adjacent nodes. We will give the node of $K_{1,n}$ that is joined to the ith node of G the number $n + i$ for $i = 1, \ldots, n$, whereas the single node of $K_{1,n}$ that is not joined to any node of G, will be numbered $2n + 1$. (See Fig. 1 where the construction is carried out for a clique $G = K_4$.) Denote by

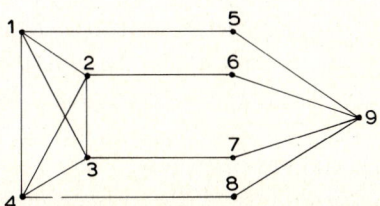

$V^* = \{1, \ldots, 2n+1\}$ the node set of G^* and by E^* its edge set. It turns out that G^* is facet-defining, and moreover, that G^* is strongly facet-producing.

Theorem 3.8. *Let $G = (V, E)$ be a graph on $n \geq 2$ nodes and let $\pi x \leq \pi_0$ be a (nontrivial) facet defined by G. Denote by $G^* = (V^*, E^*)$ the graph obtained from G by joining every node of G to the pending nodes of the claw $K_{1,n}$ as indicated above. Then G^* strongly produces the facet*

$$\pi x^{(1)} + \pi x^{(2)} + \left(\sum_{j=1}^{n} \pi_j - \pi_0\right) x_{2n+1} \leq \sum_{j=1}^{n} \pi_j \qquad (3.9)$$

where $x^{(1)} = (x_1, \ldots, x_n)$, $x^{(2)} = (x_{n+1}, \ldots, x_{2n})$ and x_{2n+1} are the variables of the node-packing problem on G^* in the numbering defined above.

The proof of Theorem 3.8 can be found in [36], where also a further construction is presented that supports the opinion that "enumeration" of facet-producing graphs is a tremendously difficult task. Even though the above is the basis for our contention that a complete linear description of hard integer programming problems will probably prove itself to be elusive, use of facetial inequalities given by a possibly small subset of *all* facets of $P_I(A, e)$ is very effective in actual computation. We will demonstrate this point by means of an example which takes up again the point of "tighter" formulations of integer programming problems mentioned in the introduction.

Example. Consider the maximum-cardinality node-packing problem on an odd antihole G with $n \geq 5$ vertices and let A_G denote the edge versus node incidence matrix of G. Denote by R the following permutation matrix:

$$R = \begin{bmatrix} 0 & 1 & 0 & \cdots & 0 \\ 0 & 0 & 1 & 0 & \cdots & 0 \\ \vdots & & & & & \vdots \\ 0 & \cdots & & & 0 & 1 \\ 1 & 0 & \cdots & & & 0 \end{bmatrix} \qquad (3.10)$$

We can write $A_G^T = (A_1^T, \ldots, A_p^T)$ where $p = [n/2] - 1$ and $A_i^T = (I + R^i)^T$ for $i = 1, \ldots, p$ with I being the $n \times n$ identity matrix. Let $P = \{x \in \mathbf{R}^n \mid A_G x \leq e, x \geq 0\}$ be the linear programming relaxation of the node-packing problem and P_I the convex hull of integer solutions. As one readily verifies, $\max\{\sum_{j=1}^{n} x_j \mid x \in P\} = n/2$ for all n. But, the integer answer is *two*, no matter what value n assumes, i.e., $\max\{\sum_{j=1}^{n} x_j \mid x \in P_I\} = 2$ for all n. Suppose now that we work with a linear programming relaxation of P_I utilizing a subset of the facets of P_I given in Theorem 3.8. Specifically, suppose that we have identified all cliques of G that are of maximum cardinality (this is in general a *proper* subset of *all* cliques of antiholes). Denote by $\tilde{A}$ the corresponding clique-node incidence matrix. Then $\tilde{A} = \sum_{i=0}^{p} R^i$. Let $\tilde{P} = \{x \in \mathbf{R}^n \mid \tilde{A}x \leq \tilde{e}, x \geq 0\}$ be the linear programming relaxation of the node-packing problem on G. Then $P_\mathrm{I} \subseteq \tilde{P} \subseteq P$. As one readily verifies, $\max\{\sum_{j=1}^{n} x_j \mid x \in \tilde{P}\} = 2 + 1/[n/2]$ and the integer optimum of 2 follows by simply rounding down.

The interesting fact exhibited by the example is that the knowledge of merely a few of the facets of P_I *in the case of odd antiholes* permits one to obtain a bound on the integer optimum that is "sharp" as compared to the bound obtained by working on the straightforward linear programming relaxation of the node-packing problem involving the edge-node incidence matrix of the antihole (a bound that is *arbitrarily bad* according to how large one chooses n). The general question raised by this example is of course, *how often* (in a statistical sense) it will be sufficient to work with only a small subset of all facets of a set packing polyhedron P_I (such as those given by cliques, holes, etc.) in order to get "close enough" to the integer optimum. As we have mentioned in the introduction, an empirical study for the symmetric travelling salesman problem [38] has provided convincing evidence for the use of facetial inequalities in large-scale combinatorial optimization.

References

[1] E. Balas, Facets of the knapsack polytope, Math. Programming 8 (1975) 146–164.
[2] E. Balas, On the facial structure of the knapsack polytope, paper presented at the NATO Advanced Study Institute on Combinatorial Programming, Versailles (September 2–13, 1974).
[3] E. Balas and M.W. Padberg, Set partitioning: A survey, SIAM Rev. 18 (1976) 710–760.
[4] E. Balas and E. Zemel, All the facets of the knapsack polytope, MSSR No., Carnegie–Mellon University (1975).
[5] E. Balas and E. Zemel, Facets of the knapsack polytope from minimal covers, Carnegie–Mellon University, Management Sciences Research Report No. 352 (December 1974).
[6] C. Berge, Graphes et Hypergraphes (Dunod, Paris, 1970).
[7] C. Berge, Balanced Matrices, Math. Programming 2 (1972) 19–31.
[8] N. Cristofides and S. Korman, A computational survey of methods for the set covering problem, Report 73/2 (1973), Imperial College of Science and Technology.
[9] V. Chvatal, On certain polytopes associated with graphs, CRM-238, University de Montreal (October, 1973). (Forthcoming in J. Combinatorial Theory.)
[10] G. B. Dantzig, Linear programming and extensions (Princeton University Press, Princeton, 1963).
[11] J. Edmonds, Covers and packings in a family of sets, Bull. Amer. Math. Soc. 68 (1962) 494–499.
[12] J. Edmonds, Path, trees and flowers, Canad. J. Math. 17 (1965) 449–467.
[13] J. Edmonds, Maximum matching and a polyhedron with 0,1 vertices, J. Res. Nat. Bur. Standards 69B (1965) 125–130.
[14] D.R. Fulkerson, Blocking and anti-blocking pairs of polyhedra, Math. Programming 1 (1971) 168–194.
[15] D.R. Fulkerson, G.L. Nemhauser and L.E. Trotter, Two computationally difficult set covering problems that arise in computing the 1-width of incidence matrices of Steiner triple systems, TR No. 203 (1973), Department of Operations Research, Cornell University.

[16] D.R. Fulkerson, A.J. Hoffman and R. Oppenheim, On balanced matrices, Math. Programming Study 1 (1974) 120–132.
[17] R. Garfinkel and G.L. Nemhauser, Integer Programming (Wiley, New York, 1972).
[18] R. Garfinkel and G.L. Nemhauser, A survey of integer programming emphasizing computation and relations among models, in: T.C. Hu and S.M. Robinson, eds., Mathematical Programming (Academic Press, New York, 1973).
[19] F. Granot and P.L. Hammer, On the use of boolean functions in 0–1 programming, O.R. Mimeograph No. 70, Technion (1970).
[20] F. Glover, Unit-coefficient inequalities for zero-one programming, University of Colorado, Management Sciences Report Series, No. 73-7 (July 1973).
[20a] M. Grötschel, Polyedrische Charakterisierungen kombinatorischer Optimierungsprobleme, Doctoral Dissertation, University of Bonn (1977).
[21] P.L. Hammer, E.L. Johnson and U.N. Peled, Regular 0–1 programs, University of Waterloo, Combinatorics and Optimization Research Report, CORR No. 73–18 (September 1973).
[22] P.L. Hammer, E.L. Johnson and U.N. Peled, Facets of regular 0–1 polytopes, Math. Programming 8 (1975) 179–206.
[23] P.L. Hammer, E.L. Johnson and U.N. Peled, The role of master polytopes in the unit cube, University of Waterloo, Combinatorics and Optimization Research Report, CORR No. 74–25 (October 1974).
[24] F. Harary, Graph Theory (Addison-Wesley, Reading, MA, 1969).
[25] A.J. Hoffman and J.B. Kruskal, Integral boundary points of convex polyhedra, in: H.W. Kuhn and A.W. Tucker, eds., Linear inequalities and related systems, Annals of Math. Study No. 38 (Princeton University Press, Princeton, 1956) 233–246.
[26] A.J. Hoffman and R. Oppenheim, On balanced matrices, Thomas J. Watson Research Center Report, IBM Research (1974).
[27] E.L. Johnson, A class of facets of the master 0–1 knapsack polytope, Thomas J. Watson Research Center Report RC 5106, IBM Research (October 1974).
[28] S. Lin and B.W. Kernighan, An effective heuristic algorithm for the travelling-salesman problem, Operations Res. 21 (1973) 498–516.
[26] J. Lorie and L. Savage, Three problems in capital rationing, Journal of Business, 28 (1955) 229–239.
[30] G.L. Nemhauser and L.E. Trotter, Properties of vertex packing and independence system polyhedra, Math. Programming 6 (1974) 48–61.
[31] M.W. Padberg, Essays in integer programming, Ph.D. Thesis, Carnegie–Mellon University (May, 1971).
[32] M.W. Padberg, On the facial structure of set packing polyhedra, Math. Programming 5 (1973) 199–215.
[33] M.W. Padberg, A note on zero–one programming, Operations Res. 23 (1975) 833–837.
[34] M.W. Padberg, Perfect zero–one matrices, Math. Programming 6 (1974) 180–196.
[35] M.W. Padberg, Almost integral polyhedra related to certain combinatorial optimization problems, Linear Algebra and Appl. 15 (1976) 69–88.
[36] M.W. Padberg, On the complexity of set packing polyhedra, Annals of Discrete Math. 1 (1977) 421–434.
[37] M.W. Padberg, A note on zero–one programming-II, T.J. Watson Research Center Report, IBM Research (1977).
[38] M.W. Padberg and S. Hong, On the travelling salesman problem: A computational study, Thomas J. Watson Research Center Report, IBM Research (1977).
[39] U.N. Peled, Properties of facets of binary polytopes, Ph.D. Thesis, Dept. of Comb. and Opt., University of Waterloo, Waterloo (1976).
[40] W. Pulleyblank, Faces of matching polyhedra, Ph.D. Thesis, Dept. of Comb. and Op., University of Waterloo, Waterloo (1973).
[41] H.M. Salkin, Integer Programming (Addison-Wesley, Reading, MA, 1975).
[42] H.M. Salkin and J. Saha, Set covering: Uses, algorithms results, Technical Memorandum No. 272, Case Western Reserve University (March 1973).
[43] M.L. Simmonard, Linear Programming (Prentice Hall, Englewood Cliffs, 1966).

[44] L.E. Trotter, Solution characteristics and algorithms for the vertex packing problem. Ph.D. Thesis, Cornell University (January 1973).
[45] L.E. Trotter, A class of facet producing graphs for vertex packing polyhedra, Technical Report No. 78, Yale University (February 1974). (Forthcoming in Discrete Math.)
[46] A.F. Veincott, Jr. and G.B. Dantzig, Integer extreme points, SIAM Rev. 10 (1968) 371–372.
[47] H. Weyl, Elementare Theorie der konvexen Polyeder, Comm. Math. Helv. 7 (1935) 290–306 (translated in Contributions to the Theory of Games, Vol. I, 3–18, Annals of Mathematics Studies, No. 24, Princeton, 1950).
[48] L.A. Wolsey, Further facet generating procedures for vertex packing polytopes, Math. Programming 11 (1976) 158–163.
[49] L.A. Wolsey, Faces for a linear inequality in 0–1 variables, Math. Programming 8 (1975) 165–178.
[50] L.A. Wolsey, Facets and strong valid inequalities for integer programming, Université Catholique de Louvain, CORE Discussion paper (April 1974).
[51] E. Zemel, Lifting the facets of 0–1 Polytopes, MSSR No. 354, Carnegie–Mellon University (December 1974).
[52] E. Zemel, On the facial structure of zero–one polytopes, Ph.D. Thesis, GSIA, Carnegie–Mellon University (September 1976).

Report of the Session on
NETWORK FLOW, ASSIGNMENT AND TRAVELLING SALESMAN PROBLEMS*

G. de GHELLINCK (Chairman)

L. Wolsey (Editorial Associate)

1. Assignments and network flow problems

The capacitated transshipment problem can be viewed as representative of most network flow problems and includes the assignment and simple transportation as special cases. They were among the first linear programmes to be solved (Hitchcock 1941, Koopmans 1951, Dantzig 1957) by methods related to the simplex. In 1961 Ford and Fulkerson proposed the out-of-kilter method which was considered the most efficient way to solve network flow problems until the early seventies. Specialized implementations of the simplex method proved then to be computationaly faster. However it appears now that the Ford–Fulkerson approach could also be substantially improved in its two basic steps. A sequence of cost-change steps can be replaced by a single application of Dijkstra's shortest path algorithm; this was proposed and implemented by Tomisawa (1971). Similarly a sequence of flow augmentations can be replaced by a single application of the new max-flow algorithm of Karzanov (1974). In both cases rather curiously, less emphasis is being given to the concept of flow augmenting paths.

A first discussant during his presentation affirmed that at present the most efficient network codes were primal codes. He suggested the explanation of this phenomenon as a research topic, and the Markov chain model of Liebling (presented in Vancouver) as a possible aid. He also suggested the further analysis of different objective functions (i.e. algebraic) as a topic for further research.

Reactions to these remarks included the warning that dual algorithms should not be written off too quickly, given that 15 years ago everyone asserted the contrary. It was suggested that there was at present strong evidence that dual algorithms were more efficient for location problems (i.e., 100×100 good solutions found in a fraction of a second), and conjectured that this was due to the fact that strategic variables were better emphasized in the dual.

Another discussant affirmed his conviction that primal codes were better. He reported that for small problems and for particular problems such as constrained regression dual codes were competitive. However for large problems he claimed that in dual algorithms all variables are needed at each iteration (i.e. 10^6 variables) which explained the advantage of the primal. The question was raised whether little changed from iteration to iteration, so that this work could be substantially reduced.

However it was suggested that "degeneracy" was the big problem, implying that the work of both Cunningham and Klingman had not substantially reduced computation times, and the number of pivots was unreduced. However the unusual behavior of convergence to the objective was pointed out namely a successive reduction in degeneracy, and hence acceleration of the convergence. This behavior, contrary to that of LP, was suggested as a subject for research.

A further important question was whether "degeneracy" was helpful or not. Perturbation techniques to avoid degenerate pivots increased computation times by a factor of three. The question was raised of whether similar experience was found with LP codes.

Several participants confirmed that in general perturbation techniques were not used, but rather row selection rules.

The expected behavior of algorithms has been a topic raised several times during the conference, in particular the idea that an average analysis might be preferred to a worst case analysis and the practical validity of reporting results based on data drawn randomly.

The chairman opened the discussion by giving computational results for the assignment problem and suggesting some of the difficulties and pitfalls of the latter approach.

In particular he reported that for integer data drawn from a uniform distribution between 1 and 100 for the (c_{ij}) matrix, a phase I procedure of searching for zero's and assigning rows allowed one to cover all but a number $r(n)$ of rows, where the expected value of $r(n)$ is proportional to the cost range and independent of n for large n, e.g. for cost range = 100, $r(n) \simeq 60$ for n larger than 500. In addition the number of cost changes required (an expensive part of assignment algorithms) tended to zero. Problems clearly become easier as n increases for data drawn in the given way. Contradicting this, for a practical problem drawn from the 120 city TSP of Grötschel, only 7 assignments were made in phase I as opposed to 90 on average for the random data. This leads to the difficulty: if uniform distributions do not lead to typical problems there appears to be no probabilistic structure that emerges as a natural alternative.

Various reactions then included comments indicating that practical assignment problems got more difficult as the dimensions increased. Among the reasons for the discrepancy between practical and random problems, it was suggested that the assumption of "independence" was at fault.

However practical problems are never totally dense — in particular manpower planning models might have 50% of the arcs infeasible, and problems having 6×10^6 arcs, where 0.5×10^6 arcs have zero cost, are still difficult i.e. relative to the phase I behavior. Problems also arise from the fact that initially the chosen arcs do not even form a connected graph.

Several other topics of research were suggested.

(1) To implement existing network codes on minicomputers so as to convince more people of the value of such codes.

(2) Computation times for networks are sufficiently faster than for LP so as to warrant the exploitation of network structure in commercial LP codes. In particular a large number of IP's such as distribution problems have such a structure, and in addition the modeller is usually able to identify it. This raises the question of knowing at what point even if the network structure were unidentified, the advantages taken of sparsity in commercial codes might balance the use of network structure. It was suggested that 25% network structure was about the break-point. The primary reason being the graphical basis representation, the avoidance of multiplication, division and testing tolerances, i.e., all logical operations.

In concluding the discussion on this topic it was pointed out that a good LP code essentially uses arithmetic while a good network code uses list processing.

2. The travelling salesman (TSP) and quadratic assignment problem (QAP)

The first discussant suggested the further study of polytopes, such as for the symmetric TSP, to aid in the design of algorithms, such as that reported by Padberg. He suggested as an open question "How to automatically find cuts or facets to add".

The next two speakers then described two recent approaches to the TSP.

It has been mentioned that even with a sophisticated method of generating facets for the travelling salesman polyhedron, ILP methods are limited by the sheer size of the problem, e.g., a 120 city symmetrical problem has over 7000 variables. However, in the same way that it is not necessary to generate explicitly even the 2^{n-1} LP constraints of the n-city problem, so it is possible to solve the problem with only a subset of the variables and confirm dual feasibility on the remaining variables. It is possible to generate sufficient constraints to find the solution and prove optimality using only $4n$ or $5n$ variables, instead of $\frac{1}{2}(n^2-n)$.

Any variable which has been suppressed must be shown to satisfy the duality conditions (and there must be a method for enlarging the problem for variables which violate the duality conditions). All the constraints which are required to prove optimality of the reduced problem can be stored in an implicit fashion. Knowing i and j of a new variable, x_{ij}, you know immediately which of the vertex-degree constraints apply. The coefficients of a whole set of k tour-barring constraints can be stored in a vector of length b, by simply identifying which set each vertex lies in for a particular partition into k sets. A new variable, x_{ij}, has a non-zero coefficient in any particular constraint if both i and j have the same vertex label in the vector.

Miliotis, in a forthcoming paper in Mathematical Programming, has demonstrated considerable success in using Gomory's method of integer forms cutting planes on travelling salesman problems, and these constraints can be recorded by storing non-negative weights of the pre-existing constraints, and hence the coefficients of new variables can be generated sequentially. However, even more simply, by generating them according to a logical analysis of the graph, they can also be recorded in terms of a subset of the vertices, plus some information about a subset of the reduced set of the variables.

This algorithm is complete, apart from the final step of closing a very small remaining duality gap by a branching procedure.

A recent extension involves comb constraints. Again these could be carried implicitly, via a list of vertices plus a limited amount of information on a set of variables. After adding combs, there is a small remaining duality gap. This is removed by Integer Forms, or Branch and Bound. The algorithm is in Fortran, using integer arithmetic. A 100-city problem was solved easily by this method.

The next discussant suggested that the TSP polyhedron be called an "especially hard polyhedron" implying that it was especially difficult to linearly describe this polyhedron.

The conjecture by Nemhauser mentioned earlier in this context ($Z^{LP}/Z^{opt} \leq 9/8$) does not contradict at all the intuition about the incredible complexity of the TSP polyhedron and in fact, it would very well explain the considerable success that the 1-tree relaxation due to Held–Karp (with refinement by Krarup and coworker H. Helbig) has in the context of symmetric TSP.

He then briefly describes some of the methodology he has used in the computational study on TSP's. Basically, it uses a hybrid approach of heuristics plus linear programming. More precisely, one proceeds almost exactly as Dantzig, Fulkerson and Johnson did in 1954: one just uses the most effective heuristic known to get a good, possibly optimal tour (Heuristic due to Lin and Kerningham). This solution is then used as a starting solution to essentially a *primal* cutting plane algorithm (à la Gomory, Glover, Young, etc.) where, however, the cutting planes used are those that we know to be facets of the TSP polyhedron. The implementation of this approach is not easy: One needs to find algorithms that identify constraints (facets) as needed in the computation.

The computational study was divided in four parts:

(a) a statistical study: 15, 30, 45, 60, 75, 100 cities.

(b) a study concerning pathological problems for TSP's due to Papadimitriou and Stieglitz which are unsolvable for local search algorithms.

(c) a study concerning all readily available test problems in the literature.

(d) a case-study of the 318-city problem published in Lin and Kerningham. The objective of the computational study was to empirically validate the usefulness of cutting planes that define facets of the TSP, i.e., to try to establish by how much the gap between the best tour and a "tight" formulation of the combinatorial problem can be reduced.

Relative to the Quadratic Assignment Problem, Burkard suggested in his presentation a study of its polyhedral structure, and different linearizations. He reported a recent linearization in $O(n^2)$ variables (Kaufman and Broeckx), and raised the problem of developing an algorithm from it. He also reported a paper of Christofides (1977) on solvable QAP's.

Another open question was a good algorithm for the three-dimensional assignment problem for applications such as timetabling, perhaps using subgradient methods.

He announced that the special version of the linear arrangement problem he has posed in Versailles has been well-solved. In addition, although the general problem was NP-complete, the case of a tree had been solved by Shilsack in $O(n^3)$.

He looked forward to the day when other efficient algorithms would be found for special cases of the TSP and QAP. He also conjectured that solving TSP will become "routine" in the years to come, and the TSP algorithm will become just a subroutine for more complex problems.

Report of the Session on
ALGORITHMS FOR SPECIAL CLASSES OF COMBINATORIAL OPTIMIZATION PROBLEMS

J.K. LENSTRA (Chairman)

U.N. Peled (Editorial Associate)

Due to recent developments in the theory of computational complexity, it is now known that many combinatorial problems are likely to require superpolynomial time for their solution (cf. [12]). On the other hand, large subclasses of problems do exist for which a polynomial-time algorithm is available or for which such an algorithm might still be found, in the sense that its existence would not imply the equality of $\mathscr{P}$ and $\mathscr{NP}$. The purpose of this report is to review the basic achievements in the latter area, to examine the prospects of possible extensions, and to mention open problems that seem worthy of further attack.

As a first example, the class of *matching* problems ought to be mentioned. It was in this context that the importance of polynomial running-time bounds was first emphasized [3]. Broadly understood, matching encompasses all integer linear programming problems with a nonnegative integer constraint matrix A whose column sums are at most 2; for example, A can be the vertex-edge incidence matrix of a graph. The polynomial-time matching algorithm is based on an efficient characterization of the convex hull of feasible solutions [7]. We return to such good characterizations below.

A closely related area is the theory of *matroid optimization*, which has been particularly fruitful [9]. A matroid, in one of many equivalent definitions, consists of a finite set E and a family of *independent* subsets of E such that

(i) every subset of an independent set is independent, and

(ii) all maximal independent subsets of any subset of E are of the same size.

Given a numerical weight for each element of E, the simplest matroid optimization problem is to find an independent set of maximum total weight. It is well known that this problem is solved by the *greedy algorithm*, in which one starts from the empty set and at each step adds a maximum-weight element such that independence is maintained [4]. In fact, given axiom (i) above, axiom (ii) is equivalent to the correctness of the greedy algorithm for all possible weights. The worst-case performance of this algorithm when applied as a heuristic in general *independence systems*, in which only axiom (i) holds, has also been investigated [8].

The *matroid intersection* problem is as follows: given two matroids on the same set and a weight for each element, find a common independent set of maximum weight [5]. This problem embodies many discrete optimization problems as special cases. An example is the *linear assignment* problem; in this case, the matroids are defined on the edge set of the complete bipartite graph K_{nn}, a subset of edges being independent in the first (second) matroid if no two edges have the same head (tail). In fact, it is an adaptation of the augmenting path technique familiar from network flow theory that has yielded a polynomial-time algorithm for the matroid intersection problem. The similar problem for k matroids is reducible to the case $k=3$, which is NP-hard. The outstanding open problem in this area is the *matroid parity* problem, which generalizes both matroid intersection and nonbipartite matching [9]. It has been under heavy attack for a long time and there are few clues as to what its complexity status might be.

The *matroid partition* problem is the problem of partitioning the set E into as few independent subsets as possible. Its polynomial-time solution originated from the solution of the Shannon switching game [11]. Matroid partition and matroid intersection are technically equivalent problems. Both have been employed to solve the *optimum branching* problem as a special case; this is the problem of finding a minimum-weight spanning directed forest in an arc-weighted graph.

Basically, it appears that the *submodular inequality* $r(S \cup T) + r(S \cap T) \leq r(S) + r(T)$ rather than the subadditive inequality $r(S \cup T) \leq r(S) + r(T)$ is what makes matroid optimization algorithms work so well; here $r(S)$ is the size of a maximal independent subset of S. Submodularity is an extremely powerful property and has allowed generalizations in such diverse areas as heuristics for warehouse location [13] and the Lucchesi-Younger theorem [6]. This theorem says that in a directed graph the smallest number of arcs that meet every directed coboundary is equal to the largest number of arc-disjoint directed coboundaries; a directed coboundary is the nonempty set of arcs joining some subset S of vertices with its complement $\bar{S}$ such that they are all directed from S to $\bar{S}$.

A lot has to be done in finding useful *applications* of matroid optimization theory. Some promising and elegant recent examples of such applications involve efficient solution methods for the following two problems: given a matrix, determine whether it is row-equivalent to a network flow constraint matrix, or find a row-equivalent matrix that is in a sense as decomposed as possible. These algorithms might be worth using as preprocessing tools for linear programming.

Another area in which spectacular progress has been made and more can be expected concerns *graphical algorithms*. Computer scientists have been particularly active in combining clever data structures such as priority queues or mergeable heaps with appropriate search techniques such as depth-first or breadth-first search to solve a large variety of problems on graphs in an extremely efficient way. A recent paper by Tarjan [16] provides a good survey of what has been accomplished.

By way of a typical example, we mention the *disjoint set union* problem. This involves the execution of a sequence of $O(n)$ UNION and FIND instructions, starting from n singleton sets. Tarjan has shown that this can be done in $O(n\alpha(n))$ time, where $\alpha(n)$ is equal to the smallest integer k such that $A(k) \geq n$, where

$A(0) = 1$ and $A(k+1) = 2^{A(k)}$; α is a very slowly growing inverse of the Ackermann function [17]. There is also strong evidence that no strictly linear algorithm exists.

Other problems that have been attacked successfully by these techniques include *planarity testing* and finding *subgraphs of maximum connectivity, shortest paths, optimum spanning trees, (weakly* or *strongly) connected components* of a directed graph and *dominators* in a rooted acyclic graph (vertex x dominates vertex y when x lies in all paths from the root to y).

More exotically, *global flow analysis* has yielded efficient methods to optimize compiled codes [18], and it is possible to *generate all maximal independent sets of an independence system* in time linear in their number for a large class of independence systems [10].

Some remarkable work has been done in *computational geometry*. Suppose we are given the coordinates of n points in the Euclidean plane and we wish to find the shortest spanning tree, the smallest circle containing all n points, or the largest circle contained in their convex hull but containing none of them. All these problems and many others are solvable in $O(n \log n)$ time through the construction of the *Voronoi diagram* [15]. This is a planar polyhedral graph which partitions the plane into n subsets, each containing all points in the plane that are closer to the corresponding given point than to any other one. For example, the first problem is solved by taking the planar dual of the diagram and finding its shortest spanning tree rather than the one of the complete graph. The Voronoi diagram can be found in $O(n \log n)$ time. It represents an important unifying principle that is currently being extended to higher dimensions.

Since many of these geometrical problems are generalizations of sorting, $O(n \log n)$ is also a *lower bound* on the running time of any method for their solution. Too little is known about similar lower bounds for other problems. As a notable exception, it should be mentioned that the *Aanderaa-Rosenberg conjecture* has recently been proved correct [14]: recognizing a certain property of a graph requires at least $O(n^2)$ inspections of the adjacency matrix if this property is monotone (maintained under addition of edges), nontrivial (satisfied by the complete graph but not by the edgeless graph) and symmetric (invariant under vertex permutation).

Even traditionally more difficult problems can be solved in polynomial time for some special cases. For instance, *graph isomorphism* can be verified in polynomial time for trees, strongly regular graphs and series-parallel digraphs, although the general problem remains open.

As a typical example of a problem which is known to be NP-hard in the general case but still yields many fascinating open problems, let us briefly discuss the problem of finding a *maximum independent set* of a graph. Polynomial algorithms exist for, among others, interval graphs and comparability graphs. However, nothing is known about the problem for perfect graphs, although the underlying polytope has a good characterization. Conversely, the problem has recently been solved by Minty for *claw-free* graphs, i.e. graphs having no induced subgraph of the type $(\{a, b, c, d\}, \{\{a, b\}, \{a, c\}, \{a, d\}\})$, but for that case the polytope has not (yet) been characterized. Minty's method generalizes Edmonds' *maximum matching* algorithm (since line graphs are claw free). Another interesting connection between these two problems was pointed out by Berge in 1957. Consider a graph with a maximum matching and let X (Y) be the set of all vertices that can be reached from some unsaturated vertex only by an even (odd) alternating path. If $X \cup Y$ contains all the vertices, then Y is a maximum independent set [1]. Approaching the problem from the other side, it would be of interest to determine the smallest ε such that it can be solved in $O((1+\varepsilon)^n)$ time, where n is the number of vertices.

An interesting variation is the problem of finding a *maximal* independent set of *minimum* size. Very little is known about this problem, even for highly structured (hyper)graphs for which the *maximum* independent set problem has been solved. For example, consider the graph whose vertices correspond to all the k-subsets of an n-element set ($k \leq n/2$), two of them being adjacent when the sets are disjoint. Then the size of a maximum independent set is known to be $\binom{n-1}{k-1}$, but the above problem remains open.

Finally, one would hope for more insight into the reasons for the existence of good algorithms. A unifying concept might be that of *obstructions*, i.e. properties that allow us to recognize efficiently the nonexistence of a combinatorially described subset of a set. We have *good characterization* if enough obstructions are known so that one of them is always available when the problem has no solutions. Examples are the good characterizations of polytopes, Kuratowski's condition for the planarity of graphs, Hall's condition for bipartite matching and the duality theorem of linear programming.

It is an intriguing problem to find the connection between the availability of a good characterization and the existence of a good algorithm. Even if no such algorithm is known to exist, obstructions can be very useful, for example in branch-and-bound algorithms.

A general method to obtain obstructions for integer linear programming problems is to get a linear relaxation and to generate cuts. This method will always work, as demonstrated by Gomory and Chvátal [2], but often very slowly, as demonstrated by Grötschel's traveling salesman example (although, in general, the facial cuts that are known for this problem appear to be algorithmically useful). However, there are other ways to obtain important obstructions and good characterizations, and methods from linear algebra, category theory, topology and probability theory may find fruitful application in this area.

Acknowledgements

C. Berge, J. Edmonds, A.J. Hoffman, E.L. Lawler, L. Lovász and A.H.G. Rinnooy Kan participated in the discussion in Banff. I have made extensive use of the notes prepared by U.N. Peled.

References

[1] C. Berge, The Theory of Graphs and its Applications (Methuen, London, 1962).
[2]. V. Chvátal, Edmonds polytopes and a hierarchy of combinatorial problems, Discrete Math. 4 (1973) 305–337.
[3] J. Edmonds, Paths, trees and flowers, Canad. J. Math. 17 (1965) 449–467.
[4] J. Edmonds, Matroids and the greedy algorithm, Math. Programming 1 (1971) 127–136.
[5] J. Edmonds, Matroid intersection, Ann. Discrete Math. 4 (1979) this volume.
[6] J. Edmonds and R. Giles, A min-max relation for submodular functions on graphs, Ann. Discrete Math. 1 (1977) 185–204.
[7] J. Edmonds and E.L. Johnson, Matching: a well-solved class of integer linear programs, in: Combinatorial Structures and Their Applications (Gordon and Breach, New York, 1970) 89–92.
[8] D. Hausmann, T.A. Jenkins and B. Korte, Worst case analysis of greedy type algorithms for independence systems, Report 7781–OR (Institut für Ökonometrie and Operations Research, Universität Bonn, 1977).
[9] E.L. Lawler, Combinatorial Optimization: Networks and Matroids (Holt, Rinehart und Winston, New York, 1976).
[10] E.L. Lawler, J.K. Lenstra and A.H.G. Rinnooy Kan, Generating all maximal independent sets: NP-hardness and polynomial-time algorithms, Report BW 87 (Mathematisch Centrum, Amsterdam, 1978).
[11] A.B. Lehman, A solution of the Shannon switching game, SIAM. J. Appl. Math. 12 (1964) 687–725.
[12] J.K. Lenstra and A.H.G. Rinnooy Kan, Computational complexity of discrete optimization problems, Ann. Discrete Math. 4 (1979) this volume.
[13] G.L. Nemhauser, L.A. Wolsey and M.L. Fisher, An analysis of approximations for maximizing submodular set functions I, Math. Programming 14 (1978) 265–294.
[14] R.L. Rivest and J. Vuillemin, A generalization and proof of the Aanderaa–Rosenberg conjecture, Proc. Seventh Annual ACM Symp. on Theory of Computing (1975) 6–11.
[15] M.I. Shamos and D. Hoey, Closest point problems, Proc. 16th Annual IEEE Symp., Foundations of Computer Science (1975) 151–162.
[16] R.E. Tarjan, Complexity of combinatorial algorithms, SIAM Review 20 (1978) 457–491.
[17] R.E. Tarjan, Efficiency of a good but not linear set union algorithm, J. Assoc. Comput. Mach. 22 (1975) 215–225.
[18] R.E. Tarjan, Iterative algorithms for global flow analysis, in: J.F. Traub, ed., Algorithms and Complexity: New Directions and Recent Results (Academic Press, New York, 1976) 71–101.

QA
402.5
D56
v.1

OCT 9 1979